T0300813

GERMAN AIRCRAFT OF WORLD WAR II

GERMAN AIRCRAFT OF WORLD WAR II

FIGHTERS BOMBERS TRANSPORTS SEAPLANES

Bing Chandler

amber
BOOKS

Published by Amber Books Ltd
United House
London N7 9DP
United Kingdom
www.amberbooks.co.uk
Facebook: amberbooks
YouTube: amberbooksltd
Instagram: amberbooksltd
X(Twitter): @amberbooks

ISBN: 978-1-83886-368-5

Project Editor: Michael Spilling
Picture Research: Terry Forshaw
Design: Mark Batley

Printed in China

CONTENTS

Introduction

The German aircraft industry bloomed under the Nazi regime as it threw off the restrictions of the Treaty of Versailles and embarked on a programme of massive rearmament. This would see new aircraft and engine designs commissioned to replace types that were themselves only just entering service.

With the experience of the Spanish Civil War to further guide development, the Luftwaffe entered World War II as the best tactical air force in the world. Its frontline fighter, the Bf 109, was the equal of the RAF's Hurricane and Spitfire; the Junkers Ju 87 dive-bomber provided them with a capability its opposition didn't possess; and the Ju 52/3m fleet gave it an airlift capability that the Allies would not gain until much later in the war. This superiority would soon peak, however, as few of the replacement programmes would produce aircraft with the required performance, the Me 210 spending years in development hell

Opposite: A pair of Messerschmitt Me 210 A-2 fighters on patrol. After a difficult gestation, the Me 210/410 emerged as a powerful *Zerstörer* ('destroyer'), but only after a major redesign. In the latter part of its career, it came to be known as the *Hornisse* ('Hornet').

Below: Focke-Wulf Fw 190A-0 fighters at Rechlin during testing. When the Focke-Wulf Fw 190 first appeared in the skies over the northern coast of France in the summer of 1941, it was certainly the most advanced fighter in frontline service in the world.

VN AT

before finally emerging as the at best adequate Me 410, while the Ju 188 would arrive too late to fully replace the Ju 88 medium bomber. The lack of a truly capable strategic bomber, although not a major shortcoming during the blitzkrieg that opened the war, would soon prevent Germany hitting back at the Allies in an effective manner. Finally, the numerous attempts to produce war-winning super-weapons, such as the Amerika Bomber or the Me 163, would merely serve to divert limited resources from the improvement and production of proven aircraft.

The Luftwaffe's fighter arm entered the war with the advantage of having operated its main type, the Bf 109, in the Spanish Civil War and proved the tactics for operating high-speed monoplanes. This would see it reign supreme through Poland, Norway and the invasion of France, while even in the opening stages of the Battle of Britain, the RAF would attempt to fight using unworkable techniques developed in interwar exercises.

Battle of Britain

Only when these were abandoned would the Luftwaffe face its first defeat as the Spitfire and Hurricane would prove capable of countering the Bf 109 and Bf 110 fighters, crucially allowing the RAF to then engage the bombers. A game of developmental superiority would follow as the principal fighter types were upgraded with more powerful engines to gain the upper hand and new types were introduced to try to leapfrog the opposition.

For the Luftwaffe, this would lead to the Focke-Wulf Fw 190 – one of the great fighters of the war and without equal until the RAF could field the Spitfire Mk IX. This success would be something of an aberration as the promised replacement for the Bf 110 heavy fighter, the Me 210, suffered a protracted development that almost broke Messerschmitt and would finally result in the Me 410, whose introduction into service had almost no effect whatsoever on the course of the war. The excellent twin-engine Do 335, meanwhile, would through initial indifference enter development far too late, with only a handful produced before the end of the conflict. Consequently, the Luftwaffe's day-fighter force would be dominated by Fw 190s and Bf 109s right up to the end.

Night-fighters

In response to night-time incursions by aircraft of the RAF's Bomber Command, Germany would produce a range of night-fighters. Although some single-seat aircraft would be used in the role, the majority were adapted from the Bf 110 heavy fighter or the Do 217 and Ju 88 medium bombers. This allowed them to carry airborne radar and the necessary additional crew to operate them, as well as the remarkably successful upwards-firing Schräge Musik system

The Junkers Ju 88 was one of the most adaptable Lufwaffe aircraft of the war. Painted in a tropical finish, this Ju 88A-11 was on strength of Lehrgeschwader 1, fighting in North Africa in 1942.

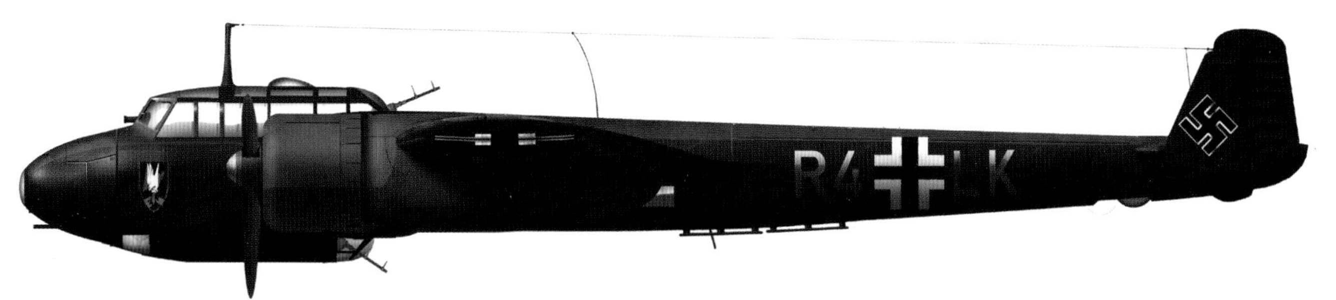

This Dornier Do 17Z-10 night-fighter was based at Gilze-Rijen in November 1941. The first of several Do 17Z-10 kills was recorded on the night of 16/17 October 1940 by Ludwig Becker of 4./Nachtjagdgeschwader 1, based at Deelen in the Netherlands, who downed an RAF Wellington bomber.

that unknowingly took advantage of a lack of ventral turret on Bomber Command's aircraft. A dedicated night-fighter, the He 219 was developed but would only enter service in limited numbers as the Third Reich struggled to prioritize its use of resources in the final years of the war.

Tactical bombers

With a doctrine based on supporting the Army as its primary task, the Luftwaffe had prioritized tactical and medium bomber development. This had resulted in the Ju 87 Stuka, Ju 88 and He 111, all of which had proven themselves in the blitzkrieg that opened the war and during Operation Barbarossa, the invasion of the Soviet Union. However, this gave them limited options for taking the war to the United Kingdom and beyond the front line on the Eastern Front. Although the medium bombers were able to strike targets across the UK, in some cases to great effect, this meant diverting aircraft that would otherwise be supporting the Army. Consequently, raids on the UK dropped off as German forces were transferred to the East. In the East, meanwhile, the Soviets were able to move much of their industry beyond the range of

A squadron of Heinkel He 111H-16s maintains a tight formation while returning from a sortie on the Russian front. The H-16 was powered by the Jumo 211F-2 engine.

the Luftwaffe. Attempts to belatedly introduce a heavy bomber were less than spectacular with the He 177 Greif being as dangerous to its crews as to the enemy, with a worrying tendency for the engines to burst into flames. The Me 264, meanwhile, was a potential strategic bomber able to reach beyond the Urals or across the Atlantic. However, its on-again, off-again development ensured that only one would fly and then never for more than a few hours at a time.

Germany pioneered the use of paratroopers through the opening battles of the war, relying on its vast fleet of Ju 52/3m transports. Various programmes would attempt to develop successors or replacements. The Ju 252 and 352 would expand on the trimotor concept while the Ar 232 pioneered many features that would be used on post-war types, such as the C-130. However, none of these types would enter series production, the difficulties of setting up new construction lines outweighing any improvement in capability. The largest transport of the war, the ungainly six-engine Me 323, would allow the Luftwaffe to move anything up to a medium tank or 200 troops in a single flight. Although widely employed, it would prove vulnerable if unescorted and was not suited to operations in contested airspace.

Maritime patrol

Maritime patrol, the Cinderella of the Luftwaffe, would use a range of cast-off land planes, such as the Fw 200 Condor airliner and He 177 bombers, for long-range patrol in search of Allied convoys. They would also use a range of seaplanes and flying boats for coastal reconnaissance and attack as well as search and rescue. A handful of Ar 196 seaplanes would embark on the Kriegsmarine's battleships and armed merchant raiders. Some of these would be the most far-ranging of the

The Blohm und Voss The BV 138 matured into a reliable and useful patroller. The majority operated in Arctic waters from Norwegian bases, but this BV 138C served with 3.(F)/SAGr 125 at Constanza, on the Black Sea coast.

This Messerschmitt Me-262A-1a was flown by Major Rudolf Sinner, who flew with III/JG-7 'Nowotny', from Brandenburg-Briest, Germany, February 1945.

Luftwaffe's aircraft, with a few reaching Japan on board their parent ships before operating out of Penang in Malaysia. It is a sign of how low the priority was for this branch that little if any thought or effort was wasted on speculative future designs despite the importance of maritime transport to the Allied war effort!

Jet aircraft

The area of German aviation development that has perhaps been the focus of most speculation is its pioneering jet aircraft. This saw them make a quantum leap in performance over anything the Allies had in the closing stages of the war. The Me 262, as well as being the first jet fighter to enter operational service, was also in terms of outright performance better than the Gloster Meteor that would fly with the RAF before the war's end.

The Me 163B could have been ripped from the pages of Dan Dare or Flash Gordon, a rocket-powered, tail-less fighter that climbed almost vertically, trailing a plume of steam. The Ju 287, meanwhile, was a six-engine bomber featuring a forward-swept wing, something that has only been seen on a handful of aircraft since. Yet again, though, these developments were more a misuse of scarce resources than practical weapons with which to turn the tide of the war. Undoubtedly the most practical, the Me 262's engines were pushing what was capable with the materials available to Nazi Germany, resulting in them needing an overhaul after 10–25 hours of use. Slow acceleration also left the early jet vulnerable to Allied piston-engine fighters during take-off and landing – a factor they took advantage of. The Me 163B fell far short of the Me 262's usability, with a flight time measured in minutes. Only a handful of USAAF bombers fell to its guns, while its propensity to explode on landing contributed to 80 per cent of its losses being unrelated to combat. Finally, the Ju 287 was a valuable aerodynamic research project, but it should have been abundantly clear that the situation in 1944 did not permit time or resources to be spent on an experimental design that was years from service.

Limitations

Ambitious in the range and scale of its plans, the Luftwaffe suffered from the failure of many of its follow-on projects to produce viable successors to the aircraft it entered the war with. It was also hindered by doctrine that led to it fighting at a strategic scale with equipment designed for a tactical role. At the same time, its seeming inability to prioritize those developments most likely to succeed in favour of a scattergun approach to commissioning aircraft hindered German industry as the war progressed.

Fighters

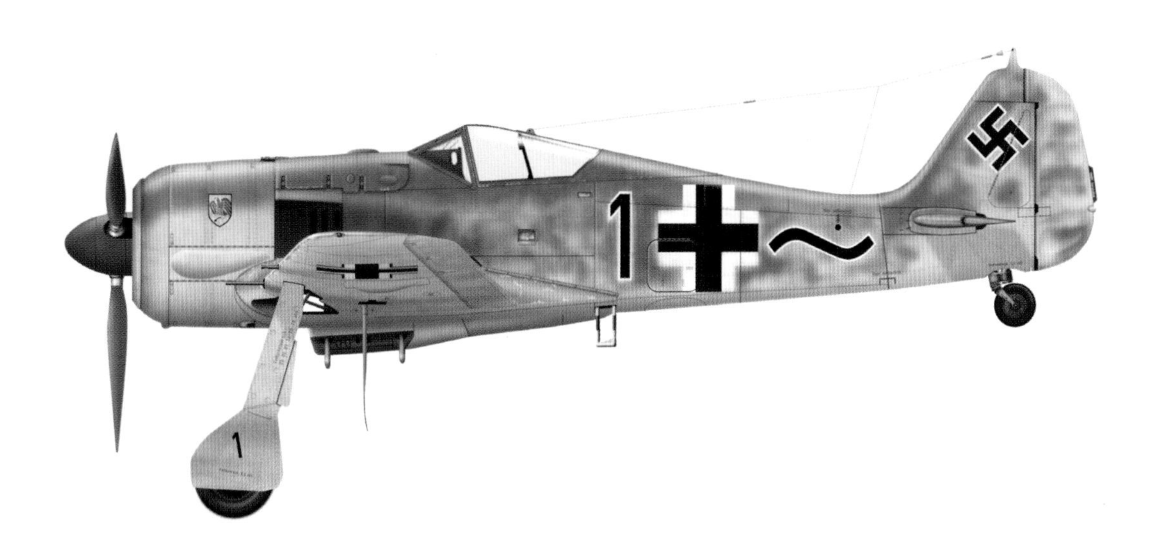

The Luftwaffe entered the war with the Bf 109 as its main single-engine fighter and the Bf 110 as its heavyweight long-range *Zerstörer* ('Destroyer'). The former would still be in service in much upgraded form in 1945, alongside the similarly improved Fw 190, which entered service in 1941. The heavy fighter concept would, however, prove something of a dead end with the Bf 110 requiring its own escort during the Battle of Britain. It would, though, find a role as a night-fighter alongside modified Do 217 and Ju 88 bombers. The only new German piston-engine fighter to be introduced after 1943, the superlative Dornier Do 335 could have been competitive with the early Allied jets if the war had continued.

Opposite: A Dornier 215 flies towards the camera. An adaptable type, the Do 215 was used as a tactical bomber and night-fighter.

Messerschmitt Bf 109E (1938)

The classic Luftwaffe fighter of World War II, the Bf 109 served throughout the conflict in a series of increasingly capable variants. On the way, it was the mount for Germany's most celebrated aces, including Hartmann, Barkhorn and Marseille.

The origins of the Bf 109 lie in the creation of the Luftwaffe in the mid-1930s and the concurrent demand for a new generation of modern fighters. As the schematic basis for a new fighter design, Willy Messerschmitt took the Bf 108, a low-wing cantilever cabin monoplane with retractable main undercarriage. In fact, preliminary work on a new fighter had begun before the Bf 108 took to the air.

The initial prototype of Messerschmitt's Bf 109 completed its maiden flight in May 1935, powered by a Rolls-Royce Kestrel engine, while the second prototype featured the planned Junkers Jumo 210A. In a fly-off competition, the Bf 109 saw off competition from the Arado Ar 80 and Focke-Wulf Fw 159, and together with the rival Heinkel He 112 prototype was selected for further development, an initial batch of 10 examples of each being ordered. The Bf 109A pre-production prototypes tested a variety of armament installations, and served to refine the design of the initial production model, the Bf 109B, which received improved Jumo 210B, 210D, 210E or 210G engines.

Deliveries of the fighter commenced in early 1937, and initial machines went to the Luftwaffe's premier fighter

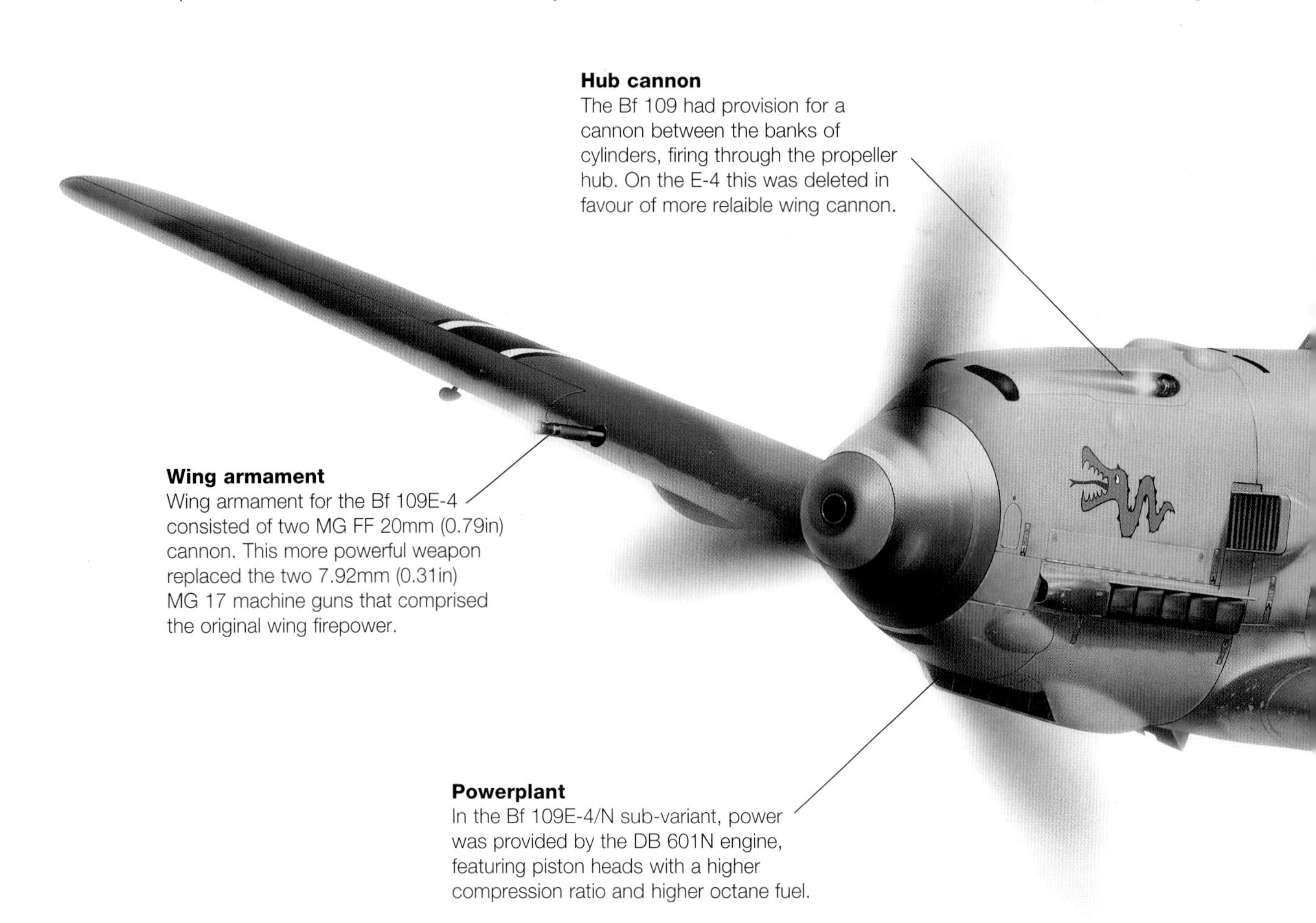

unit, Jagdgeschwader 132 'Richthofen'. In the summer of 1937 the type was subject to combat trials in Spain, as part of the Legion Condor. Before the end of the year a specially prepared Bf 109 took the world land plane speed record, at 610.55km/h (379.38mph), making use of a boosted DB 601 engine. Next of the production versions was the Bf 109C, with a Jumo 210Ga engine and armament increased from three to four machine guns. By September 1938 almost 600 examples had been completed, using production facilities at Arado,

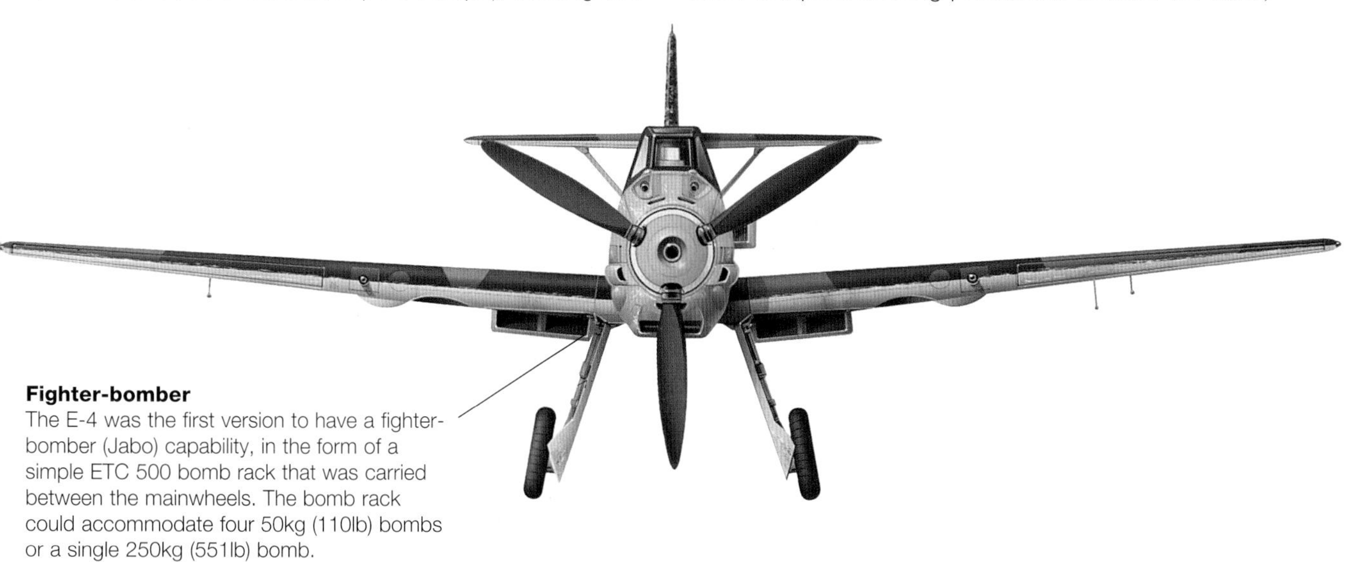

Fighter-bomber
The E-4 was the first version to have a fighter-bomber (Jabo) capability, in the form of a simple ETC 500 bomb rack that was carried between the mainwheels. The bomb rack could accommodate four 50kg (110lb) bombs or a single 250kg (551lb) bomb.

A Bf 109E-4, the mount of Hans von Hahn, Gruppenkommandeur of I. Gruppe, Jagdgeschwader 3, based at Grandvillier, France, in August 1940. The 'Tatzelwurm' emblem on the cowling was used throughout I. Gruppe of JG 3.

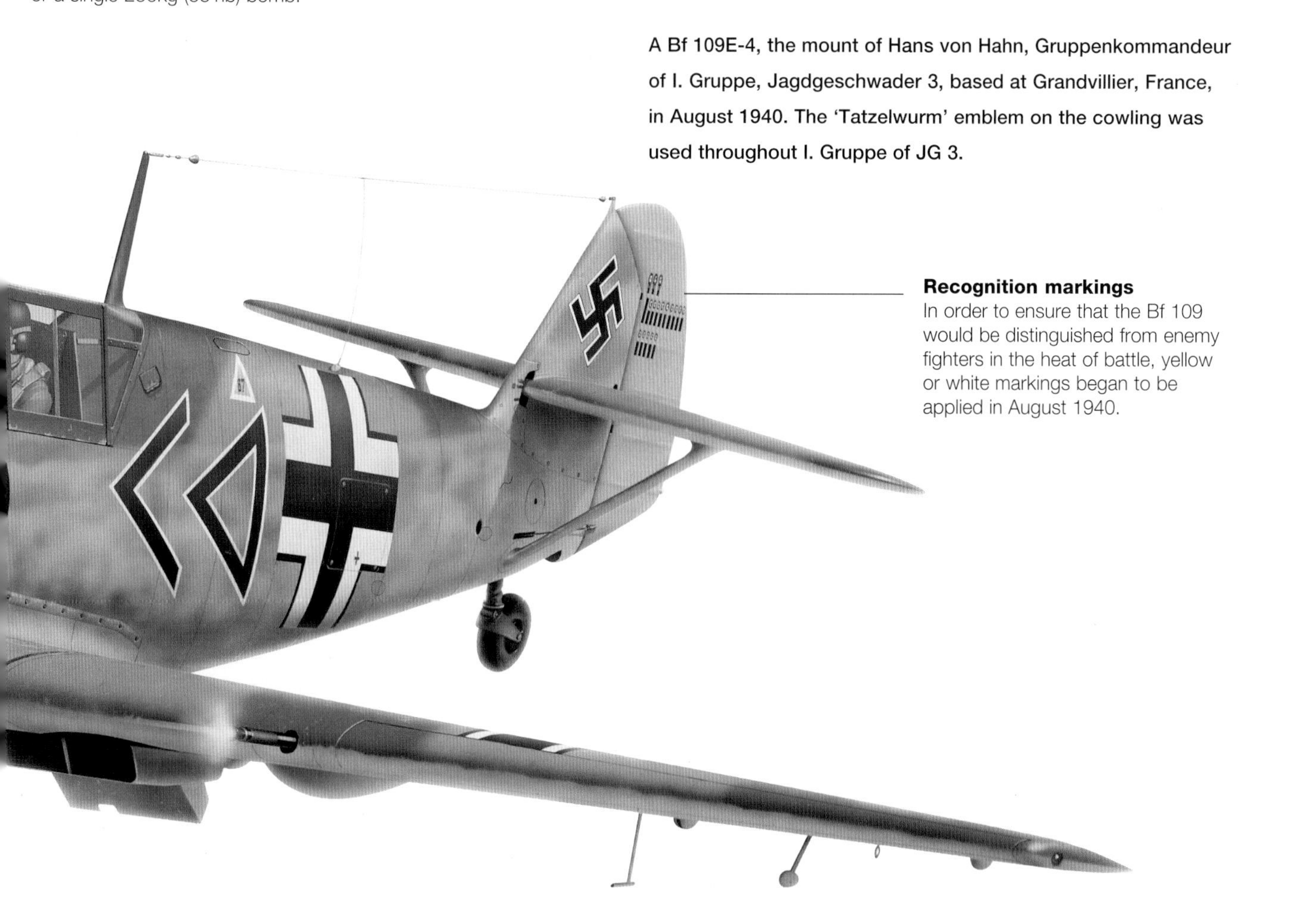

Recognition markings
In order to ensure that the Bf 109 would be distinguished from enemy fighters in the heat of battle, yellow or white markings began to be applied in August 1940.

This Battle of Britain Bf 109E-4 was flown by Helmut Wick, Gruppenkommandeur of I./Jagdgeschwader 2 'Richthofen', based at Beaumont-le-Roger, France in October 1940. Wick was the highest-scoring Luftwaffe ace at the time of his death.

Erla, Focke-Wulf and Fieseler. The Bf 109D featured the Daimler-Benz DB 600 engine and a hub-firing cannon. It was produced in 1938–39. The first of the 109s to be built in significant quantities was the Bf 109E, powered by a DB 601D engine with direct fuel-injection. Armament options for the 'Emil' were based around two machine guns, four machine and a single hub cannon. Representing the versatility of the basic design, the Bf 109E was fielded in specialized fighter-bomber and reconnaissance sub-variants.

By making continual improvements to the basic design, the Bf 109 remained viable right until the end of World War II, and despite the appearance of the more capable Fw 109, the Bf 109 remained the backbone of the Luftwaffe fighter arm. The Bf 109 remains associated, therefore, with the legendary aces of the *Jagdverband* (fighter units). Top-scoring ace of all time, Erich Hartmann achieved his tally of 352 victories in the space of three and a half years, all while at the controls of a Bf 109. The highest-scoring German ace in North Africa, Hans-Joachim Marseille similarly scored all his 158 victories flying the Messerschmitt fighter.

Specifications: Bf 109E-4

Type:	Fighter
Dimensions:	Length: 8.76m (28ft 7in); Wingspan: 9.87m (32ft 4in); Height: 2.28m (7ft 5.5in)
Weight:	2505kg (5523lb) maximum take-off
Powerplant:	1 x 894kW (1200hp) DB 601N 12-cylinder inverted-V engine
Maximum speed:	570km/h (354mph)
Range:	720km (447 miles)
Service ceiling:	10,500m (34,450ft)
Crew:	1
Armament:	2 x 20mm (0.8in) MG FF/M wing cannon, two 7.92mm (0.31in) MG 17 machine guns

Africa service

While the Bf 109E was the primary Luftwaffe fighter during 1939–40, it was superseded by the Bf 109F, powered initially by the DB 601N and later by the DB 601E, and with nitrous-oxide power boost, faster-firing guns and provision for underwing gun pods. Tropicalized versions of both the Bf 109E and F spearheaded the fighter arm in North Africa during 1941–42. Numerically the most important version was the Bf 109G, built from 1942–45 and active on all fronts. The 'Gustav' featured a DB 605 engine and a variety of armament options, including 30mm (1.18in) cannon. The similar Bf 109K featured boosted versions of the DB 605.

The Bf 109 had also been intended for service aboard the abortive aircraft carriers of the Kriegsmarine, and a

Lieutenant Steindl, headquarters adjutant of Jagdgeschwader 54, overflies the Stalingrad sector in summer 1942 in a Bf 109E-4.

Record-breaking Me 209

Officially the world's fastest piston-engined aircraft between 1939 and 1969, the Messerschmitt Me 209 V1 was loosely based on the Bf 109 airframe, with the specific aim of capturing speed records for Nazi Germany. Based around a specially prepared DB 601 engine that had a peak power output of 1715kW (2300hp), the one-off Me 209 was flown by Fritz Wendel on 26 April 1939, when it took the speed record at 755.136km/h (469.22mph). In an effort to increase the propaganda value of the achievement, the German authorities described the record-breaking aircraft as the Me 109R, suggesting a version of the production Bf 109 fighter. Wendel and the Me 209 retained the speed record for piston-engined aircraft until April 1969, when Darryl Greenamyer flew the Grumman Bearcat Conquest I to an average speed of 769.23km/h (477.98mph).

batch of navalized Bf 109T versions were completed with folding wings, arrester gear and tailhook. In the event, these aircraft saw service from land bases.

Production numbers

By the outbreak of World War II, a total of over 1000 Bf 109s had been delivered, production increasing in 1942 when Messerschmitt completed close to 2700 examples, and additional aircraft began to be completed at production facilities in Hungary the following year. In 1944 German industrial output accounted for almost 14,000 aircraft. Although no comprehensive records survive, it is estimated that around 35,000 Bf 109s were ultimately built, making it the most prolific fighter of all time.

Alongside Germany, the Bf 109 was operated by Bulgaria, Finland, Hungary, Japan, Romania, Slovakia, Spain, Switzerland, the USSR and Yugoslavia. Beginning in 1945 it was also built under licence in Spain as the Hispano HA-1109 and improved HA-1112. Bf 109s were also built post-war in Czechoslovakia, as the Avia S-99 and S-199, and both Czech and Spanish derivatives remained in service into the 1950s.

Junkers Ju 88C & Ju 88G (1938)

A lower priority than its bomber sibling, the Ju 88C's numbers would be slow to build up. Nevertheless, it developed into one of the pre-eminent night-fighters of the war due to its speed, armament and range.

The Junker Ju 88 was designed against a Luftwaffe requirement for a multi-role combat aircraft, although this was refined to focus on the bomber variant that first flew in 1936, while a heavy-fighter, or *Zerstörer*, version remained under development. The seventh prototype was modified accordingly: the glazed nose was replaced with a sheet metal fairing housing two 20mm (0.79in) MG FF cannon and two 7.92mm (0.31in) machine guns. Meanwhile, the blister under the starboard side of the fuselage that housed the bomb aiming site and a rearward-firing machine gun was removed. The crew now consisted of three – pilot, flight engineer and radio operator, the bombardier being surplus to requirements. Powered by two Junkers Jumo 211B-1 engines producing 895kW (1,200hp), the prototype first flew in 1938 and demonstrated a maximum speed of 500km/h (311mph) – around the same as the Messerschmitt's Bf 110 but with a range of 2,900km (1,802 miles), three times that of the smaller aircraft.

A Ju 88C-4 of II./NJG 2 flown by Hptm. Horst Patuschka while operating from Comiso, Sicily, in 1942–43. Although most markings have been overpainted to reduce the aircraft's conspicuity his tally of kills is displayed on the tail.

Development

A handful of Ju 88C-0s were produced by modifying Ju 88A-1 bombers. These gained the solid nose of the prototype but retained the ventral gondola, as would most subsequent C series models. The C-1 was intended to be produced with BMW 801 radial engines but as these were

Nose armament
The fighter variants of the Ju 88 received a solid metal nose to hose the forward-firing armament, initially comprising a 20mm (0.79in) MG FF cannon and three 7.92mm (0.3in) MG 17 machine guns the C-4 gained two more MG FF cannon.

Gondola
The Ju 88C retained the bomber version's ventral gondola, which housed a rearward firing machine gun for self-defence. Although some units would remove it to improve performance others would add a further two forward-firing MG FF cannon.

With national markings obscured, this Ju 88C-6 Zerstörer served with V./Kampfgeschwader 40 against Allied anti-submarine aircraft and as an escort fighter for the Focke-Wulf Fw 200 Condor maritime patrol aircraft.

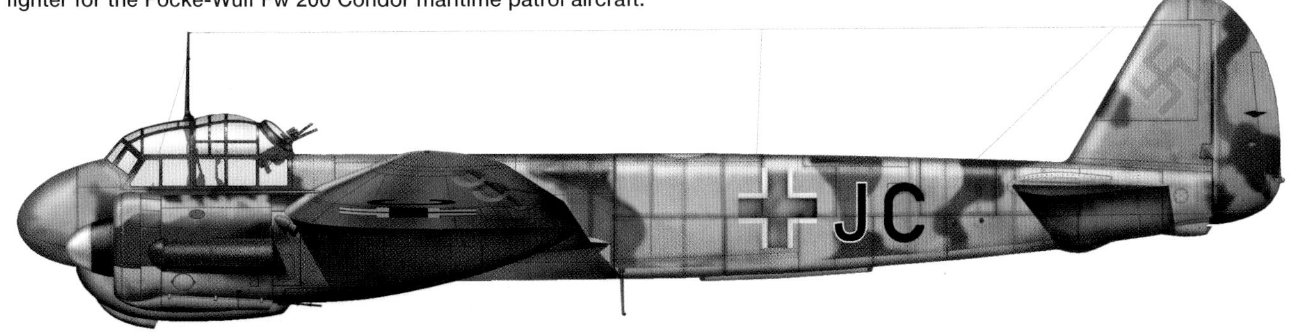

prioritized for the Fw 190, the next variant produced was the Ju 88C-2, which retained the Jumo 211 liquid-cooled engines. In common with many other German aircraft, these used a circular radiator that gave the installation the look of a radial engine but with a nacelle extending far forwards of the wing leading edge. The C-2 had only one 20mm (0.79in) cannon, swapping the other for an additional 7.92mm (0.31in) machine gun. As with the C-0, only a small number were produced.

Some Ju 88Cs saw service in the Battle of Britain, on one occasion acting as long-range escorts to the Ju 88As of KG 30 for a raid on Driffield in the north of the UK. This produced a mixed result for the escorts, who lost five of their 12 aircraft, two more barely making it back to occupied Europe. However, only two of the 40 bombers were lost to the Spitfires and Hurricanes that had opposed the raid, the Ju 88Cs preventing them from attacking the main force.

A more fruitful role for the Ju 88C arose in late 1940. With RAF Bomber Command conducting night raids over Germany, Nachtjagdgeschwader 1 was formed with a Gruppe each of Bf 110s, Bf 109s and Ju 88Cs. Redesignated I./NJG 2 in September of that year, the Ju

Tail
The Ju 88C retained the rounded tail of the Ju 88A bomber. Hptm. Dr Patuschka would ultimately have 23 kill markings on his aircraft before being lost over Tunisia due to an engine failure.

Rear fuselage
The antenna for the FuBl 2 blind landing system was mounted under the rear fuselage. This received morse dots if the aircraft was left of track and dashes if to the right, once on track a continuous tone would be heard.

4R

Junkers Ju 88G-1 of 7./NJG 2 based at Gilze-Rijen in July 1944. After becoming lost while searching for mine-laying Stirling bombers, the crew inadvertently homed on to the radio beacon of RAF Woodbridge and landed there, gifting the Allies their first look at the SN-2 radar system.

88C Gruppe was tasked with conducting night-intruder operations over the UK. This involved attacking the bombers as they were making their recovery to their home bases. Slow and draggy in their landing configuration, the British aircraft were highly vulnerable. In early 1941, while the night-intruder missions were still ongoing, the Ju 88C-4 joined I./NJG 2. Based on the A-4 bomber version, this featured wider wings, more powerful Jumo 211G engines, deletion of the bomb bay and additional armour plating around the crew compartment.

Multi-role fighter

Up to this point, relatively few of the fighter variant had been built, the C-0 and C-2 being conversions of existing Ju 88A airframes while only 65 Ju 88Cs in total were built in 1941, the majority C-4s. The C-6, however, would be built in quantity and serve in a range of roles. Based on the C-4, the C-6 gained more armour protection and a revised armament of three MG FF 20mm (0.79in) cannon and three MG 17 7.92mm (0.31in) machine guns.

The year 1942 saw Ju 88Cs increasing the scale and range of their activities. Aircraft involved in operations on the

Specifications: Ju 88G-1

Type:	Night-fighter
Dimensions:	Length: 16.5m (54ft 2in); Wingspan: 20.0m (65ft 7in); Height: 4.85m (15ft 11in)
Weight:	14,690kg (32,386lb)
Powerplant:	2 x 1,268kW (1,700hp) BMW 801D-2 air-cooled radial engine
Maximum speed:	575km/h (357mph)
Range:	2,195km (1,364 miles)
Service ceiling:	8,840m (29,003ft)
Crew:	3
Armament:	4 x 20mm (0.79in) MG 151 cannon in the ventral gondola; 1 x 13mm (0.51in) MG 131 machine gun firing aft from the cockpit

Eastern Front were used as long-range fighters, for low-level strafing, and as bombers. They were regularly used to target the Soviet lines of communication, attacking lorries and trains. At the opposite extreme of Nazi-occupied Europe, V./KG 40 was using Ju 88C-6s over the Bay of Biscay to intercept aircraft of RAF Coastal Command, which themselves were attempting to intercept U-boats based on the French Atlantic coast. Meanwhile, NJG 1 would fit four of their Ju 88C-6s with the FuG 212 Lichtenstein C-1 radar, significantly expanding the Junker's capabilities as a night-fighter.

Although the radar antenna array reduced the aircraft's top speed by around 8km/h (5mph), its utility on a dark night was soon shown to compensate. The German night-fighter control system involved a chain of radars running north to south through Europe. Known as the *Himmelbett* ('canopy bed') system, a search radar would be paired with two pencil beam control radars, one of which would track a target while the other was used to track a night-fighter and direct it to an intercept. The Lichtenstein radar allowed the aircraft to take over control of the final few kilometres of the approach to the target, permitting attacks to be made in conditions of much poorer visibility. The major weakness of this system was that each radar cell could only track one target at a time.

In May 1942, RAF Bomber Command concluded that directing the bombers in a stream over one cell would saturate the defenders, rendering them unable to direct an attack. Overnight bomber losses dropped dramatically as Himmelbett failed to cope with as many as 1,000 aircraft routing through a narrow area. In response to this, the Luftwaffe developed a new tactic, known as 'Zahme Sau' or 'Tame Boar'. In this, the night-fighters were directed to the bomber stream and then used their own radars to guide themselves in running battles until they were short of fuel or ammunition. Faster and longer-ranged than the Bf 110, the Ju 88C-6 was better suited to this new form of night warfare and gradually took over from the smaller aircraft in the Nachtjagdgeschwader. During the second half of 1943, Zahme Sau would see Bomber Command's losses climbing to 8 per cent during the 'First Battle of Berlin' – well above the 5 per cent considered acceptable.

Operation Gisela

The final major operation for the Ju 88 was Operation Gisela. This was a return to the night-intruder role attacking aircraft of Bomber Command as they returned to their bases. It was hoped that without the chaff and jamming now regularly experienced over Germany the work of the night-fighters would be easier. Three waves of Ju 88Gs, numbering around 100 aircraft in total, struck on the night of 3 March 1945. However, pre-warned by intelligence and detecting the raid on radar, the RAF managed to divert most of its aircraft to bases beyond the range of the attackers. Twenty-two heavy bombers that had not diverted were shot down by the intruders. However, 24 Ju 88s were also lost. These were either shot down by Mosquito night-fighters or ground fire, crashed into the ground while strafing targets of opportunity, or simply ran out of fuel on the return journey. Considered a failure, no further attempts were made and barely two months later the war would be over.

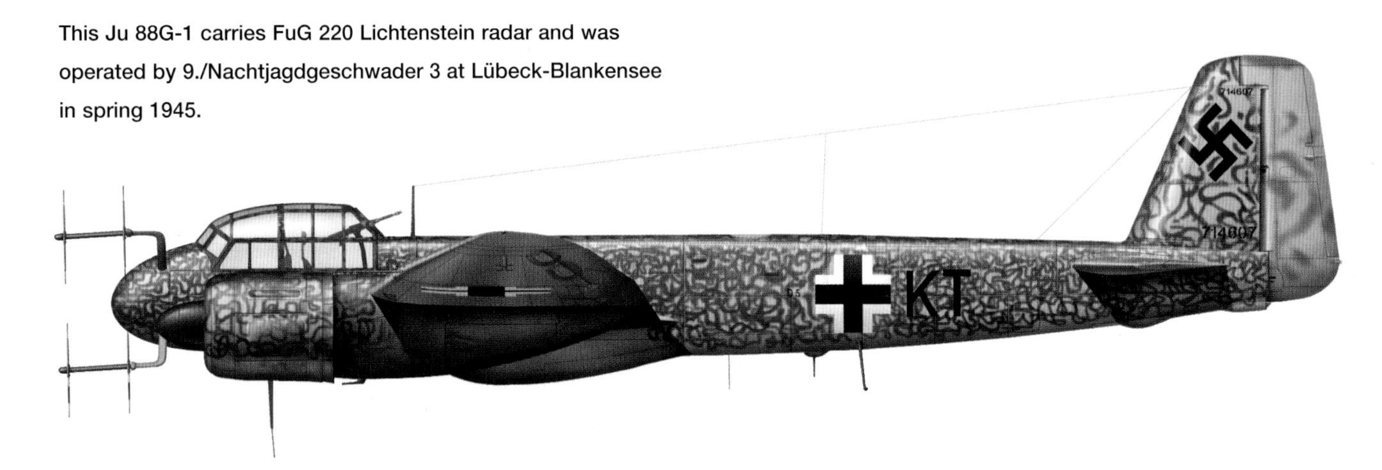

This Ju 88G-1 carries FuG 220 Lichtenstein radar and was operated by 9./Nachtjagdgeschwader 3 at Lübeck-Blankensee in spring 1945.

Junkers Ju 88G

Early 1944 saw a new night-fighter variant introduced into service, intended to compensate for the additional equipment that was beginning to affect the Ju 88C's handling. Powered by BMW 801D air-cooled radial engines producing 1,270kW (1,703hp), the Ju 88G also had the tail of the Ju 188 but with a much squarer vertical stabilizer and could reach a top speed of 537km/h (334mph) – a significant improvement over the 470km/h (292mph) of the Ju 88C-6. The Gustav also received a revised armament package consisting of four MG 151/20 cannon in a gondola on the left-hand side of the aircraft's belly, while a 13mm (0.51in) MG 131 could be fired aft from the radio operator position. The G-4, -6 and -7 variants also had the *Schräge Musik* system of upward-firing guns in the aft fuselage. These allowed a simple zero-deflection shot to be made once a position was taken, formatting underneath the target, and took advantage of the RAF's lack of downward-firing gun turrets, or even observation windows.

The Ju 88G carried a range of radar systems. The G-1 used the FuG 220 Lichtenstein SN-2, an improvement on the C-1 set used in the Ju 88C-6, and many also carried the C-1 as it had a lower minimum range, although was more affected by chaff. The FuG 218 Neptun was used by some Ju 88C-6s and as well as air interception capabilities, had a tail-mounted antenna to warn of aircraft approaching from behind. Towards the end of the war, around 10 Ju 88C-7 were fitted with the FuG 240 Berlin radar.

This operated in the centimetric band, having been developed from an RAF H2S navigation radar that had been captured. The much higher frequency of the Berlin would have provided the crews with a far superior radar picture than the earlier systems and greater resistance to the effects of chaff. Ju 88Gs were also fitted with the Naxos and Flensburg systems, which homed on the RAF bombers' own radar transmissions.

By this time, the night-fighter force was fighting a desperate battle. Although early 1944 had seen Zahme Sau achieving great success, the bombing campaign's shift of focus for the invasion of Normandy had removed most of the opportunities to intercept aircraft over occupied Europe. After the landings took place in June, Ju 88Gs would be called upon to carry out night-strafing runs as part of Operation Heidelberg.

These tactics would be used again during the Battle of the Bulge with Ju 88Gs and Bf 110s conducting night-interdiction strikes behind the front line. As on the Eastern Front, this saw road traffic and trains targeted and caused significant damage and disruption when the conditions were in the night-fighters' favour. Learning from operations over Normandy, some Ju 88Gs had also been fitted with bomb racks, enabling them to carry 250kg (551lb) and 500kg (1,102lb) weapons.

A Ju 88G-1, the first night-fighter to enter production with BMW 801D radial engines, earlier aircraft employing inline engines with annular radiators. The G series also introduced the squared-off tail developed for the Ju 188.

Dornier Do 217 (1938)

Intended for use as a heavy bomber, the Do 217 fulfilled many roles admirably. Efforts to take advantage of its range and payload in the night-fighter role would, however, not see a similar level of success.

The Dornier Do 217 was a successor to the Do 17 and was visually similar despite being a completely new design. First flying in 1938, it had a protracted development as handling issues took time to be resolved. By the time the Do 217E entered service in late 1940, it was a generally adequate aircraft. Following the example of the Do 17, a night-fighter version was proposed.

The Do 17Z-10 Kauz II night-fighter had replaced the glazed bombardier's position with a solid nose containing four MG 15 machine guns and a pair of MG FF 20mm (0.79in) cannon. Aft-facing armament was retained and fired by the radio operator while the engineer was responsible for reloading the MG FF's magazines in flight. Although not fitted with radar, the Do 17Z-10 were equipped with a Spanner-Anlage IR detector to find the heat from the exhausts of RAF bombers. Also used on Bf 110 night-fighters, the device was primitive and not particularly successful in its intended role. The Do 17Z

Dornier Do 217N-1 of II./NJG 4 3C+DV was fitted with FuG202 Lichtenstein radar and operated from Rheine in late 1943.

did, however, achieve a number of kills, the first over the Zuider Zee in Holland in October 1940 when an aircraft of 4./NJG 1 shot down a Wellington bomber.

The nine Kauz IIs built remained in service until the beginning of 1942. They were followed by the Do 215B, which was essentially a Do 17Z powered by DB 601A engines. The Do 215B-6 Kauz III night-fighter version used the nose from the Do 17Z-10 but also carried the FuG 202 Lichtenstein BC airborne interception radar. Despite the increased capability, this was really paving the way for the radar's use in the Bf 110 and Ju 88, and the Do 215s were all retired by mid-1942 shortly after the Do 17Z-10.

The Do 217J-1 night-fighter followed the example of the Kauz II, replacing the bombardier's position with a solid nose, this time containing four 7.92mm (0.31in) MG 17 machine guns and four 20mm (0.79in) MG FF cannon. The aft-firing armament was again retained, and the J-1 entered service at the beginning of 1942. Although heavy and sluggish, the Do 217 had the firepower and endurance required of a night-fighter. Without a radar, however, the J-1 would struggle to find its target on a dark night. This came with the J-2, featuring the FuG 202 Lichtenstein BC radar and deleting the bomb bay that had been retained on the J-1. This reduced the aircraft's all-up mass, which just about countered the reduction in performance caused by the radar antenna array on the nose. Due to this, the Do 217J were less popular than the Bf 110 and Ju 88 night-fighters, which were lighter and more agile. However, the Do 217J were vital in maintaining the mass of the night-fighter force in the face of slow deliveries of the Ju 88C, and was used by NJG 1, 2, 3 and 4.

The Do 217J's service life was short-lived as by January 1943 the Do 217N-1 was joining the Nachtjagdgeschwader (NJG – the night-fighter wings of the Luftwaffe). This variant used an inline engine, as opposed

Communications
The aerial for the FuG 10 HF aerial was held between the port tailplane and a fuselage mounted mast. After the RAF began jamming communications a FuG 16 VHF radio was also fitted.

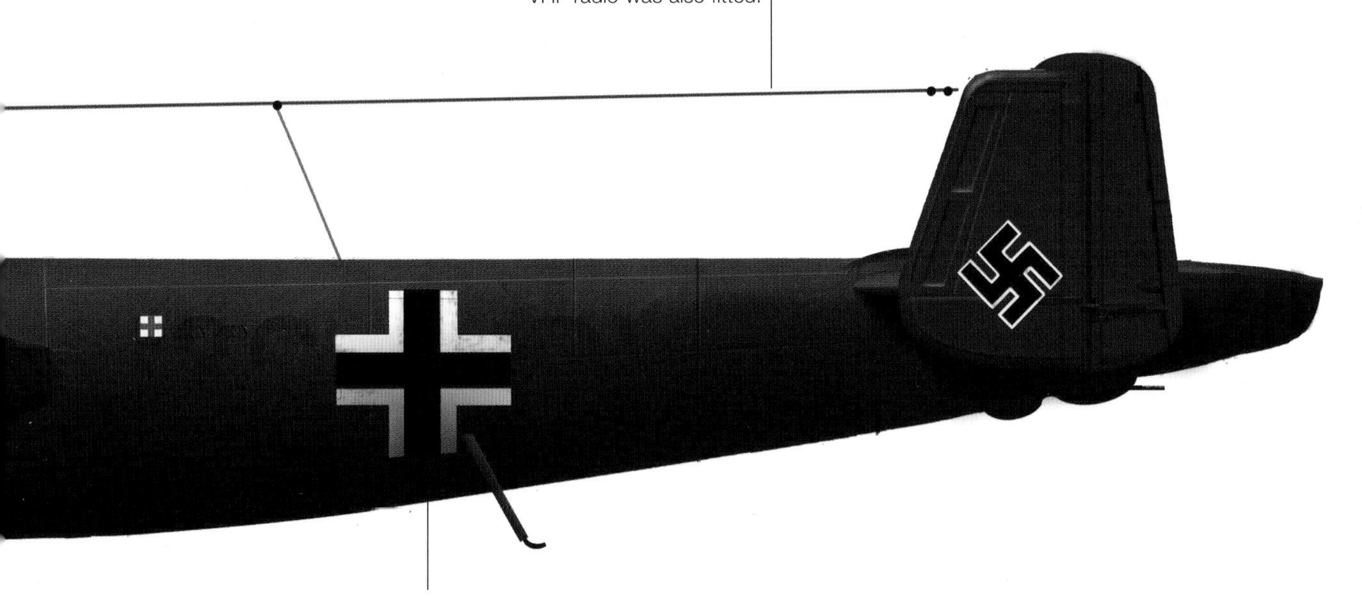

Radio equipment
The Do 217N carried a FuG 25 beacon that allowed the ground-based Himmelbett controllers to identify the aircraft as friendly.

Specifications: Do 217N-2/R22

Type:	Night-fighter
Dimensions:	Length: 18.9m (62ft 0in); Wingspan: 19.0m (62ft 4in); Height: 5.0m (16ft 4in)
Weight:	13,700kg (30,203lb)
Powerplant:	2 x 1,380kW (1,851hp) Daimler-Benz DB 603A inverted V liquid-cooled engines
Maximum speed:	500km/h (311mph)
Range:	1,755km (1,091 miles)
Service ceiling:	8,400m (27,559ft)
Crew:	4
Armament:	4 x 7.92mm (0.31in) MG 17 machine guns and 4 x 20mm (0.79in) MG 151 cannon in the nose; 4 x 20mm (0.79in) MG 151 cannon in the central fuselage aimed 70 degrees upwards

to the BMW radials used on the Do 217Js, in this case two Daimler-Benz DB 603As providing 1,380kW (1,850hp) each. In most other respects, however, it was the same as the Do 217J-2. Although the night-fighter crews had suggested a number of improvements there had not been time to incorporate them without an unacceptable reduction in production levels. They were instead included in the Do 217N-2, the definitive night-fighter version of the Do 217. On these aircraft, and subsequently modified Do 217N-1/U1s, the dorsal turret and ventral gondola were removed and faired over. Along with other weight savings, around 2 tonnes (2.2 tons) was lost from the all-up weight, allowing for a maximum speed of 500km/h (311mph) in level flight.

In addition to the four machine guns and cannon in the nose, modification Rüstsätz 22 added a *Schräge Musik* system in the rear fuselage. This featured a further four MG 151 20mm (0.79in) cannon angled upwards at 70 degrees from the horizontal. With this, the crew could make their attack by formatting underneath the enemy bomber and

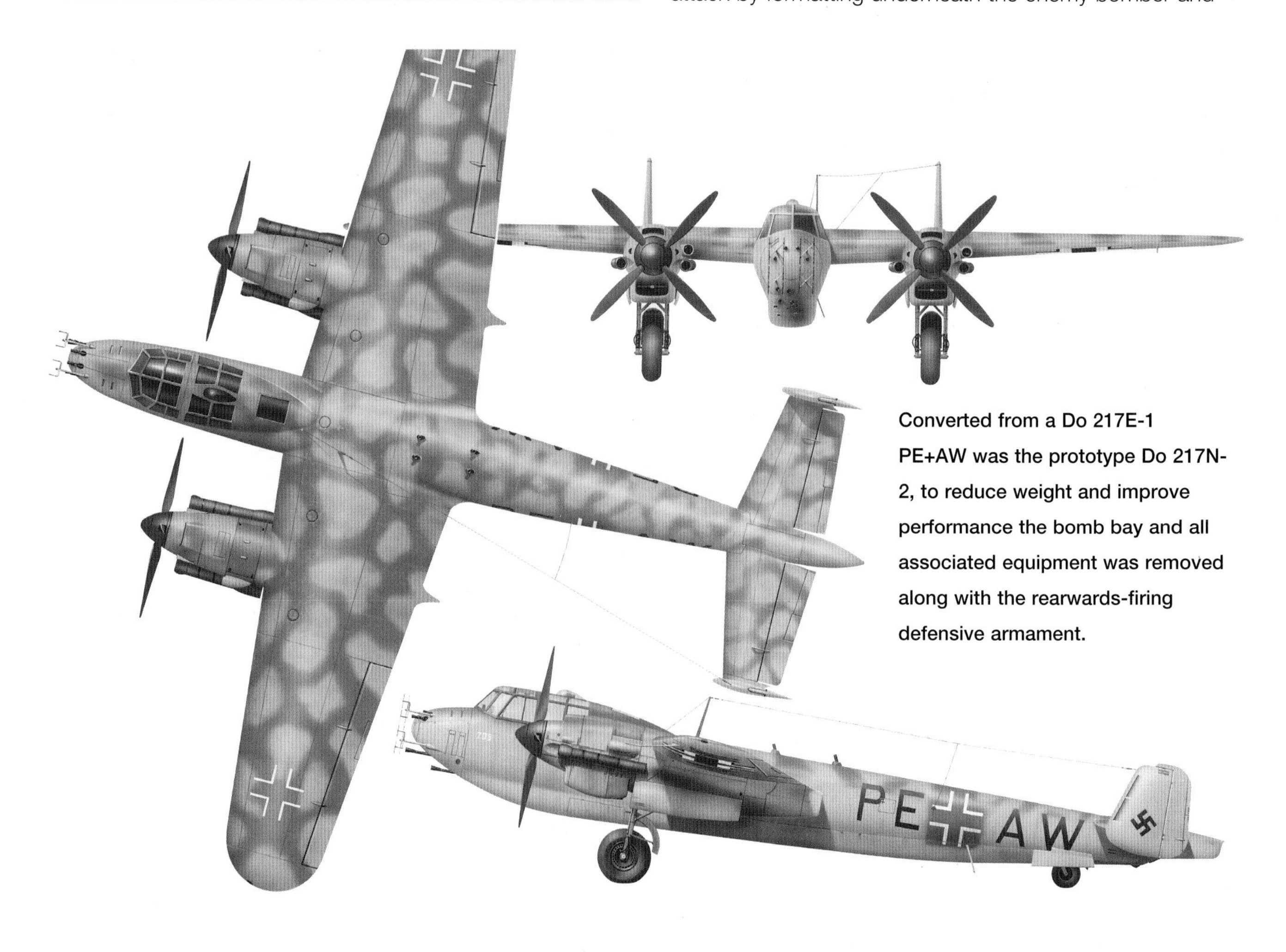

Converted from a Do 217E-1 PE+AW was the prototype Do 217N-2, to reduce weight and improve performance the bomb bay and all associated equipment was removed along with the rearwards-firing defensive armament.

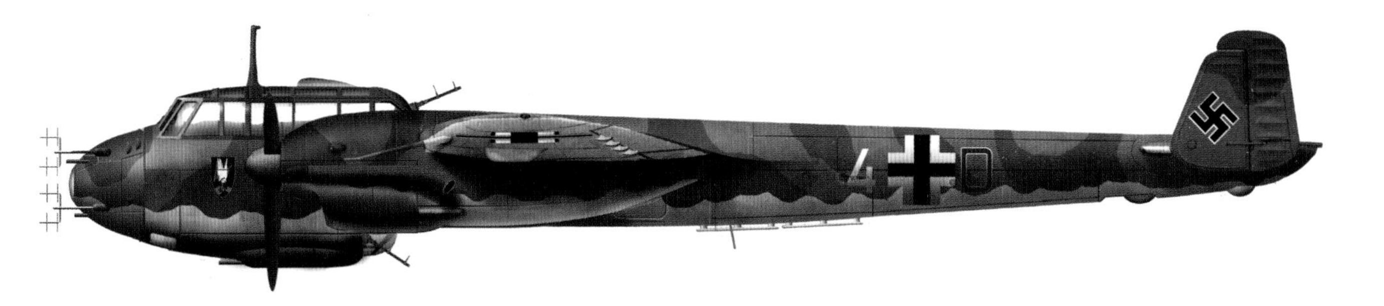

A Dornier Do 215B-5 Kauz III. R4+DC was an aircraft of the Stab. II/Nachtjagdgeschwader 2 based at Leeuwarden in the Netherlands in summer 1942. Visible under the forward fuselage is a weapons tray that increased total forward-firing armament to three MG 17s and three MG/FF cannon.

opening fire. This was easier than closing from astern as the pilot could judge the closing rate more accurately and had time to place the aircraft in the optimum position. The extra 500kg (1,102lb) mass of the system would, however, have an impact of around 10km/h (6.2mph) on the top speed. As well as the FuG 202 or the improved FuG 212 Lichtenstein C-1 radars, the Do 217N-2 would gain the FuG 227 Flensburg and FuG 350 Naxos systems, which could home on RAF bombers' tail warning and navigation radar emissions respectively. To aid the fighter controllers on the ground in their work, Do 217Ns would also be equipped with the FuG 25 IFF (identification friend or foe) system, which allowed them to be clearly identified on radar scopes.

Despite this, the Do 217's career as a night-fighter was brief, its low top speed and sluggish handling preventing it from being truly effective, with some pilots even refusing to fly it operationally. By the beginning of 1944, there were sufficient Bf 110s and Ju 88s available for the Dornier to be withdrawn from the role. In total, 364 Do 217Js and 217Ns were built along with a handful of Do 17Zs. The only other operator was the Italian Air Force, who received six in 1943, but these saw very little action.

A Do 215J-2 banks away from the camera. Built in relatively small numbers, the BMW 801-powered J-2 improved on the J-1 with the installation of the FuG 202 radar. The deletion of the bomb bay compensated for the increased drag from the antenna.

Seeing in the dark

The Lichtenstein airborne radar was developed by the Telefunken company at the request of Generalmajor Josef Kammhuber, head of the Luftwaffe's night-fighter force. Developed by modifying their existing Lichtenstein A radio altimeter, the operating wavelength of 61cm (24in) required a large antenna array. Inside the aircraft, three cathode-ray tubes would indicate the target's range, bearing and elevation. A trial installation was made on a Do 215B-5 to be flown by Oberleutnant Becker, which was ready by the middle of 1941. The first successful interception was of an RAF 301 (Polish) Squadron Wellington bomber on the night of 9 August near the German–Dutch border. Over the next few months, Becker would refine his tactics with the radar operator guiding him in from astern and remaining below the target. Once underneath and matching its speed, he would pull up into a climb and open fire. As his own aircraft slowed, the bomber would fly through the stream of shells, with devastating effect.

Messerschmitt Bf 110C (1938)

Designed to sweep all before it as a long-range fighter clearing the skies for the Luftwaffe's bomber force, the Bf 110 achieved mixed results due to the contradictory nature of its role.

The Messerschmitt Bf 110 was designed to fulfil a unique Reichsluftfahrtministerium (RLM) requirement for a Zerstörer or destroyer, long-range heavily armed fighter. This would clear a path through the enemy's fighter force allowing the following bomber stream to operate unmolested. Ignoring those areas of the RLM's requirements that he considered unachievable, engineer Willy Messerschmitt produced a design for an all-metal monoplane powered by two Daimler-Benz DB 600A engines capable of speeds higher than those of contemporary single-seat fighters.

The first Bf 110 flew in May 1936, barely a year after work had started on the design, and with a maximum

S9+FP flew with II Gruppe, Erprobungsgruppe 210, deployed in raids across the English Channel in the summer of 1940. It features the *Wespen* ('wasp') nose marking of Zerstörergeschwader 1.

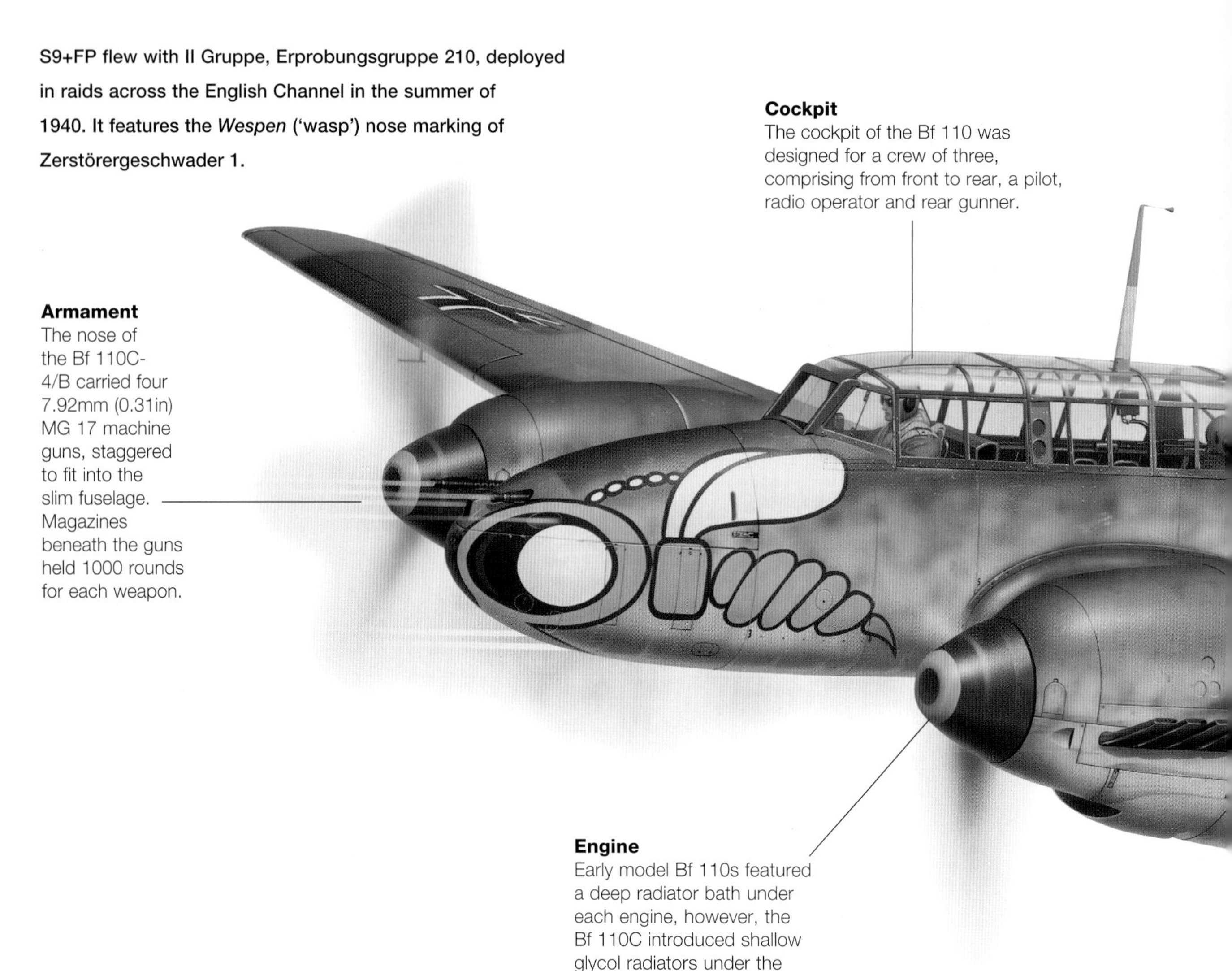

speed of 505km/h (314mph), it was faster than its stablemate, the Bf 109B-2. The test and evaluation pilots did note, though, that the 110 was slower in acceleration and less manoeuvrable than the smaller fighter. Despite this, WWI ace Ernst Udet flying a Bf 109B found it difficult to hold the larger Messerschmitt in his gun sight long enough to guarantee a hit and struggled to stay with the more powerful aircraft in a steep turn. Ignoring the misgivings, Hermann Goering ordered the type into production and the first Bf 110B-1 entered service in late 1938.

Due to a shortage of Daimler-Benz DB 600s, the B variant Bf 110s were powered by Jumo 210 engines, inverted V-12s producing 508kW (681hp) to the DB 600A's 735kW (986hp). With the addition of two 20mm (0.79in) MG FF cannon and four 7.92mm (0.31in) MG 17 machine guns for the forward-firing armament, and a single MG 15 to the rear, the B-1's maximum speed was only 455km/h (283mph).

Heavy fighter wings

With no other option, the Bf 110B equipped some Schweren Jagdgruppen (heavy fighter wings) while Messerschmitt worked on the aircraft's design, until the supply of Daimler-Benz engines improved. This resulted in the Bf 110C, which featured a modified fuselage, and improved lower drag radiators that were moved from the engine nacelle to under the wings. Powered by the DB 601A-1 producing 820kW (1,100hp), the maximum

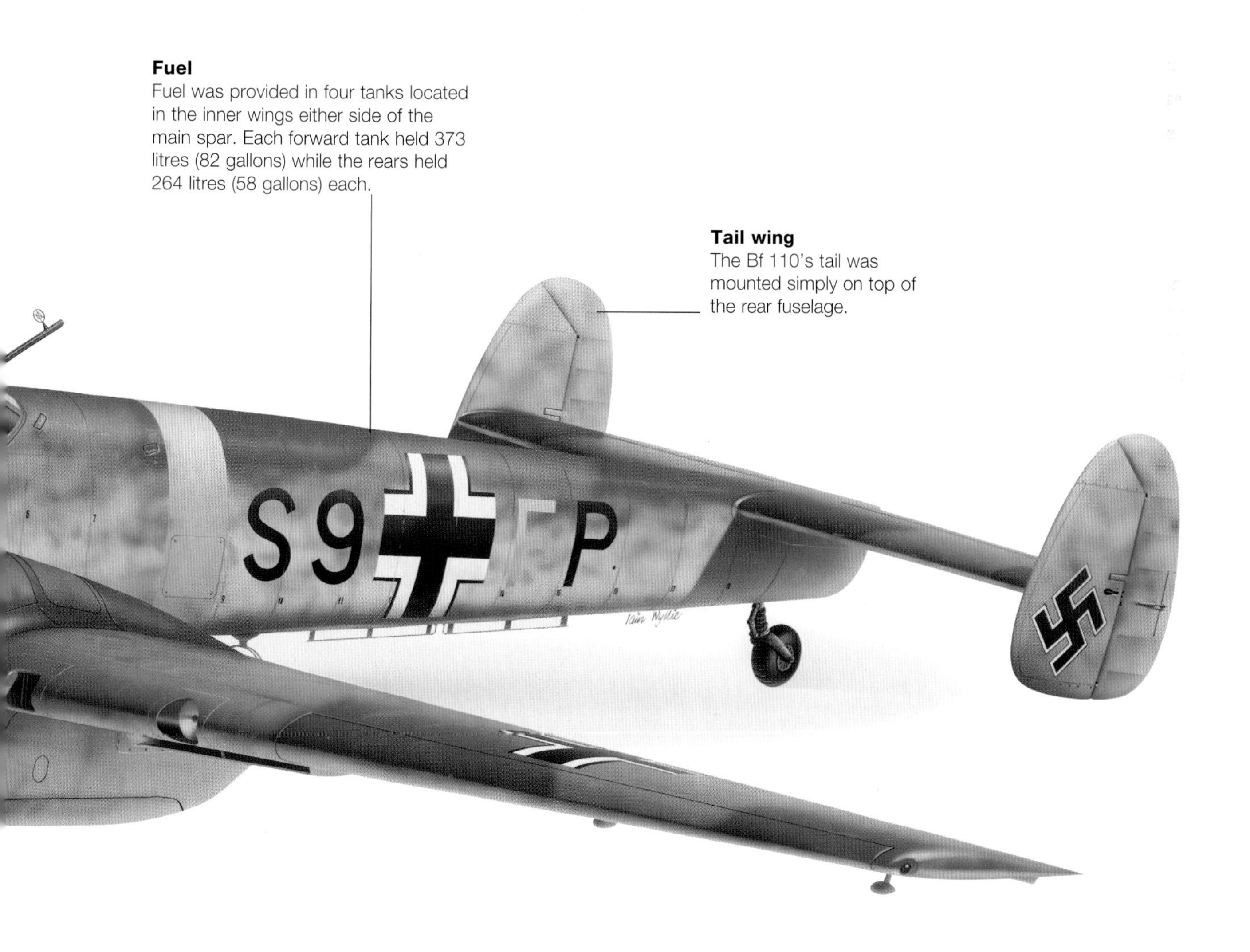

Specifications: Bf 110C-4/B

Type:	Heavy fighter
Dimensions:	Length: 12.65m (41ft 6in); Wingspan: 16.27m (53ft 5in); Height: 3.5m (11ft 6in)
Weight:	6,750kg (14,881lb)
Powerplant:	2 x 821kW (1,116hp) Daimler-Benz DB 601A inverted V-12 piston engine
Maximum speed:	560km/h (348mph)
Range:	775km (482 miles)
Service ceiling:	10,000m (32,808ft)
Crew:	2
Armament:	2 x 20mm (0.79in) MG 151 cannon and 4 x 7.92mm (0.31in) MG 17 machine guns firing forwards; 1 x 7.92mm (0.31in) MG 15 firing aft

speed increased to 560km/h (348mph), restoring the heavy fighter's key strength of speed. Operations over Poland in the opening days of the war demonstrated that the Bf 110C was a capable fighter when used correctly. Escorting Heinkel and Dornier bombers, the *Zerstörer* operated at 6,000m (19,685ft) and used their speed advantage to dive on the slower but more manoeuvrable PZL P.11 fighters of the Polish Air Force. By avoiding being drawn into a turning fight, the Bf 110 Geschwader soon racked up an impressive tally, a two-second burst from their battery of guns being all that was required to down an enemy aircraft.

On one occasion, I.(Z)/Lehrgeschwader 1 downed five Polish fighters over Warsaw while covering a bombing raid by He 111Ps. In the west, meanwhile, Bf 110s helped to destroy 11 of the 22 Wellington bombers that raided Wilhelmshaven on 18 December 1939. Although the raid had reached its target relatively unscathed, the loss of half the force on the return leg helped drive the RAF to adopt night-time bombing raids in an effort to minimize

Bf 110C-3 D5+DT was operated by 9./NJG 3. It retains the *Haifischmaul* ('shark's mouth') marking originally used when the unit had been part of II./ZG 76 from which III./NJG 3 was formed.

future losses. Improvements to the Bf 110C came with the Bf 110C-3, which featured better MG FF cannon, and the Bf 110C-4, which added 9mm (0.35in) armour plate to protect the pilot and gunner. The Bf 110C-4/B gained ETC 250 racks under the central fuselage to enable two 250kg (551lb) bombs to be carried. The increased weight of armour and weaponry saw the Bf 110C-4/B fitted with the DB 601N, which with a greater compression ratio and higher-octane fuel could provide 894kW (1,199hp) for take-off.

Battle of Britain

Retaining the tactics used in Poland and France would initially serve the Bf 110C well during the opening stages of the Battle of Britain. Sweeps were carried out at heights above 6,700m (21,982ft) and with the ability to out-climb the Spitfire Mk I and generally outperform the Hurricane Mk I, the heavy fighter was essentially invulnerable. That would change, though, as Luftwaffe bomber casualties mounted and increasing pressure was applied for the fighters to fly close escort to the bomber squadrons. ZG 2, 26 and 76 would suffer devastating losses operating at medium level and cruising at the speed of the bombers. Unable to outmanoeuvre the Spitfires and Hurricanes, they were also unable to out-accelerate them and became sitting ducks. Further difficulties were caused by the automatic leading-edge slats, which could deploy asymmetrically, causing the aircraft to oscillate laterally when trying to turn with an opponent, throwing the pilot's aim off. Consequently, in August 1940 alone, some 120 Bf 110s were lost during operations, around 40 per cent of the number that had been available at the start of the month. A further 83 were lost through September as the hard-pressed Luftwaffe had no choice but to continue using the twin-engined Zerstörer despite its unsuitability for the task of escort fighter.

Bf 110C-4/B and the similar C-7 with strengthened undercarriage and uprated ETC 500 bomb racks participated in the Battle of Britain in the fighter-bomber, or Jagdbomber, role. These were operated by 1 and 2 Staffel of Erprobungsgruppe (E.Gr.) 210, which had been formed as the service trials unit for the Me 210, although delays to that aircraft saw them initially operate the Bf 110. With their high speed, the Jagdbomber eschewed conventional bomber tactics for low-level precision strikes. Operating in small formations, they were less likely to attract the attention of the RAF's fighter controllers, who concentrated their efforts on directing Spitfires and Hurricanes to the large waves of bombers coming over at medium level.

Initially, E.Gr. 210 carried out attacks on shipping around the south-east of the UK. With their heavy forward-firing armament and bombs, the Bf 110Cs were able to sink 80,000 tons (72,575 tonnes) of shipping in the second half of July 1940. They would then transition to attacks on land targets, primarily airfields and radar installations. The 12 August saw them carrying out three missions. In the morning, they attacked four radar stations along the south coast of England. By midday, E.Gr. 210 had re-launched to attack Manston air base together with Do 17s of KG 2. Their approach at low level being undetected, the force was able to escape unscathed apart from some slight damage to one of the Bf 110Cs. Finally, in the afternoon, an attack was made on Hawkinge airfield, causing considerable damage, again for no losses. E.Gr. 210 would continue to carry out

Middle East adventure

On 1 April 1941, a coup d'état in Iraq overthrew the pro-British Regent, Prince Abd al-Ilah, and installed a pro-Axis regime. To take advantage of the situation, 4./ZG 76 with Bf 110Es and a Staffel of He 111s were dispatched by Germany along with some transport aircraft. Painted with Iraqi markings and routing from Crete via Syria to Mosul airfield, the force, less five He 111s lost on the way, arrived by mid-May. Operations were primarily against British ground positions and the RAF base at Habbaniyah, which was equipped with training and second line aircraft. The Bf 110s shot down two Gloster Gladiators during these raids, however with a weak logistics chain and continued counterattacks by British forces it proved near impossible for the Luftwaffe to keep its aircraft serviceable. By 25 May, the last two operational Bf 110s were lost on a mission and by the end of the month, German forces were evacuated by Ju 90s.

Bf 110C M8+EP of 6./ZG 76 over the English Channel in August 1940. Fighting here would be the first real challenge to the *Zerstörer* concept.

A former II./ZG 76 Bf 110D captured in North Africa and put back into service with the Allies as a communication aircraft bearing the name 'Belle of Berlin' across the nose.

precision raids with its Bf 110s for the rest of the Battle of Britain, although not without loss. Attrition would reduce their numbers and four commanding officers would be killed by the end of operations against the UK. By 27 September, only 10 aircraft were available to take part in a raid near Bristol – half the number involved in earlier sorties.

As German focus shifted from the UK to other theatres, the Bf 110 followed, with ZG 26 taking its C-7 and extended-range D-3 aircraft to Sicily and then Africa and the Eastern Mediterranean. In June 1941, Operation Barbarossa, the invasion of the Soviet Union, saw Bf 110s operating with I. and II./Schnellkampfgeschwader 210, a renamed E.Gr. 210 in the fighter-bomber role, and I. and II./ZG 26 operating as Zerstörer. Here, the big Messerschmitt's poor performance in a turning fight at low level would again prove a weakness as Soviet fighters refused to engage them at altitude, and by April 1942, ZG 26 had been withdrawn to Germany to convert to night-fighters. SKG 210, meanwhile, would predominately operate the Bf 110E, which had two additional bomb racks under each wing and DB 601N-1 engines for increased power.

Limited numbers of aircraft would be a problem for the Bf 110 units in the East with only 51 available for the Zerstörer role at the start of Barbarossa. This situation was not helped by production being paused for the first half of 1942 as an attempt was made to switch production to the Me 210.

As the tide of the war turned against Germany, Bf 110 units were withdrawn to defend the Reich against the daylight bomber threat. These were re-equipped with Bf 110G-2s armed with two 30mm (1.18in) MK 108 cannon in place of the MG 17 machine guns and which could also carry 21cm (8.3in) Werfer-Granate rocket mortars for use against formations of aircraft. Initially, these aircraft achieved considerable success against the waves of B-17s and B-24s. However, as longer-ranged escort fighters entered service for the Allies, the Bf 110 would again be decimated by more manoeuvrable and faster single-seat aircraft. By April 1944, it had been withdrawn from the day anti-bomber role apart from one Staffel in Austria, although it still saw service as a fighter in areas such as Norway, where there was limited fighter opposition, accounting for several RAF Coastal Command long-range patrol aircraft.

Day-fighter

The Bf 110 had a mixed career as a day-fighter; when employed in a way that leveraged its strengths it had shown great promise, able to destroy enemy aircraft in a single high-speed pass. However, when forced to fight against single-seat, single-engine fighters on their own terms it had been found lacking, unable to outmanoeuvre them and struggling to outpace them. As a fast bomber, it was able to carry out pinpoint raids at low level with a degree of defence from its speed but never in sufficient numbers to prove decisive. In total, 6,170 Bf 110 were produced although it's likely this number would have been lower if its successor, the Me 210, had not suffered a protracted development.

Messerschmitt Me 210/410 (1939)

As early as 1938, the Reichsluftfahrtministerium (German Ministry of Aviation, RLM) were planning for a replacement for the Bf 110 heavy fighter. Although tenders were received from both Messerschmitt and Arado Flugzeugwerke, there was little doubt that with the ongoing success of its aircraft the former company would win the contract.

The Me 210 was a considerable advance on the design of its predecessor with the cockpit right at the front of the fuselage sitting above the gun armament and a small bomb bay. The Daimler-Benz DB 601F engines were mounted on the front of the low-set mainplane, the propellers sitting forwards of the fuselage. On either side of the rear fuselage was a MG 131 13mm (0.51in) machine gun, controlled remotely by the gunner.

Impressive on paper, an order was placed for 1,000 Me 210 straight from the drawing board. This would prove optimistic as the first flight of the prototype would demonstrate serious flaws with the design. Taking to the air

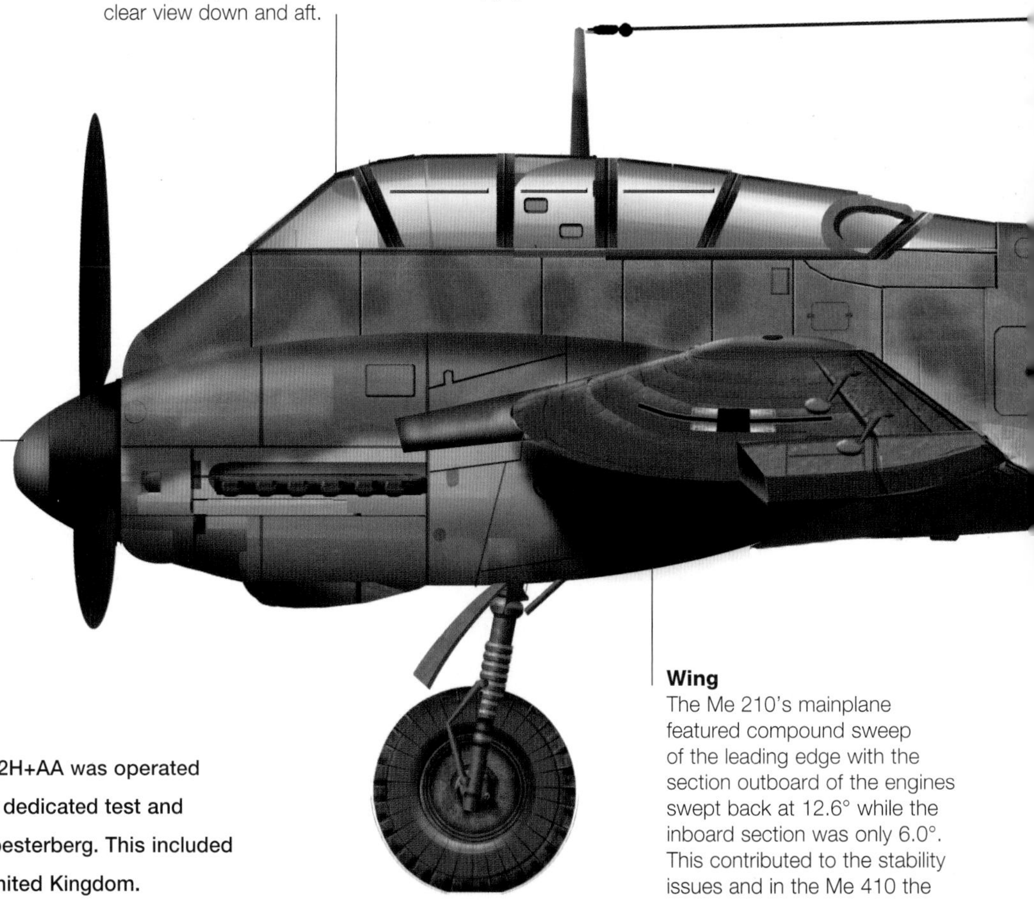

A Messerschmitt Me 210A-1, 2H+AA was operated by Erprobungsstaffel 210, the dedicated test and development unit based at Soesterberg. This included bombing missions over the United Kingdom.

The Me 210 was plagued with vicious and unpredictable handling qualities. The aircraft shown in this photograph was fitted with the longer fuselage, which largely cured the design's major faults.

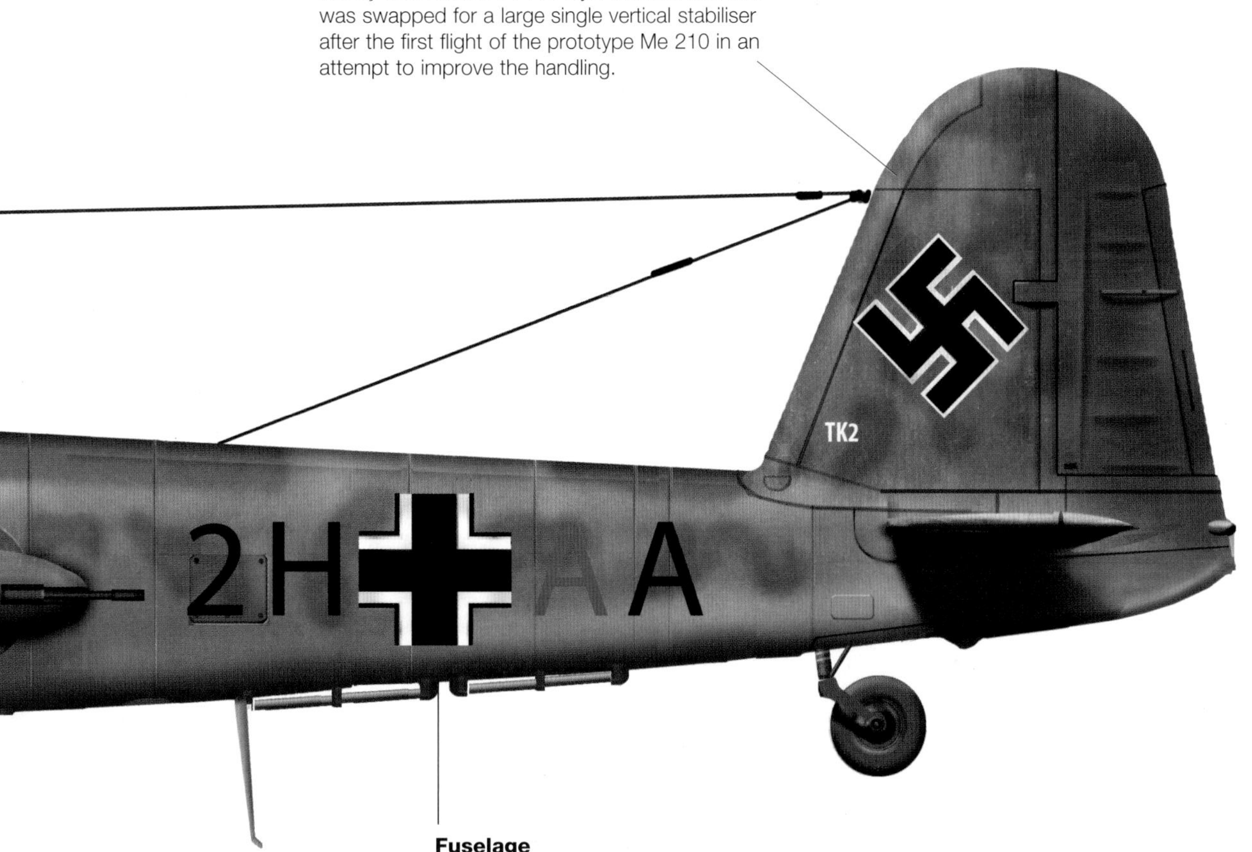

Tail
Initially fitted with a Bf 110 style twin tail unit, this was swapped for a large single vertical stabiliser after the first flight of the prototype Me 210 in an attempt to improve the handling.

Fuselage
Many of the Me 210's handling problems were cured by fitting a longer and deeper rear fuselage that increased the damping effect of the tail surfaces.

Me 410B-1/U4 3U+CC was operated by Zerstörergeschwader 26 in 1944. The aircraft was stationed at Königsberg-Neumark, where it was captured and then evaluated by the RAF.

on 5 September 1939, the test pilot found the Me 210 to be unstable in pitch and yaw to a dangerous degree. With mass orders for an essentially unusable aircraft, a desperate effort was made to rectify the situation. Initially, the twin tail was replaced with a large single tailplane with a tapering horizontal stabilizer, which was a marginal improvement. However, after 90 examples of the Me 210A had been completed the decision was made to cancel the order in early 1942.

Ironically, shortly after this, on 14 March 1942, a much-modified Me 210A-0 flew with a 92cm (36.2in) longer fuselage, leading edge slats on the outer wing sections and numerous detail changes. These transformed the handling, making the 210 stable and removing the tendency to depart into a spin at the first opportunity. The slats had in fact been a feature of the original design but due to its instability had a tendency to deploy asymmetrically as the aircraft entered a spin, exacerbating the situation.

The extended rear fuselage and leading-edge slats were retrofitted to the existing Me 210A, which primarily saw service in the Mediterranean with 16. Staffel of KG 6 and III. Gruppe of ZG 1. Meanwhile, the Me 210 Erprobungsstaffel, or Operational Evaluation Squadron, carried out a series of reconnaissance and bombing operations over the UK, often being intercepted by Typhoons. Further examples were built with the modifications and the 1,070kW (1,455hp) DB 605B engine as the Me 210C by Messerschmitt, and under licence as the Me 210Ca in Hungary, which used it extensively on the Eastern Front.

Specifications: Me 410

Type:	Heavy fighter
Dimensions:	Length: 12.48m (40ft 11.3in); Wingspan: 16.35m (53ft 7.7in); Height: 4.2m (13ft 9.3in)
Weight:	9,651kg (21,277lb)
Powerplant:	2 x 1,290kW (1,730hp) Daimler-Benz DB 603A inverted V-12 piston engine
Maximum speed:	624km/h (388mph)
Range:	1,200km (746 miles)
Service ceiling:	10,000m (32,808ft)
Crew:	2
Armament:	2 x 13mm (0.51in) MG 131 machine guns and 2 x 20mm (0.79in) MG 151 cannon firing forward; 2 x 13mm (0.51in) MG 131 firing to the rear

Me 410 development

Further improvements to the design included a revised main wing with a consistent 5.5-degree sweepback to the leading edge rather than the kinked design of the original, and DB 603A engines producing 1,360kW (1,850hp). This refined model was designated the Me 410 in an attempt to shake off some of the stigma now associated with the 210.

A range of armament options were available for the Me 410, in addition to the standard pairs of MG 151s and MG 17s in the forward fuselage as carried by the Me 210. Thus, a number of additional cannon choices were produced. Aircraft with a /U2 suffix carried an additional two MG 151s in the bomb bay while the /U2/R2 instead carried a pair of 30mm (1.18in) MK 103s or 108s for an even greater punch. Meanwhile, the /U2/R5 carried a total of six MG 151 20mm (0.79in) cannon with the additional four again in the bomb bay. The /U4 variant carried a single 50mm (1.97in) cannon, the BK 5, in the bomb bay. Initially, this was the only armament on the /U4 but later conversions carried the usual fuselage quartet of forward-firing weapons. Aside from these Zerstörer heavy fighter variants there was also the /U1, which carried a single vertical camera in the rear fuselage, losing the MG 17s, followed by A-3 and B-3 dedicated reconnaissance versions with an array of cameras fitted where the bomb bay would normally be. The B-5 anti-shipping aircraft, meanwhile, could carry a variety of torpedoes under the left side of the fuselage and was also generally fitted with the FuG 200 search radar.

Surrounded by Soviet troops after its capture, this Me 410A-2/U4 is armed with a BK 5 50mm (1.97in) cannon. The formidable BK 5 was intended to be used to destroy Allied bombers, with the Me 410 being able to carry 21 rounds for the weapon.

The Me 410 saw the majority of its service engaging with the USAAF's heavy bombers over Europe – a role it was suited to with the various choices of high-calibre forward-firing weaponry. Together with its high speed, this led to many B-17s and B-24s falling to the 410 and through 1943 the aircraft developed a good reputation. Thanks to its service with II/.ZG 26 the 'Hornissengeschwader', it gained the unofficial name of 'Hornisse' or 'Hornet'. However, as long-range escorts became more common in 1944, the Hornet itself fell prey to the more agile North American P-51 Mustang and Republic P-47 Thunderbolt. Although there were still occasional successes, such as on 11 April 1944, when a sortie by the Hornissengeschwader took out 10 B-17s for no losses, the Me 410 was gradually withdrawn from the heavy fighter role to be replaced by single-engined aircraft. It would, however, continue to find a niche in reconnaissance and testing, including of a six-barrelled rotary launcher for the Werfer-Granate 21 210mm (8.3in) anti-bomber rocket. This was placed in the bomb bay with the lowest launcher exposed and could launch all six rockets in around two seconds.

Although more powerful and revised designs were proposed for the Me 410, none of these plans would come to fruition. Total production was 1,189 Me 410s, plus 90 Me 210s and 267 Me 210Cas.

Rear defensive armament

The Me 210 featured a novel rear defensive armament. A pair of MG 131 machine guns was positioned in a teardrop-shaped fairing on either side of the fuselage, and they moved together through +/- 70 degrees of elevation. Each gun could also be trained through 90 degrees of azimuth on its respective side. The second crew member operated the guns via remote control, which gave the Me 210 effective rear firepower without the additional drag of a traditional turret. An electrical interlock prevented the guns firing through the aircraft itself. The rear canopy design incorporated bulges to either side, which gave the gunner an effective field of view of the rear hemisphere and downwards in order to make use of the weapons' full arcs of fire. As with many advanced weapon systems, there were teething problems with the Me 210's rearward-firing guns, but much like the rest of the aircraft it would ultimately perform as intended and be incorporated on the Me 410.

Focke-Wulf Fw-190A (1940)

Focke-Wulf's 'Butcher Bird' was a nasty surprise for the Allies when it appeared over the skies of France and would remain a thorn in their sides until the end of the war.

Although happy with the performance of the Bf 109, the Luftwaffe were conscious of the rapid advances being made in aviation in the late 1930s. Consequently, in early 1938, they asked Focke-Wulf to submit a design for a fighter to supplement Messerschmitt's single-engined racehorse. With his experience in World War I, Focke-Wulf's chief designer, Kurt Tank, saw the need for a cavalry horse better suited to ill-prepared airstrips, maintenance by conscripts, and readily able to absorb the inevitable modifications and upgrades. This would become the Fw 190, one of the most potent fighters of World War II.

First flying in June 1939, the Fw 190 featured several design choices to fulfil Tank's goals without unduly sacrificing performance. An air-cooled radial engine was chosen for its greater resilience to battle damage. The wide-track undercarriage would provide much better

In February 1944, this Fw 190A-4 was serving with 11./JG 5, Luftflotte 5, and was tasked with operations in defence of German installations in occupied Norway.

Kommandogerät
The engine was controlled via the Kommandogerät (command device), which automatically adjusted the manifold pressure, compressor gear shift, fuel mixture and propeller pitch. This allowed the pilot to operate the engine with a single throttle lever.

Engine
The Fw 190A-4 used the BMW 801D-2 engine and was the first version to feature water-methanol injection. This worked to prevent detonation allowing higher boost pressure to be used increasing the power output.

Bomb rack
For use in the fighter bomber role the A-4, and earlier A-3, could be fitted with a centreline rack to carry an SC 250 or SC 500 bomb. This would ultimately lead to the ground attack optimised Fw 190F and G models.

ground handling than the fuselage-mounted, narrow-track set-up used by the Bf 109 and Spitfire. To provide the pilot with the best all-round vision, the Fw 190 featured one of the first bubble canopies to enter service. The design, meanwhile, eschewed the typical control linkages of cables and pulleys for rigid rods, which eliminated maintenance issues caused by the cables stretching.

The first Fw 190A-0s entered service in March 1941 with the test unit Erprobungsstaffel 190 with II./JG 26, the first operational Gruppe, equipped with Fw 190A-1s by October of that year. These featured the BMW 801C engine producing 1,194kW (1,601hp), allowing a top speed of 624km/h (388mph). Armed with four 7.92mm (0.31in) MG 17 machine guns, two of which would soon be replaced by 20mm (0.79in) MG FF cannon in the wings to produce the A-2, the Fw 190 was at this stage almost a complete mystery to the RAF. It was, however, becoming apparent that the Luftwaffe had a new fighter that was at least the equal of the Spitfire Mk V. This would be hammered home on 1 June 1942 when a force of eight Hurricanes escorted by 11 squadrons of Spitfires attacked a target near Bruges in Belgium. Fw 190s of I. and III./JG 26 attacked during the withdrawal and shot down at least eight Spitfires for no losses. The following day saw another seven Spitfires fall to JG 26, again for no losses. It had become obvious that the Fw 190 was superior to anything the Allies had available at that stage of the war.

The main criticism against the Fw 190 was its limited armament. The A-3 would address this by adding an additional pair of 20mm (0.79in) cannon in the outer wing panels, giving a total of four cannon and two machine guns. At the same time, the engine was changed to the BMW 801DG with 1,268kW (1,700hp) of power. The Fw 190A-4 would receive a further boost with the BMW 801D-2 which, with water injection to prevent detonation, could produce 1,567kW (2,101hp) for short periods at low and

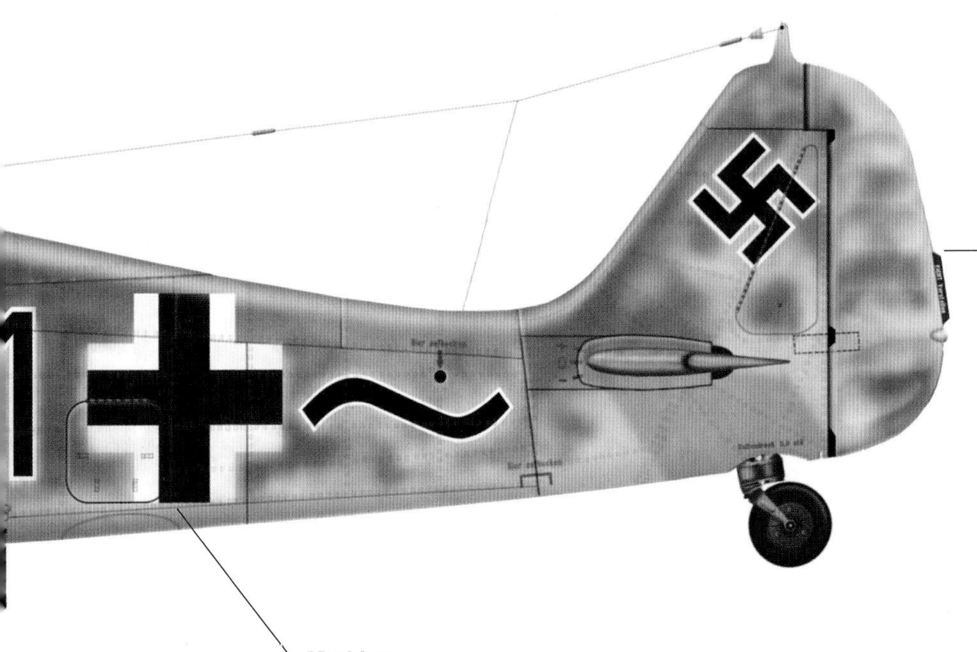

Tailplane
The Fw 190 had a simple fixed tab for trimming the rudder that could only be adjusted on the ground. The horizontal stabiliser meanwhile was a variable incidence unit driven by an electric motor allowing it to be trimmed between -3° and +5°.

Markings
The single wave indicates that this aircraft is from IV Gruppe of the Geschwader while the use of black for the aircraft number denotes the second staffel in that Gruppe. The other two staffeln in IV Gruppe were 10 and 12. From August 1944 10, 11, and 12 staffeln were renumbered 13, 14, and 15 respectively.

medium level. These could also be modified for the fighter-bomber role carrying a 500kg (1,102lb) bomb under the fuselage. 1942 would see the first Fw 190s being issued to the Eastern Front with JG 51 on the Central Sector and JG 5 in the Arctic. Here, the Fw 190A-3 and A-4 would regularly operate in the Jabo fighter-bomber role against the Soviets on the Eastern Front, and British shipping convoys around the North Cape.

With the equally capable Spitfire Mk IX entering widespread service and the US Eighth Air Force's bomber force beginning to make its presence felt, the Fw 190 would start to face greater challenges as 1943 began. With the B-17 and B-24 able to absorb a significant amount of damage, the Fw 190s would need to make multiple passes – a perilous undertaking when attacking a formation. Tactics were soon modified to attack from ahead, where the US bombers' defensive power and armour were weaker.

A further development was the Fw 190A-8/R8, nicknamed the 'Sturmbock' or 'Battering Ram'. This replaced the outer-wing 20mm (0.79in) cannon with MK 108 30mm (1.18in) cannon. A powerful weapon, the low muzzle velocity made it necessary to close to around 183m (600ft) to ensure hits were made. To withstand the intense defensive fire, the Sturmbock featured extensive protection around the engine and pilot, including 30mm (1.18in) armoured glass quarter panels on the canopy. The extra weight, however, made it highly vulnerable to the Eighth's long-range escorts and although some initial success was had, downing 11 Liberators of the 492nd Bomb Group on 7 July 1944, this was seldom repeated.

Operation Airthief

With its obvious superiority over the Spitfire in the spring of 1942, the RAF were desperate for any information they could get on the Fw 190. To provide this, Captain Philip Pinckney, a British Commando, proposed an audacious plan to steal one. This would involve him and Jeffrey Quill, chief Spitfire test pilot, being inserted into northern France by gunboat, then running to hide near a Fw 190 base to fine-tune their next move. As the ground crew carried out an engine run in the early morning, the pair would attack, Pinckney overpowering the engineers and holding off attempts to interfere while Quill took off and flew back to England. Pinckney, meanwhile, would make his way back to the coast, evading attempts to catch him, and row out by night to another gunboat. Fortunately for all involved, on the evening Pinckney submitted his detailed plan for the raid, a lost German pilot landed his Fw 190A-3 in South Wales.

Specifications: Fw 190A-4

Type:	Fighter
Dimensions:	Length: 8.8m (28ft 10.5in); Wingspan: 10.51m (34ft 5.8in); Height: 3.96m (13ft 0in)
Weight:	3,225kg (7,110lb)
Powerplant:	1 x 1,268kW (1,700hp) BMW 801D-2 air-cooled radial engine
Maximum speed:	673km/h (418mph)
Range:	805km (500 miles)
Service ceiling:	13,500m (44,291ft)
Crew:	1
Armament:	2 x 7.92mm (0.31in) MG 17 machine guns; 2 x 20mm (0.79in) MG 151 cannon in the wing roots; 2 x 20mm (0.79in) MG FF cannon in the outer wings

Focke-Wulf Fw 190D

Late 1944, meanwhile, would see the introduction of the Fw 190D. This replaced the BMW radial engine used up to this point with a Junkers Jumo 213 inverted V12 producing 1,670kW (2,240hp) at take-off and 1,491kW (2,000hp) at 3,400m (11,155ft). Cooling was provided by a circular radiator mounted between the propeller and engine and at first glance it could be assumed the Dora was still equipped with a radial engine. Performance was now comparable to the North American P-51D Mustang and the Griffon-powered Spitfire Mk XIV and superior to anything encountered on the Eastern Front. Optimized for the fighter role, the Fw 190D-9 was seen as a stepping stone to the ultimate Jumo-powered Ta 152 version.

Around 19,500 Fw 190s were built. One of the best fighters of WWII, it had done far more than merely supplement the Bf 109.

Belonging to the Geschwaderstab (wing staff) of Jagdeschwader 2 'Richthofen' based at Merzhausen in December of 1944, this Fw 190D-9 was tasked with home defence, primarily the interception and destruction of American daylight bombers.

Messerschmitt Bf 110G (1940)

With an airframe large enough for a crew of three, heavy radar equipment and long endurance, the Bf 110 was an obvious choice when the Luftwaffe found itself in need of a night-fighter.

Although night-fighter was not a role originally planned for the Bf 110, it was one in which it would come to excel. With the Bf 110 making day bombing untenable, the RAF had switched to nocturnal operations by May 1940. Goering responded by forming a night-fighter force to intercept them. I./Nachtjagdgeschwader 1 was formed in July 1940, at first equipped with Bf 110C-4s. Apart from training in night flying, the force initially had little in the way of specialist equipment; tactics involved waiting for bombers to be illuminated by searchlights before making a visual attack. Although this had some success, with I./NJG 1 claiming a Whitely near Saerbeck on 19 July, only 42 British bombers were claimed that year. To address this, a chain of radar stations was developed, extending from Denmark, through occupied Europe to the Swiss border. Each station had one surveillance and two Würzburg pencil-beam radars, one of

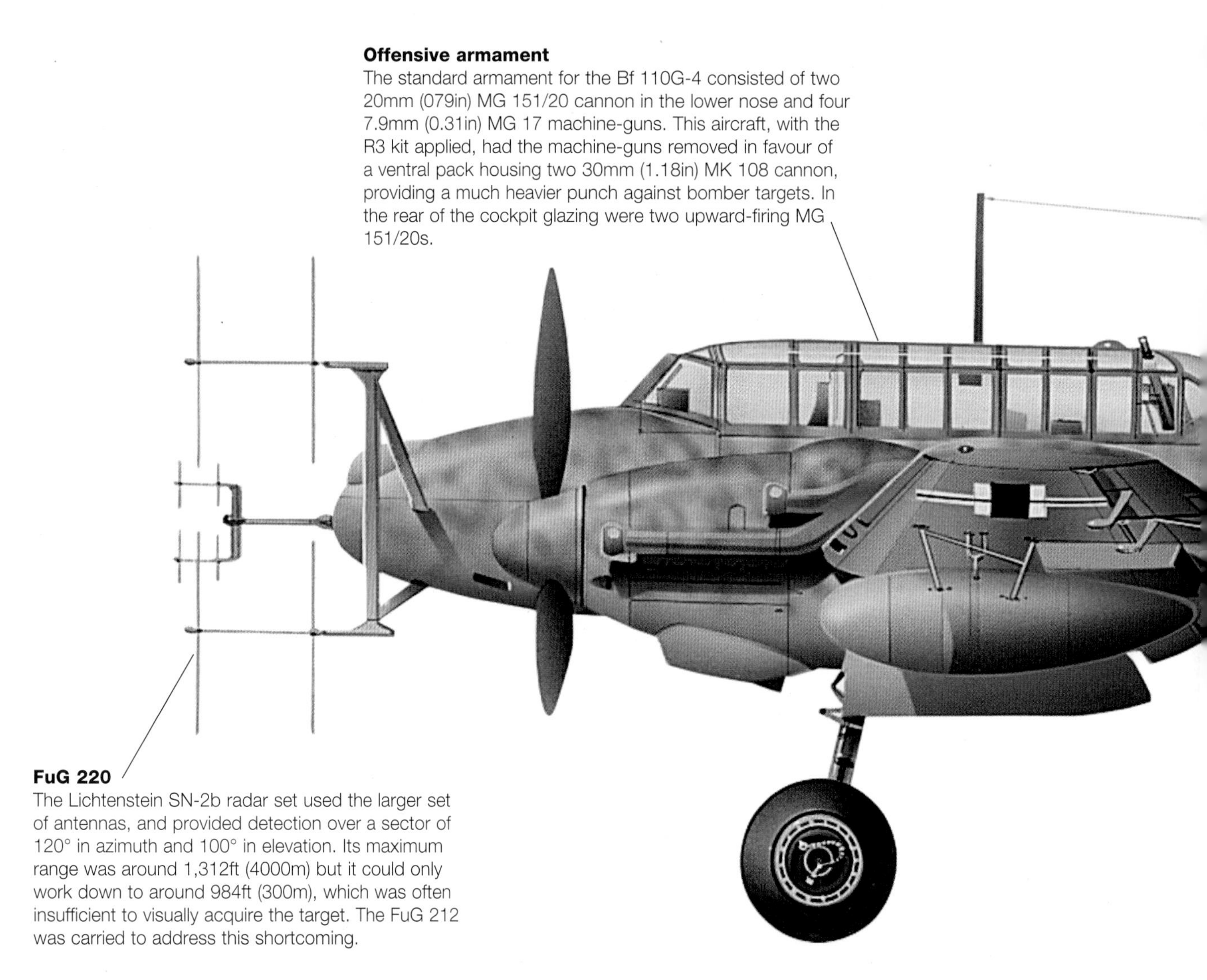

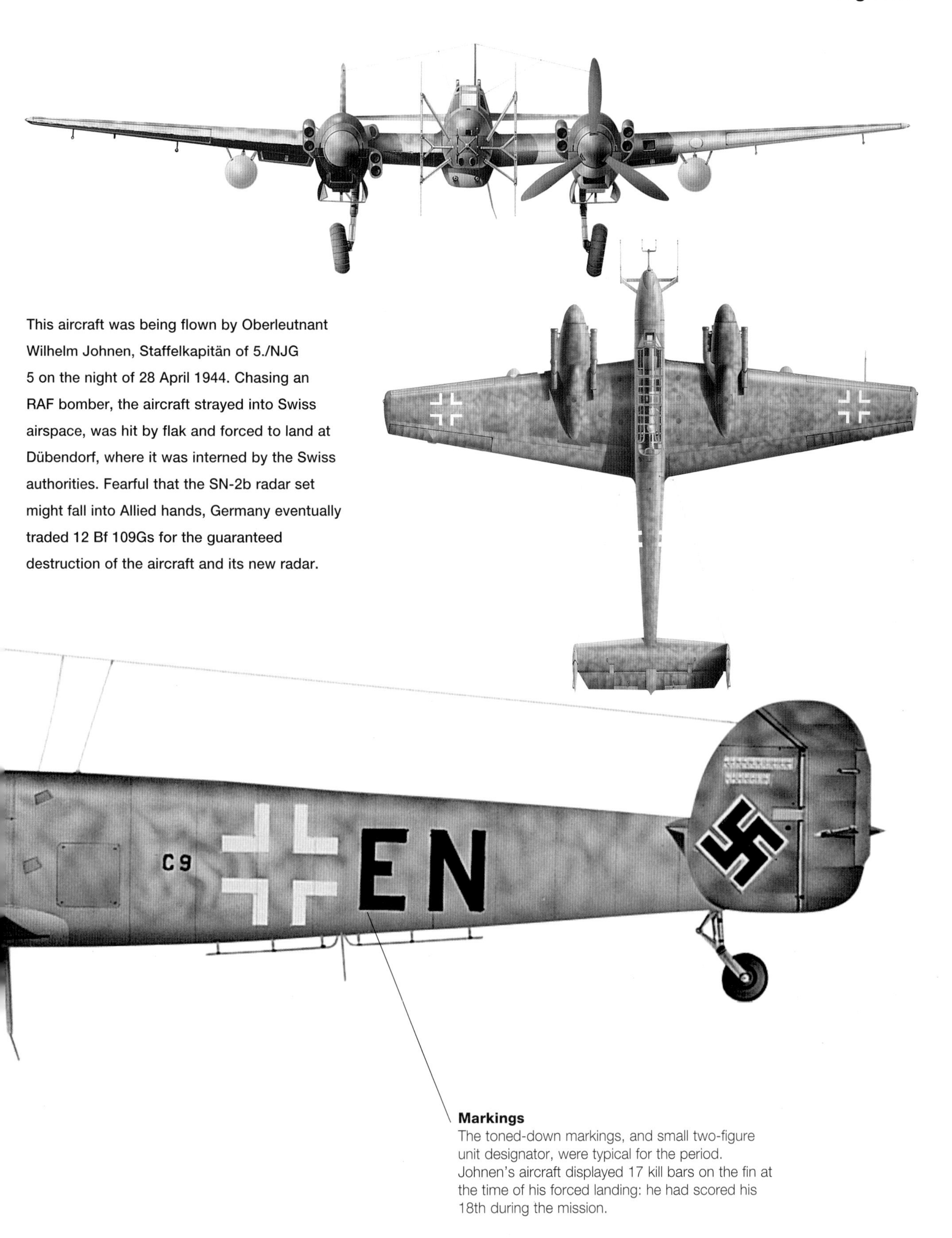

This aircraft was being flown by Oberleutnant Wilhelm Johnen, Staffelkapitän of 5./NJG 5 on the night of 28 April 1944. Chasing an RAF bomber, the aircraft strayed into Swiss airspace, was hit by flak and forced to land at Dübendorf, where it was interned by the Swiss authorities. Fearful that the SN-2b radar set might fall into Allied hands, Germany eventually traded 12 Bf 109Gs for the guaranteed destruction of the aircraft and its new radar.

Markings
The toned-down markings, and small two-figure unit designator, were typical for the period. Johnen's aircraft displayed 17 kill bars on the fin at the time of his forced landing: he had scored his 18th during the mission.

which would track the target and the other a friendly night-fighter. The radar operator would direct the fighter in an attempt to gain visual contact with the bomber. With good handling and a speed advantage over the RAF's heavies, the Bf 110 was preferred to the Do 17s and Ju 88s by most night-fighter pilots.

New night-fighter

A dedicated night-fighter version of the Bf 110 would emerge in 1941 with the F-4. As well as more powerful DB 601F engines producing 970kW (1,301hp), this featured improved night-flying instruments, while two 30mm (1.18in) MK 108 cannon could be fitted in a ventral tray in place of the nose machine guns. By the middle of 1942, the Bf 110F-4a was entering service. This introduced the Telefunken Lichtenstein B/C intercept radar. Using this, the aircraft's crew could now guide themselves for the final stages of an intercept once the target was inside its 3.5km (2.2-mile) range and hold it down to 200m (656ft) – close enough for visual contact on all but the darkest nights.

These improvements did not come without cost, however: the extra weight of the heavier weapons and radar system, together with drag from the antenna array sprouting on the nose, reduced the Bf 110F-4a's top speed to 500km/h (311mph). This made it all but impossible to intercept the RAF's fastest bombers, the Lancaster and Halifax. Additionally, Bomber Command had realized the weakness of the German's fighter-control system, in that each station could only conduct one intercept at a time. Sending all their bombers through one link in the chain, they could saturate its capacity and significantly reduce the number of aircraft that were intercepted. First tried on the night of 30 May 1942 during a raid on Cologne, they experienced a loss rate of 3.8 per cent – significantly lower than during previous raids.

To address the lost performance, the Bf 110 was re-engined again, this time with the 1,100kW (1,475hp) DB 605B-1, over double the power installed on the first Bf 110s. As such, the Bf 110G-4 was the ultimate night-fighter variant although the maximum speed was not much better than the F-4's as more firepower, fuel and electronics were added.

To overcome the RAF's new tactic, night-fighters were vectored to the bomber stream by the ground controller and then engaged in a running battle to the target, guiding themselves with their Lichtenstein radars. In July 1943, the RAF started deploying the technique codenamed 'Window', dropping bundles of aluminium foil to jam

Schräge Musik

Finding the target by night was one problem, shooting it down was another. With limited visual references, it was a much more difficult challenge than by day, with the most desirable option – an attack from astern – complicated by the presence of rear gunners in all the Allies' heavy bombers. *Schräge Musik* was an answer to this. On the Bf 110, this generally featured a pair of MG FF 20mm (0.79in) cannon mounted in the rear of the cockpit and elevated 70 degrees. The fighter could now approach the bomber from underneath, formatting on its silhouette against the stars before opening fire in relative safety. The first successful use of the upward-firing guns is believed to have been by a Bf 110 of II./NJG 5 flown by Hauptmann Schönert in May 1943. It would go on to have a devastating impact on the RAF's Bomber Command, not least because for several months it wasn't even understood how the attacks were taking place, preventing a countermeasure from being developed.

Specifications: Bf 110G-4b

Type:	Night-fighter
Dimensions:	Length: 13.05m (42ft 10in); Wingspan: 16.27m (53ft 5in); Height: 3.5m (11ft 6in)
Weight:	9,900kg (21,826lb)
Powerplant:	2 x 1,100kW (1,475hp) Daimler-Benz DB 605B-1 inverted V-12 piston engine
Maximum speed:	500km/h (311mph)
Range:	1,300km (808 miles
Service ceiling:	11,000m (36,089ft)
Crew:	3
Armament:	2 x 20mm (0.79in) MG 151 cannon and 4 x 7.92mm (0.31in) MG 17 machine guns firing forwards; 2 x 7.92mm (0.31in) MG 15 firing aft

the Luftwaffe's radar. With both ground and airborne radars operating in the same frequency range it became significantly harder for Bf 110 crews to identify targets. The solutions to this were the Lichtenstein SN-2, an airborne radar operating in a lower frequency range; Flensburg, which homed on the RAF bomber's tail warning radar; and Naxos, which did the same for their H2S bombing radar. Although the SN-2 could see through 'Window', some fighters still carried the earlier Lichtenstein C-1 as well due to its lower minimum range, which gave a better chance of getting close enough to engage the target. The Bf 110G-4b had both radars fitted while the Bf 110G-4a only had the C-1.

Taking on the RAF

Although the new equipment gave the Bf 110G-4b a much better chance of finding its target, the performance was now so degraded that an unfortunate pilot could find himself falling out of the sky in a stall as he was outmanoeuvred by a heavy bomber. Despite this, the night-fighter force, which was still over 60 per cent Bf 110s, continued to extract a heavy toll on the RAF. On the night of 30 March 1944, during a raid on Nuremberg, 795 heavy bombers were attacked by 145 Bf 110G-4s and 60 other twin-engined night-fighters, resulting in the loss of 94 aircraft.

This would be the zenith for the night-fighter force, and the Bf 110. As the Allies retook France, the Luftwaffe's radar chain was broken open. At the same time, the RAF developed a jammer for the SN-2 radar, making it almost useless, and from the autumn of 1944, with fuel supplies being targeted, it became increasingly hard to even get aircraft airborne. Worse still, the Bf 110 night-fighters were pressed into acting as day-fighter-bombers, a role the draggy Bf 110F-4s and G-4s were suicidally unsuited for.

Between 1936 and late March 1945, 6,170 Bf 110s were built. Over a third of those were Bf 110G-4s – an indication of how well-suited it had been to the night-fighter role.

Two Bf 110G-4a/R1 of NJG 6, which formed in August of 1943 by redesignating elements of NJG 4 based in Mainz-Finthen in south-west Germany.

Messerschmitt Bf 109G (1942)

By 1942, the Bf 109 had been in service for five years and performance was starting to lag behind that of its contemporaries and adversaries. Designed to address this, the Bf 109G-6 would become the most produced 109 version.

The Bf 109G, known as the 'Gustav' by its Luftwaffe pilots, replaced the earlier model's Daimler-Benz DB 601 engine with the larger DB 605. This increased power by around 224kW (300hp), although it also added 156kg (344lb) to the weight and required a heavier structure to support it.

'Red 13' was a 'Kanonenboote' flown by Feldwebel Heinrich Bartels of 11./JG 27, from Kalamaki, Greece in September 1943. The rudder displays 56 kills of Bartels's eventual tally of 99, most of which had been scored on the Russian Front. It also records his award of the Knight's Cross.

Consequently, although outright speed and high-altitude performance were improved, handling and manoeuvrability took a hit, especially below 3,000m (9,843ft), where the controls became almost unmovable in dives above 645km/h (400mph).

Entering production in late 1942, the Bf 109G-6 was the first to introduce the 13mm (0.51in) MG 131 machine gun to the Bf 109, replacing the 7.92mm (0.31in) MG 17 that had previously been used in the above-engine mountings. This required a bulge either side of the cowling for the spent casing return feed, giving the type a distinctive

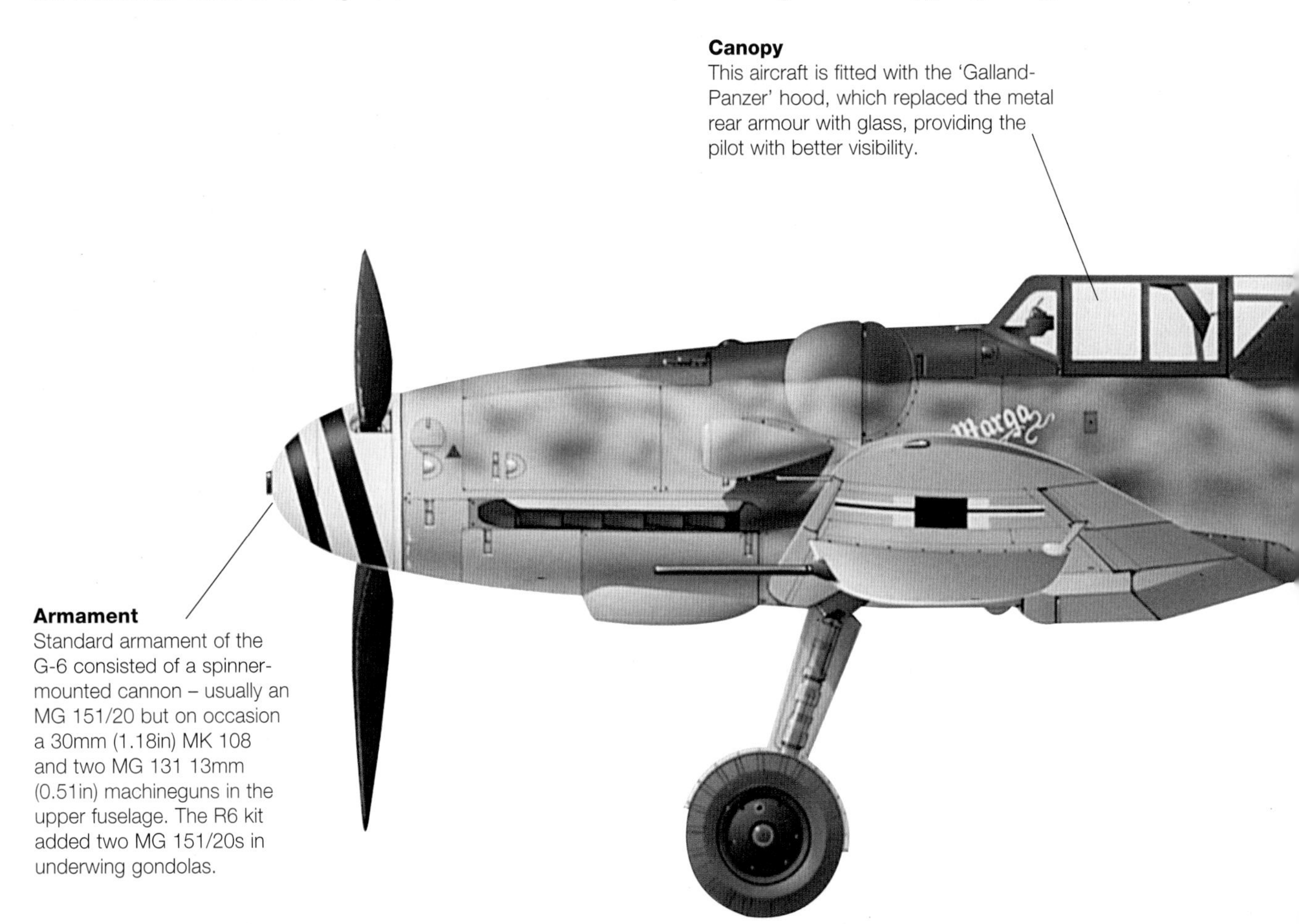

Canopy
This aircraft is fitted with the 'Galland-Panzer' hood, which replaced the metal rear armour with glass, providing the pilot with better visibility.

Armament
Standard armament of the G-6 consisted of a spinner-mounted cannon – usually an MG 151/20 but on occasion a 30mm (1.18in) MK 108 and two MG 131 13mm (0.51in) machineguns in the upper fuselage. The R6 kit added two MG 151/20s in underwing gondolas.

profile and the nickname 'Beule' or 'Bump'. Together with the standard 20mm (0.79in) MG 151/20 cannon firing through the propeller hub, this gave the G-6 better hitting power against the medium and large bombers starting to appear over Germany, although again at the expense of increased weight.

Bf 109G-6s began to be delivered to the front line in February 1943 with deliveries to JG 27, 51, 53, and 77 in the Mediterranean. For service, a tropical filter was fitted to the engine in a lengthened intake extending forwards above the exhaust pipes on the left of the fuselage. In a nod to pilot comfort, mountings for a sunshade were provided to protect them from the sun while waiting to be scrambled. With the Allied landings in Morocco having taken place at the end of 1942, the Jagdgeschwader in North Africa were facing a battle on two fronts. Although initially able to take advantage of the relative inexperience of the USAAF's pilots, the Luftwaffe

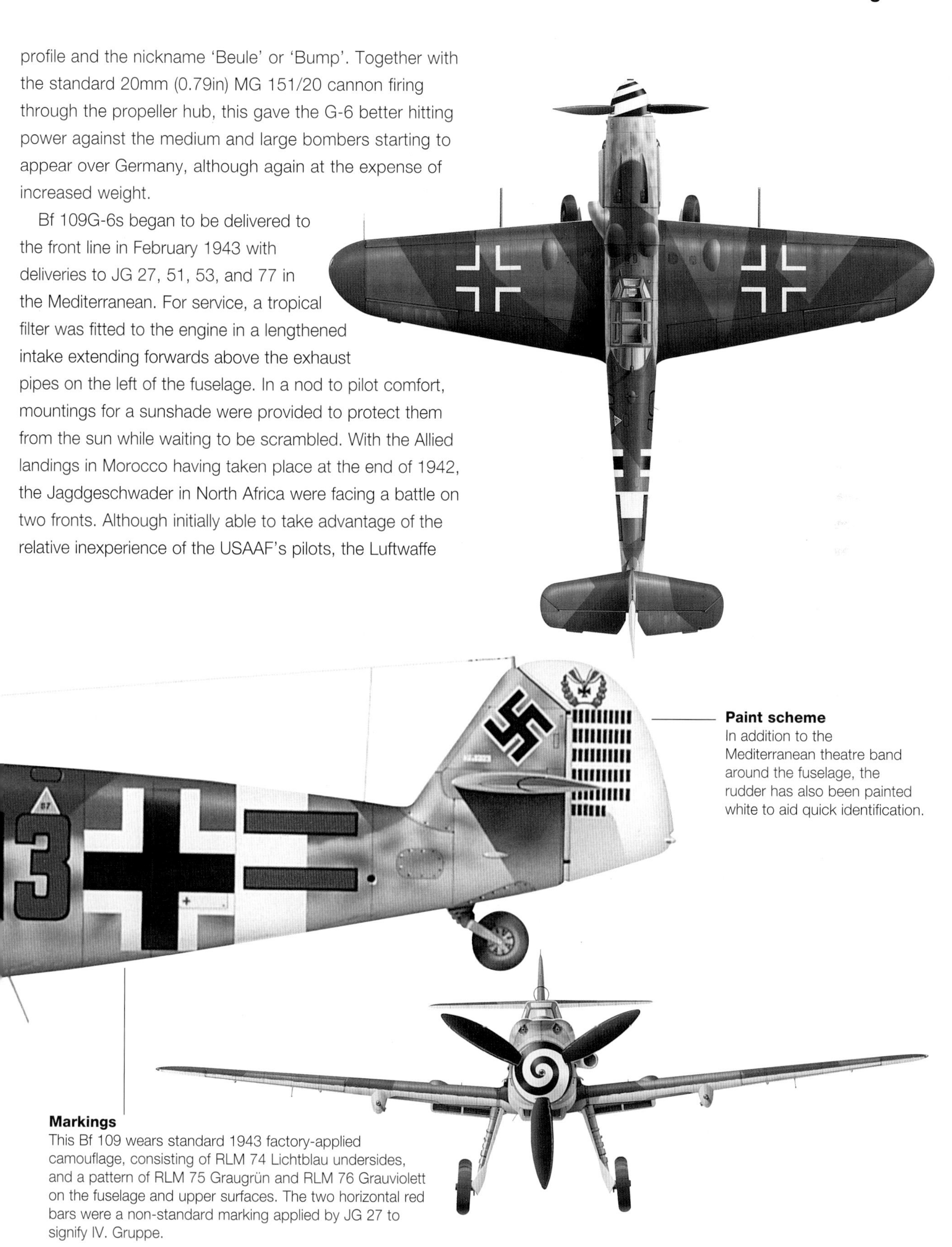

Paint scheme
In addition to the Mediterranean theatre band around the fuselage, the rudder has also been painted white to aid quick identification.

Markings
This Bf 109 wears standard 1943 factory-applied camouflage, consisting of RLM 74 Lichtblau undersides, and a pattern of RLM 75 Graugrün and RLM 76 Grauviolett on the fuselage and upper surfaces. The two horizontal red bars were a non-standard marking applied by JG 27 to signify IV. Gruppe.

Bf 109G-6 flown by Hauptmann Ludwig Franzisket, Gruppenkommandeur of I./JG 27 based in Wagram, Austria in 1944. The aircraft carries the Africa badge of the Gruppe on the nose and a green Reichs Defence Band around the fuselage.

units were soon worn down by the large numbers of better-performing Lockheed P-38 Lightnings and Spitfire Mk IXs.

With some units facing near 100 per cent attrition rates and the situation firmly in the Allies' favour, II./JG 27, II./JG 51 and II./JG 77 retreated to Sicily in advance of the Afrika Korps' surrender in May. There, they faced massed air attack by medium and heavy bombers and on 10 July, Operation Husky saw the Allies carrying out amphibious landings to recapture the island. Within days, the Luftwaffe's fighters were again retreating, this time to the Foggia Airfield Complex in Italy, with the evacuation complete by the end of the month.

Specifications: Bf 109G-6

Type:	Fighter
Dimensions:	Length: 9.02m (29ft 7in); Wingspan: 9.92m (32ft 6.5in); Height: 3.37m (11ft 1in)
Weight:	3,400kg (7,496lb)
Powerplant:	1 x 1,085kW (1,455hp) Daimler-Benz DB 605A inverted V-12 piston engine
Maximum speed:	621km/h (386mph)
Range:	998km (620 miles) (with drop tank)
Service ceiling:	11,735m (38,500ft)
Crew:	1
Armament:	2 x 13mm (0.51in) MG 131 machine guns and 1 x 20mm (0.79in) MG 151 cannon

New cannon

By mid-1943, a further upgrade to the G-6's armament was available in the form of the Umbausatz-4 (U4) modification, which replaced the Mauser MG 151/20 with the Rheinmetall-Borsig-produced MK 108 30mm (1.18in) cannon. Although this only had 60 rounds to the MG 151's 150, the increased explosive charge of the larger round was sufficient to take out any fighter with a single hit. Alternatively, a field modification was available to add a fairing under each wing containing a MG 151/20, earning them the nickname 'Kanonenboot' or 'Gunboat'. Both modifications further reduced the aircraft's manoeuvrability but made it a more effective bomber destroyer, a role that would become more significant as the tide turned against the Axis powers.

Another G-6 armament option intended exclusively for anti-bomber operations was the Pulk-Zerstörer (formation destroyer), which mounted a launch tube under each wing for a 210mm (8.3in) rocket with a 40kg (88lb) warhead. This allowed attacks to be made from outside the range of the bombers' defensive fire and even if direct hits were not obtained it could be enough to break up the bomber formations, making them easier prey for follow-on attacks. However, the lack of suitable weapon site and low speed of the rocket required a skilled operator to successfully employ it.

With the Allied landings in Salerno and the Italian surrender happening near simultaneously in September 1943, the Luftwaffe started to withdraw Bf 109 units to Germany to defend against bomber attacks or to the Eastern Front. By mid-1944, however, the Aeronautica Nazionale

Black Thursday

The US Eighth Air Force's raid on the ball bearing factories in Schweinfurt on 14 October 1943 became known as 'Black Thursday' due to the losses inflicted by the Luftwaffe. With no fighter escort, the B-17s were exposed to the attentions of German fighters for 320km (200 miles) to and from the target area in southern Germany. I./JG 27 flew from Austria to support attacks on the 291 Flying Fortresses with at least six falling to its Bf 109G-6s while a further nine fell to those of JG 2 as they were returning to the UK over the Somme. G-6s of JG 3 and 11, meanwhile, employed Pulk-Zerstörer rockets to attack the massed formations with considerable success. In total, 77 B-17s were lost with a further 121 damaged. Ultimately, this ended deep raids into Germany by the USAAF until the beginning of 1944, when long-range escorts became available, and was in many ways the final strategic victory for the Luftwaffe.

Repubblicana, Fascist Italy's continuation air force, had received a squadron of Bf 109G-6s and formed II° Gruppo Caccia. These operated over northern Italy and into Austria, where combined operations were carried out with the Luftwaffe. The Eastern Front, meanwhile, saw operations by both standard and tropical model G-6s, the latter being employed in the south where the summer dust had a similar effect on aircraft engines as the sands of Africa. Here, the G-6 was also employed in the fighter-bomber, or Jabo, role using the R1 modification that added an ETC 500 /IXb shackle to the centreline, allowing a single 250kg (551lb) bomb to be carried.

Further modifications to the G-6's basic form continued to be introduced through late 1943 and into 1944. To improve visibility to the rear, the armour plate behind the pilot that had been a long-running feature of the Bf-109 was replaced with a revised design that featured a section of armoured glass. Subsequently, the Erla-Haube canopy was introduced by Erla, one of the companies licence-building 109s. This replaced the heavily framed hinged canopy with a lighter section with large curving transparencies, improving the view in all directions bar directly ahead. As an indication of the worsening situation for Germany, 1944 also saw the

Operating from an airfield in Greece, Bf 109G-6s of 7./JG 27 patrol over the Adriatic Sea. The two aircraft to the rear are equipped with tropical filters and underwing cannon gondolas.

introduction of a wooden tailplane to the G-6, distinguished by its taller profile and straight vertical hinge. The new tailplane reduced the demand on strategic materials and was 20 per cent cheaper, but it was heavier overall than the metal unit it replaced. More worryingly, there were difficulties with quality control identified after a Bf 109G-6 shed its wooden tail in July 1944; the subsequent investigation found sub-standard wood and poor assembly techniques. This was likely deliberate in the case of Czech-assembled units. Another modification that began to be introduced with the later G-6 was a taller non-retracting tailwheel. This marginally improved the view forwards on the ground but, more importantly, raised the rudder and elevators into the propwash, giving better controllability during take-off.

While the G-6/U2 and U3 sub-variants had employed nitrous oxide and methanol-water boost respectively to provide a short boost in power, a more permanent increase was needed to counter the growing numbers of USAAF heavy bombers operating at high altitude. This resulted in the Bf 109G-6/AS being equipped with the DB 605AS, featuring the more powerful supercharger from the DB 603 to counter the effect of reducing air pressure on the engine's output. Housing the enlarged supercharger required a redesign of the engine cowling, which also faired in the bulges for the MG 131 machines guns, giving a cleaner look to the airframe that would be retained on later models. The first examples of the 6/AS were delivered to the front line in April 1944, seeing service with III./JG 1, I./JG 5 and II./JG 11 initially, with most assigned to the defence of the Reich.

Night-fighter duties

While the G-6/AS provided a response to the daylight raids by the USAAF there were still night raids by the RAF to contend with. Standard G-6s were the main aircraft used for this along with some FW-190s, all using the *Wilde Sau* ('wild boar') tactic whereby the fighters roamed freely at altitude waiting until they sighted a bomber silhouetted against the searchlights and fires below. Tested by JG 300, the tactic came into wider prominence with the RAF's introduction of chaff in July 1943, which prevented the Luftwaffe's radars being used to control its night-fighters. JG 301 and 302 were soon formed to widen its use, all three units falling under 30. Jagddivision. These three JGs achieved a great deal of success over the summer months before Germany introduced radar capable of seeing through the chaff.

In early 1944, the first of a small number of Bf 109G-6/Ns began to enter service, designed specifically for night-fighting. As well as flame-dampers for the exhausts, a tactical radio and beacon homing equipment, they were also equipped with the FuG 350 Naxos Z, which could home on to the signal of the RAF's H2S bombing radar. This was mounted in a distinctive transparent dome on the spine of the aircraft with a circular indicating unit in the cockpit.

Although used by the specialist NJG 11, the G-6/N's service life was cut short as 30. Jagddivision was dissolved in March 1944. The winter months had proved single-seat fighters poorly suited for night flying in bad weather with pilots becoming lost or unable to land, sometimes bailing out to avoid the near certainty of crashing.

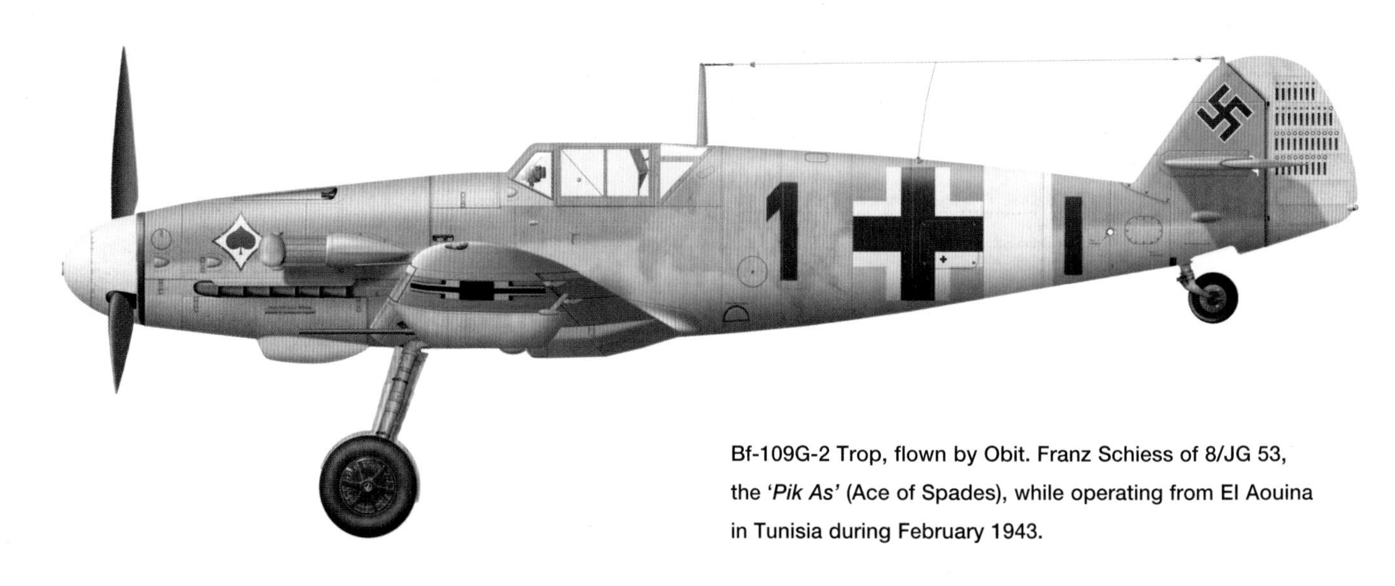

Bf-109G-2 Trop, flown by Oblt. Franz Schiess of 8/JG 53, the '*Pik As*' (Ace of Spades), while operating from El Aouina in Tunisia during February 1943.

A Bf 109G-6/R1 of JG 3 'Udet', armed with a centreline 250kg bomb. Bulges at the rear of the cowling covered the larger breeches of the 13mm (0.51in) MG 131 machine guns adopted on the BF 109G-5 and subsequent versions.

By the end of 1944, the G-6 was being phased out in favour of later G and K models. It had been the mount of multiple aces, including the three highest-scoring of all time: Erich Hartmann, Gerhard Barkhorn and Günther Rall. It had seen service on all the fronts in which Germany was engaged in fighter, fighter-bomber and fighter-reconnaissance units. The G-6 sub-variant was not only the most produced of the Gustavs, but of all Bf 109 production, with around 12,000 built between 1943 and 1944 – over a third of all 109s built. As well as the Luftwaffe and the Aeronautica Nazionale Repubblicana of Italy, the G-6 saw service with Bulgaria, Croatia, Finland, Hungary, Romania, Slovakia and Switzerland, with both Romania and Hungary also building Gustavs, including G-6s.

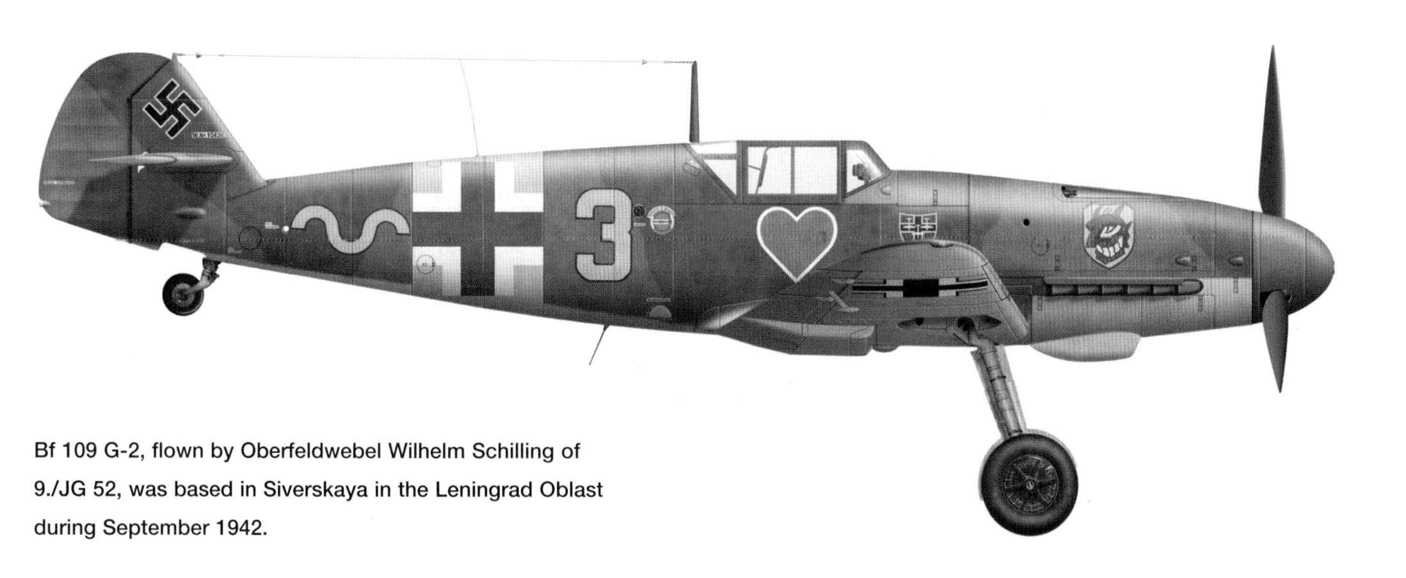

Bf 109 G-2, flown by Oberfeldwebel Wilhelm Schilling of 9./JG 52, was based in Siverskaya in the Leningrad Oblast during September 1942.

Focke-Wulf Fw 190F (1942)

Showing its adaptability, the Fw 190A was easily modified into a fighter-bomber able to support ground troops and carry out lightning raids. With a need to replace the ageing Stuka, dedicated ground-attack models were a natural progression.

The Fw 190 was first used in a ground-attack role in mid-1942 for hit-and-run raids along the English south coast. These used modified Fw 190A-2s and 190A-3s, which had the cannon in the outer wings removed to compensate for the extra weight of a centreline bomb rack carrying either a 250kg (551lb) or 500kg (1,102lb) bomb. Initially operated by 10./JG 2 and 10./JG 26, they attacked shipping in the English Channel during July before moving on to targets inland. These raids could involve more than 60 Fw 190 fighter-bombers with a similar number of escorting Fw 190 fighters.

Late 1942 saw Fw 190s deployed to Tunisia in response to British successes in the North African theatre; they arrived

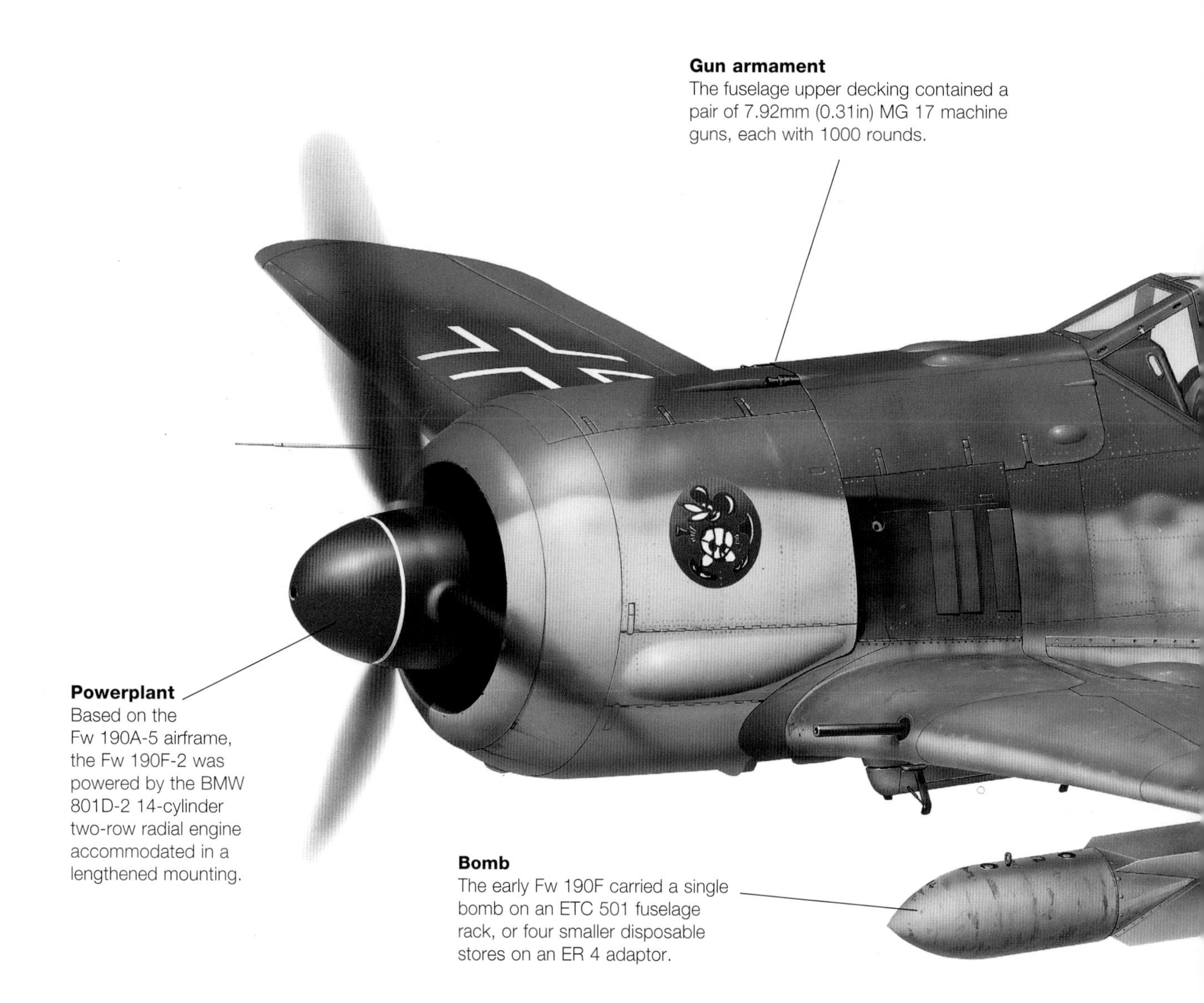

Gun armament
The fuselage upper decking contained a pair of 7.92mm (0.31in) MG 17 machine guns, each with 1000 rounds.

Powerplant
Based on the Fw 190A-5 airframe, the Fw 190F-2 was powered by the BMW 801D-2 14-cylinder two-row radial engine accommodated in a lengthened mounting.

Bomb
The early Fw 190F carried a single bomb on an ETC 501 fuselage rack, or four smaller disposable stores on an ER 4 adaptor.

shortly after the Allied landings in Morocco and Algeria. As well as air-to-air combat, the 190s were again used in the Jabo role, bombing fixed positions and attacking tanks and shipping. In April 1943, III./SKG 10 were issued Fw 190A-5/U8s. These had additional bomb racks under the wings that could also be used to carry 300-litre (79-US gallon) drop tanks and was the forerunner of the dedicated ground-attack Fw 190s.

These were the Fw 190F and Fw 190G, the first of which to enter service was the Fw 190G-1 in 1943. This was originally developed as the Fw 190A-4/U8, which had a centreline ETC 501 bomb rack. The undercarriage was strengthened while fixed armament was limited to the wing root pair of MG 151/20 cannon. In place of the outer cannon were wing racks that could accommodate two 300-litre (79-US gallon) drop tanks. The removal of the nose-mounted guns allowed a larger oil tank to be fitted, increasing the endurance of the BMW 801D-2 used to power the G-1 and allowing it to fulfil the role of Jagdbomber grosser Reichweite or long-range fighter-bomber. Excessive drag from the underwing racks led to the Fw 190G-2, based on the A-5/U8, which had the 18cm (7in) longer nose common to all A-5 models for improved cooling. These were fitted with a Messerschmitt-designed wing rack system that only knocked 3.2km/h (2mph) off the top speed once the tanks had been dropped. A night

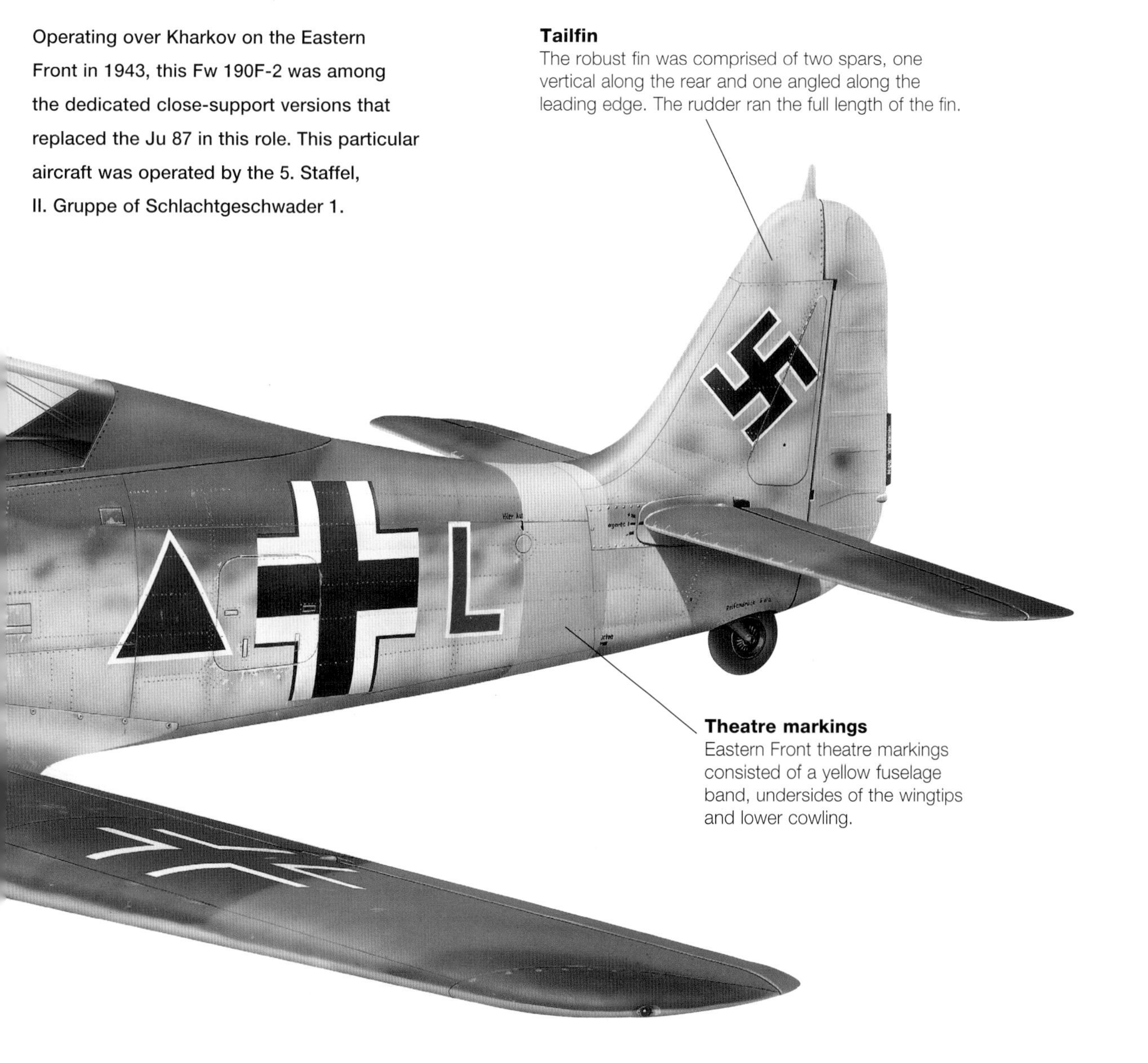

Operating over Kharkov on the Eastern Front in 1943, this Fw 190F-2 was among the dedicated close-support versions that replaced the Ju 87 in this role. This particular aircraft was operated by the 5. Staffel, II. Gruppe of Schlachtgeschwader 1.

capability came with the G-2/N, which essentially added shields and covers to the exhausts to prevent the flames from blinding the pilot or giving away the aircraft's position. For extra hitting power, Fw 190Gs were field-modified with a strengthened undercarriage to allow an increased bomb load to be carried – in some cases up to 1,800kg (3,968lb) of disposable stores.

Operation of the Fw 190A-5/U8 in Tunisia had demonstrated the vulnerability of the ground-attack aircraft's drop tanks to ground fire. At the same time, the success of the A-4/U3 and A-5/U3, with additional armour protection, had pointed the way to the replacement for the Ju 87 as the Luftwaffe's dedicated ground-attack aircraft. Developed into the Fw 190F-1 and F-2 respectively, they carried 360kg (794lb) of additional armour with 6mm (0.24in) plating around the engine, 5mm (0.20in) under the forward fuselage, 8mm (0.31in) behind the fuel tanks, 5mm (0.20in) either side of the cockpit, and on the undercarriage doors. The centreline bomb rack gained the ability to carry four SC50 50kg (110lb) bombs as an alternative to a single large bomb up to 1,800kg (3,968lb) in weight. Meanwhile, each wing carried an ETC 50 rack designed for up to four 50kg (110lb) bombs. Around 270 F-1s and F-2s were produced, including those redesignated from A4/U3 and A5/U3.

Specifications: Fw 190F-8

Type:	Fighter
Dimensions:	Length: 8.95m (29ft 4.4in); Wingspan: 10.51m (34ft 5.8in); Height: 3.96m (12ft 11.9in)
Weight:	3,225kg (7,110lb)
Powerplant:	1 x 1,567kW (2,101hp) BMW 801D-2 air-cooled radial engine
Maximum speed:	635km/h (395mph)
Range:	750km (466 miles)
Service ceiling:	10,600m (34,777ft)
Crew:	1
Armament:	2 x 13mm (0.51in) MG 131 machine guns;
	2 x 20mm (0.79in) MG 151 cannon in the wing roots; up to 1,800kg (3,968lb) of bombs dependent on sub-variant

Night ground-attack

With Allied air dominance in the final year of the war, the Luftwaffe was forced to operate by night in some theatres. To qualify for a Nachtschlachtgruppen – night ground-attack group – pilots had to be able to navigate in the dark to and from the target below 300m (984ft), then pull up to 2,500m (8,202ft) before rolling into a dive as the target was illuminated by the ground liaison officer using flares. With the propeller in fine pitch acting as a brake, the dive was held at around 40 degrees, speed building up to around 615km/h (382mph), until the bomb was released at around 1,000m (3,280ft). The aircraft would then continue its dive through the night back to 300m (984ft) for the egress from the target. Accurate knowledge of the terrain was needed to ensure the level-off wasn't carried out too late. In actual operations, this was generally complicated by a lack of ground liaison officers to illuminate the target. Most operational attacks were therefore made on moonlit nights when the pilot had a chance of spotting the target without assistance.

Fw 190F-8

The majority of the 4,000 F models built were Fw 190F-8s based on the A-8, which was the Fw 190 version produced in the greatest number overall. The main change was the addition of a fuel tank behind the pilot, which required the centreline bomb rack to move forwards to compensate for the change in position of the centre of gravity. As the ground-attack role involved greater time spent flying at low level, the F-8's engine was modified to provide more power in that regime. An array of modifications was developed for the F-8, allowing heavier bombloads to be carried under the wings, various aborted attempts at a torpedo bomber, and mounting 30mm (1.18in) cannon under the outer wings. The most common version, however, was the Fw 190A-8/R1, which had the ETC 50 racks under each wing.

After the initial success of the Fw 190 Jabo in the West, the main focus of the F and G models would be on the Eastern Front, where it was planned to replace the Ju 87. Fw 190As had already been operating in the fighter-bomber role against the Soviets, with good results. The dedicated fighter-bombers, however, arrived in Russia in mid-1943 as the tide of the war was turning against the Axis with

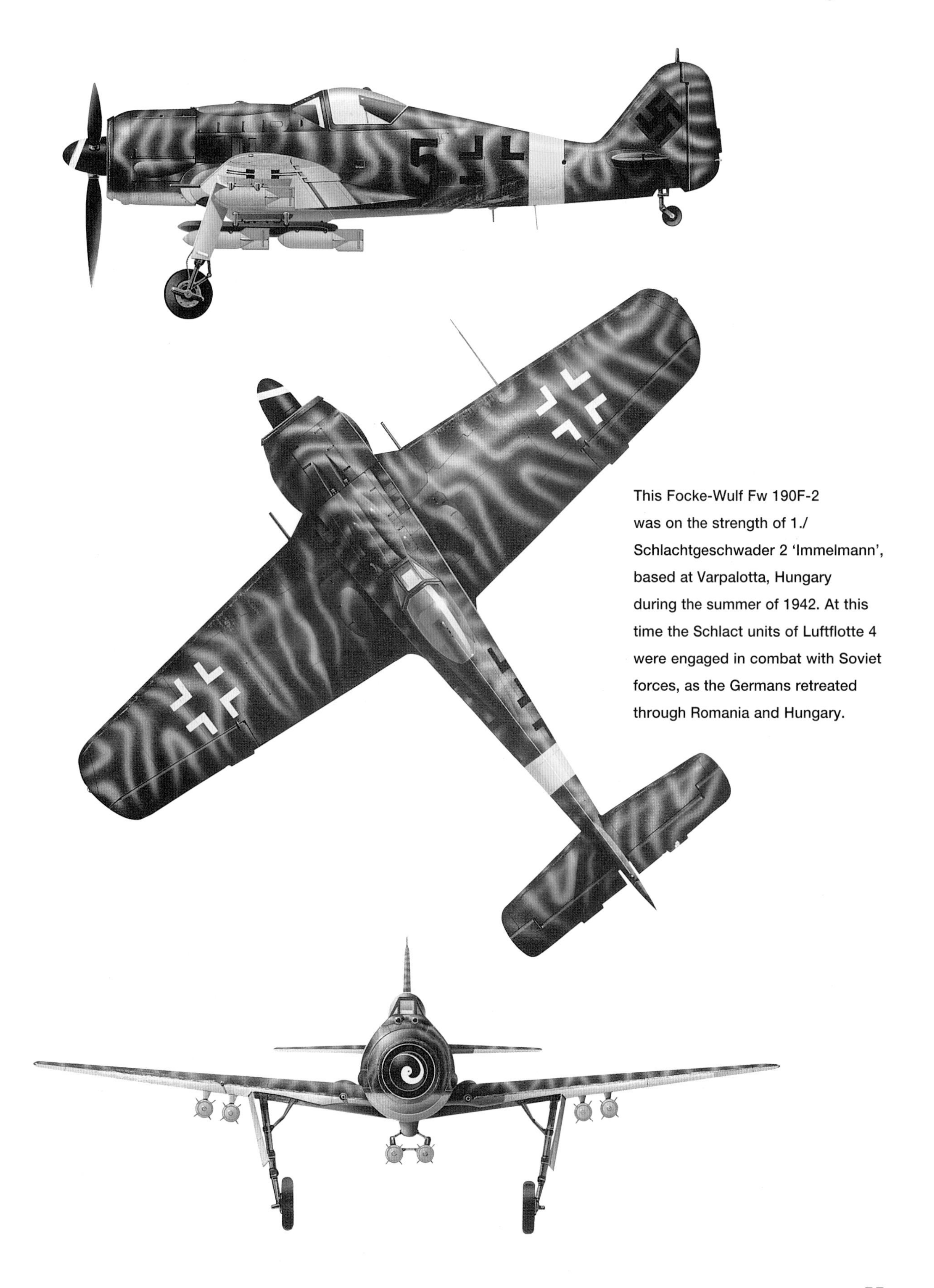

This Focke-Wulf Fw 190F-2 was on the strength of 1./ Schlachtgeschwader 2 'Immelmann', based at Varpalotta, Hungary during the summer of 1942. At this time the Schlact units of Luftflotte 4 were engaged in combat with Soviet forces, as the Germans retreated through Romania and Hungary.

A Fw 190F-8/R3 apparently abandoned on the Eastern Front. Many of the Luftwaffe's aircraft were found in similar conditions at the end of the war as fuel shortages had prevented them from flying.

An Fw-190F-8, as flown by Obfr. Heinrich Zotlotterer, 8/SG-4; Koln-Wahn, Germany; December 1944.

their first major operation being during Operation Citadel, the German attempt to remove the Soviet salient at Kursk, which would become the largest tank battle in history. The Soviet counterattack and subsequent advance saw the two Fw 190-equipped ground-attack Schlachtgeschwader working to stave off troops and armoured vehicles while the German Army prepared defensive positions to fall back to. Early morning reconnaissance sorties would locate the enemy's new positions before Gruppen were allocated their targets for the day. When armour was located it would become the priority, with the Luftwaffe pilots dropping down to around 10m (32.8ft) to make their attack runs at 480km/h (298mph). From this position, as the target vehicle disappeared below the nose, the bomb would be released, either hitting it directly or ricocheting off the ground first.

A one-second delay in the fuse allowed the aircraft to avoid self-damage from the subsequent explosion. Soft-skinned vehicles could be attacked with machine guns and cannon, helping to break the enemy's logistics chain and saving the 250kg (551lb) bombs for tanks. With the front line less than half an hour away, pilots could fly as many as eight sorties a day.

Maintenance in the field became difficult during the harsh Russian winter with a variety of measures being taken to keep the aircraft warm enough to work on, or even just start without damaging the engine. These included heated tents to cover the front of the aircraft and petrol-powered heaters that blew hot air directly into the engine bay. In extremis, small fires were even lit under the engines. The air-cooled radial engines in the Fw 190s at least had an advantage over the liquid-cooled ones of the Bf 109s, which had several gallons of fluids to prevent from freezing in the block and rendering it unusable.

As the Soviet advance gained momentum into 1944, the number of Fw 190F- and Fw 190G-equipped Gruppen on the Eastern Front increased significantly from four in May 1943 to nine in Schlachtgeschwader (SG) 1, 2, 5, 10 and

77 some 12 months later. By this point, there were around 700 Fw 190s serving with combat units, almost 400 of which were assigned to ground-attack Gruppen. A further five Gruppen were converting from the Ju 87 Stuka to the Fw 190G as the ground war in the East became ever more bitterly contested.

As the second front opened in the West, the Fw 190 fighter-bombers would be called into a desperate defence there as well. On D-Day, III./SG4, equipped with Fw 190Fs, moved to forward operating bases at Laval and Tours. Before even arriving, they were intercepted by USAAF fighters for the loss of five aircraft. Despite this setback, 9./SG4 was able to sortie 13 aircraft to attack the landing beaches that afternoon. However, with 3,000 Allied fighters operating in the area, only two brief passes were made on the target area. The pattern would be repeated the following day with the Fw 190Fs unable to push home their attacks and losing more aircraft to marauding Allied fighters. This would be a running theme for the rest of the war as the ground-attack Fw 190s' performance was hampered by the weight of weapons and armour, while even the fighter variants were overwhelmed by the sheer number of Allied fighters.

With the war entering its final phase, most Fw 190 fighter-bomber units were transferred to the Eastern Front in an attempt to hold back the Soviets for as long as possible. Fighting on this front remained fierce until the end of the war with Army Group North holding out in Latvia and cut off from the rest of Germany from July 1944 until the surrender in May 1945. With them were II. and III./SG 3 with Fw 190Fs and I. and II./JG 54 with Fw 190As, who fought valiantly, with the latter accounting for 31 Soviet aircraft on one day in October alone. However, as the end became inevitable, the Fw 190s were used to evacuate personnel, up to four ground crew being squeezed into the rear fuselage for a 900km (559-mile) flight to Schleswig-Holstein.

Despite the Allies' accelerating advance there were more than 1,600 Fw 190s in service in the final month of WWII, almost equally divided between fighter and fighter-bomber variants. This was in part due to improvements in factory output the previous year building up a healthy reserve, but also to a lack of fuel making it impossible for some units to actually fly their aircraft. Consequently, while JG 3, and I. and II./SG 1 were involved in the final struggle for Berlin, many Fw 190s simply remained hidden in their dispersals awaiting the final surrender.

American soldiers examine a captured 'Mistel' composite aircraft at Bernberg, Germany, in May 1945. The Fw 190 has been mated with a Ju 88; the lack of warhead indicates this combination was in fact for training the pilots of the piggyback aircraft.

Piloted by Oberleutnant Walter Nowotny, among others, this Fw 190A-5 was part of 1./Jagdgeschwader 54. The bulk of Nowotny's 258 victories were gained in the Focke-Wulf.

This Focke-Wulf Fw 190A-6 was flown by Leutnant Emil Lang, who served with 5./Jagdgeschwader 54 'Grünherz' ('Green Hearts'). The aircraft was operated on the Eastern Front in October 1943.

Named 'Christl', this Focke-Wulf Fw 190D-9 was the mount of Major Gerhard Barkhorn of the Stab./Jagdgeschwader 6, flying from Sorau in February 1945.

Heinkel He 219 (1942)

Developed from an internal study, the He 219 was, along with the Northrop P-61 Black Widow, one of only two dedicated night-fighters developed in World War II. Its great potential was, however, squandered by in-fighting and production difficulties.

The Heinkel 219 started life in 1940 as a design project by the head office with the title 'Projekt 1064'. Intended as a multi-role aircraft capable of fighter-, attack-, reconnaissance- and torpedo-bomber roles, it featured several then novel features, including a tricycle undercarriage, a high-mounted wing and a pressurized tandem cockpit. Despite being almost exactly what the Luftwaffe needed, concern over these novel features, combined with a lack of strategic planning, led the High Command to reject it. There matters would have lain if it weren't for General der Nachtjägd Josef Kammhuber, who was struggling to obtain an advanced night-fighter for his force. After a meeting with Hitler in late 1941, he was given carte blanche to proceed with fulfilling his requirement and shortly after, a contract was signed with Heinkel for development of the He 219.

The He 219 prototype first flew just over a year later, in November 1942. An all-metal stressed-skin monoplane, few changes had been made to the original Projekt 1064 design. The Daimler-Benz DB 603 engines were mounted in under-wing nacelles that also housed the main gear, while an integral boarding ladder provided access to the cockpit,

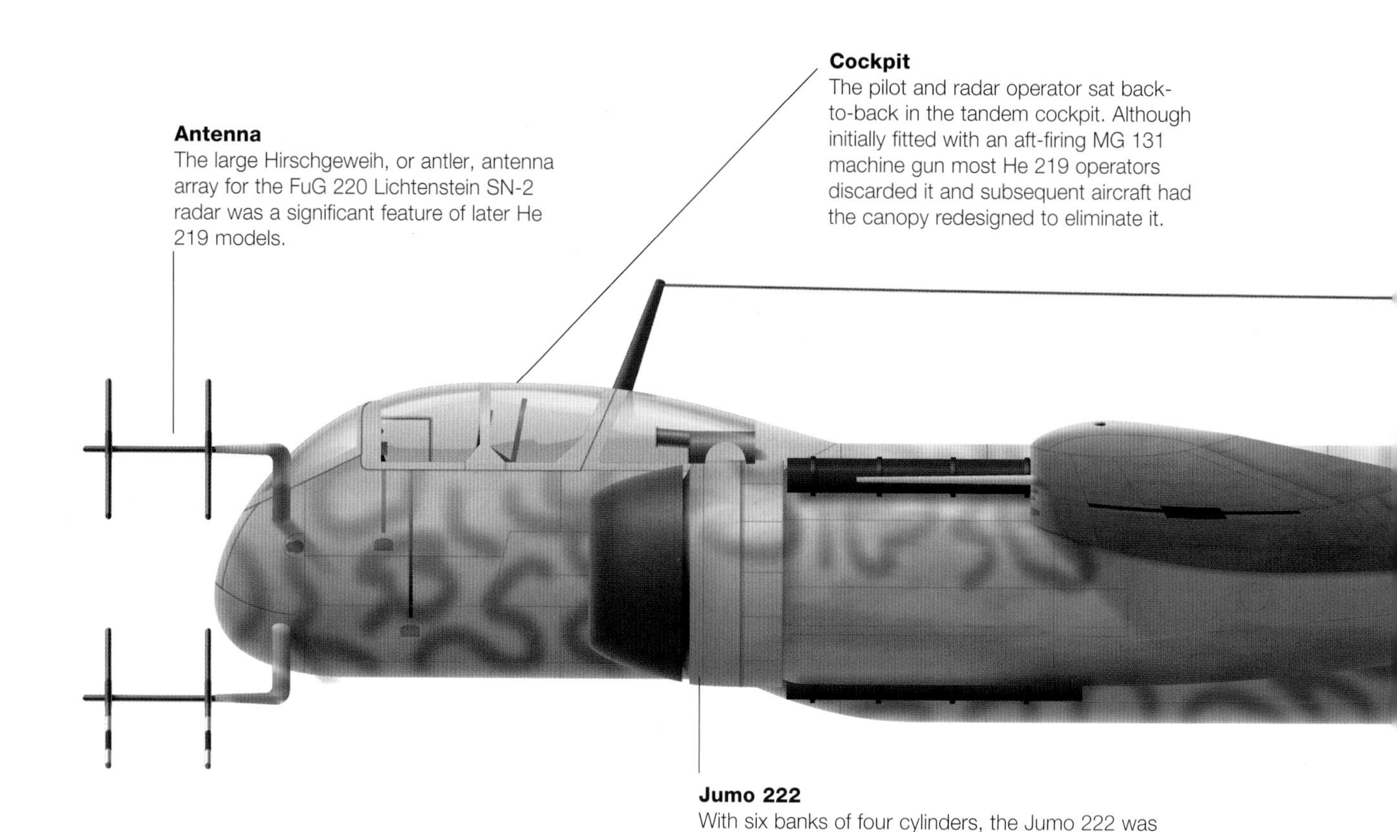

where the pilot and radar operator sat back-to-back under a large bubble canopy. Armament initially consisted of two wing root-mounted 20mm (0.79in) MK 151 cannon and a further two in a ventral pack, where the original design had had a bomb bay. There were also twin MG 131 machine guns in remotely controlled barbettes firing to the rear, one above and one below the fuselage aft of the mainplane. However, these would not be included on later production aircraft as the hydraulic system struggled to move them against the airflow at operational speeds.

Top performance

Performance was found to be outstanding apart from some issues with roll and yaw stability, which were rectified from the third airframe onwards by extending the tail and enlarging the vertical stabilizers. Although urgently needed, the He 219 then entered a protracted development phase that explored 29 different armament variations in an attempt to meet constantly changing diktats from the Luftwaffe. Progress was further delayed by air raids destroying the majority of the production drawings, and the actions of Generalfeldmarschall Milch, who was in charge of aircraft production and bitterly opposed to the He 219's development as a distraction from his priorities. Consequently, the 'Uhu', or 'Owl', didn't begin to enter service until May 1943. These were He 219A-0/R1 and / R2 models, which differed from the prototypes by housing four 30mm (1.18in) cannon in the ventral pack for greater hitting power and in at least some aircraft, compressed air-powered ejection seats for the two crew. It also now carried the FuG 212 Lichtenstein C-1 radar as standard.

I./NJG 1 based at Venlo in the Netherlands was the first unit to receive He 219A-0s and the type had an auspicious entry into service. The first operational sortie on the night of 11 June 1943 claimed five RAF heavy bombers, the only blot on this impressive opening record being when the pilot misjudged the landing, writing off the aircraft in the process, although both he and the radar operator

An artist's interpretation of the He 219B Uhu with Jumo 222 engine and extended wingspan. Note the large, ducted spinner and numerous exhaust pipes to accommodate the engine's 24 cylinders.

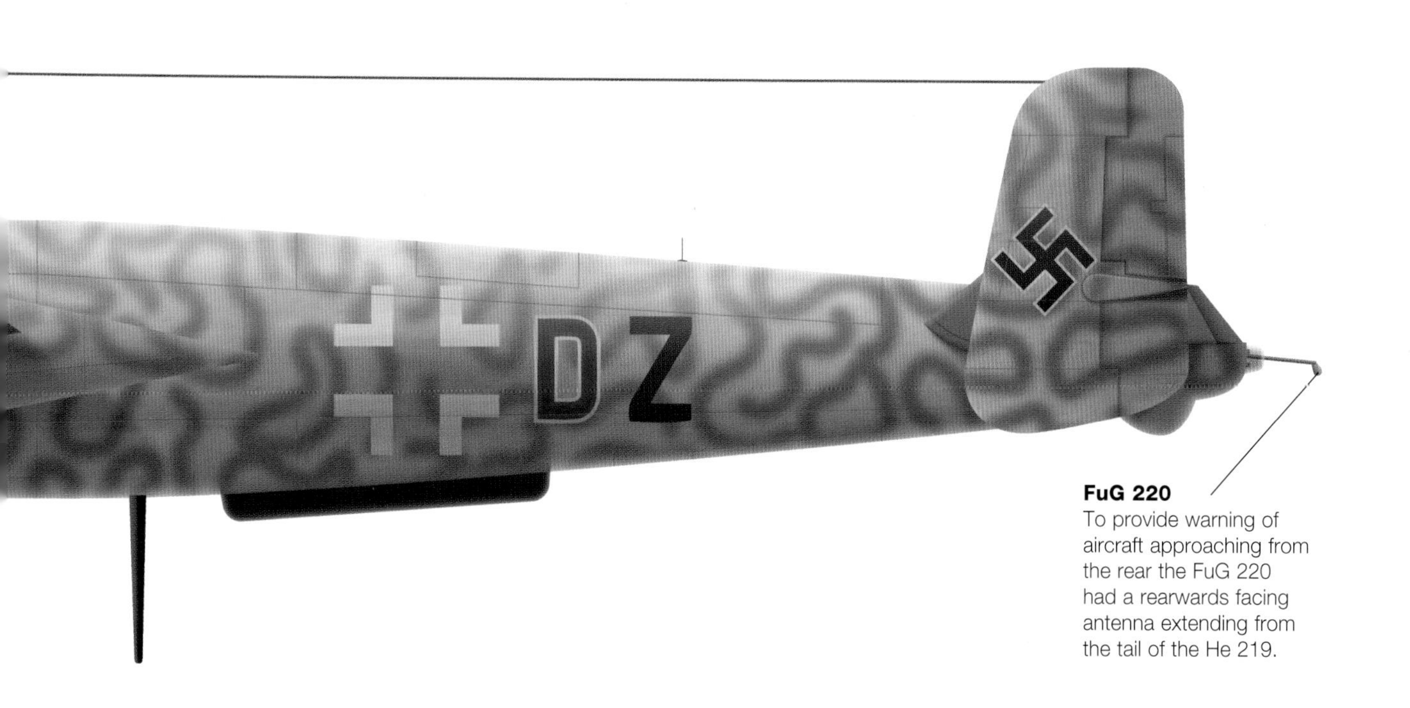

Seen at Freeman Field, Indiana, this He 219A-2 was captured in Denmark and transported to the USA onboard HMS *Reaper*.

Specifications: He 219A-7

Type:	Night-fighter
Dimensions:	Length: 16.34m (53ft 7.3in); Wingspan: 18.5m (60ft 8.4in); Height: 4.10m (13ft 5.4in)
Weight:	15,100kg (33,289lb)
Powerplant:	2 x 1,416kW (1,899hp) Daimler-Benz DB 603G 12-cylinder liquid-cooled engines
Maximum speed:	585km/h (363mph)
Range:	1,850km (1,150mph)
Service ceiling:	9,800m (32,152ft)
Crew:	2
Armament:	2 x 20mm (0.79in) MG 151 cannon in the wing roots; 2 x 20mm (0.79in) MG 151 cannon in the ventral bay; 2 x 30mm (1.18in) MK 108 cannon at 65 degrees in the rear fuselage

escaped unscathed. Over the next six sorties, a further 20 RAF bombers were claimed and Kammhuber pushed for full-scale production to commence. However, a shortage of parts, including engines, meant this was slow to build up and only just hit a tenth of the 100 aircraft per month originally planned.

As a result, I./NJG 1 would be the only unit fully equipped with He 219s, with a handful reaching other Nachtjagdgeschwader. The production models featured longer engine nacelles, each containing: 390 litres (103 US gallons) of fuel in addition to the 2,590 litres (684 US gallons) in the three fuselage tanks; a FuG 220 Lichtenstein SN-2 radar in addition to the C-1 to provide a measure of anti-chaff capability; a tail warning radar; and in common with other twin-engined night-fighters, *Schräge Musik* obliquely firing guns mounted in the rear fuselage. The production aircraft continued to demonstrate the He 219's capabilities as a night-fighter with I./NJG 1 pilots often claiming multiple aircraft in a single sortie – Oberfeldwebel Wilhelm Morlock going as far as six kills and one probable in the space of only 12 minutes.

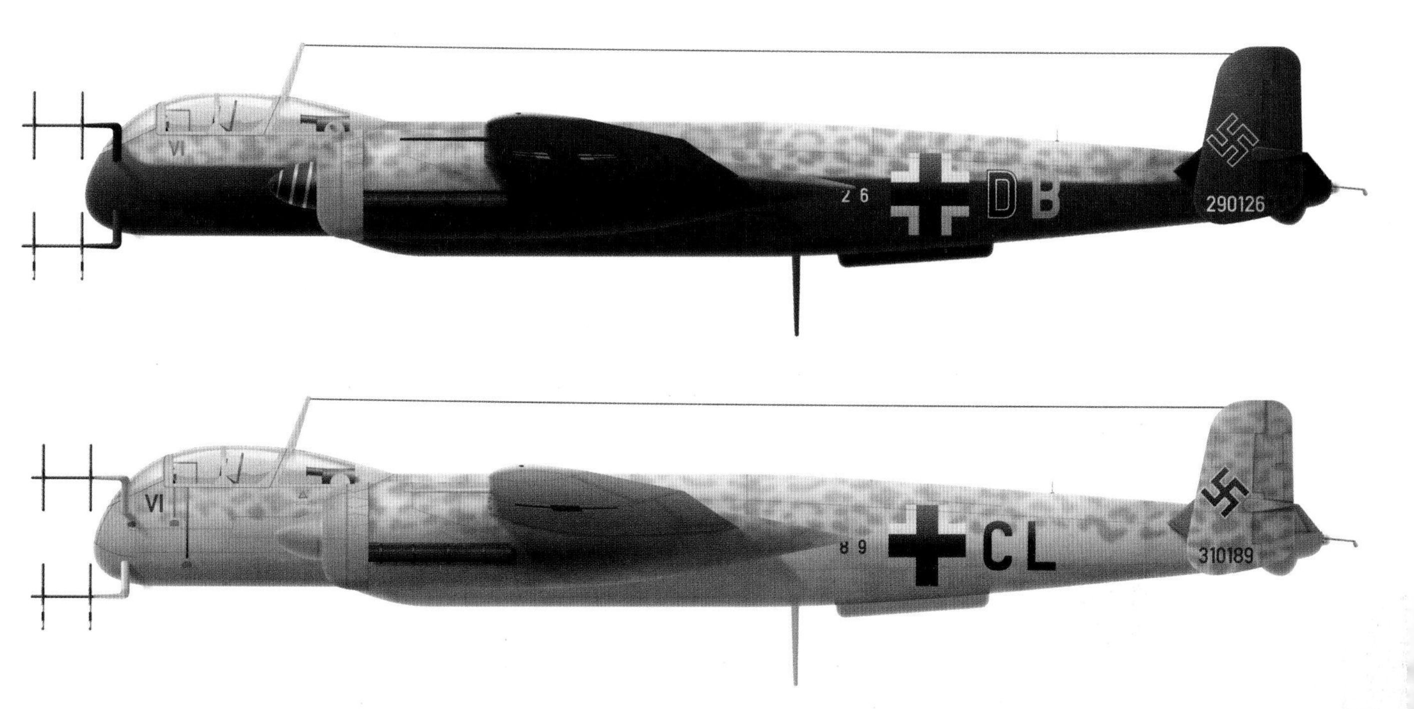

Both captured at Gove, Denmark, in May 1945, D5+BL was a He 219A-2 of Nachtjagdgeschwader 3, whereas D5+CL was an example of the He 219A-7 variant.

The extra weight and drag from the additions and modifications did however have an impact on the Uhu's performance and left it vulnerable to Mosquito night-intruders. The final production variant, the He 219A-7, featured the DB 603G producing 1,416kW (1,899hp) to compensate. This led to mixed claims for the handling of the He 219, with some German reports claiming it could in extremis take off on only one engine, while Allied post-war test pilots considered it underpowered with single-engined landings requiring careful handling. Certainly, the He 219A-7 weighed more empty than a Mosquito fully loaded while only having a few hundred more kW of power. Despite its limited power, the Uhu was found to be a good all-weather aircraft with effective cabin-heating and anti-icing systems and a reliable, easy-to-operate autopilot, while the cockpit was well laid out with good all-round visibility.

Meagre production

In total, 268 production He 219s were built while around 20 of the development aircraft were also brought up to a production standard. I./NJG 1 are also believed to have manufactured a further six from spare parts. Considering the Uhu's obvious capabilities against the heavy bomber threat, the choices that led to such meagre production can only be considered to have been to the Allies' advantage.

Night-fighter ace

The highest-scoring He 219 ace was Major Ernst-Wilhelm Modrow, a pre-war airline pilot who initially operated Do 26s and BV 222 flying boats for the Luftwaffe after the war started. Modrow would not train as a night-fighter pilot until late 1943, after which he was assigned to I./NJG 1. Initially assigned to 2 Staffel, he scored his first kill on the night of 7 March 1944, unverified as a Mosquito. By the end of May, he had claimed a further 17 aircraft, including three heavy bombers on both 23 April and 28 May. Moving to 1 Staffel, his final claim total reached 34 aircraft by 5 January 1945. This included four Lancasters on the night of 22 June 1944 – his highest claim for a single night – and a confirmed Mosquito of 571 Squadron, one of only eight aircraft lost by them during their 18-month existence. Modrow's record over such a short time shows what the He 219 was capable of and how lucky the Allies were that its production was mismanaged.

Dornier Do 335 (1943)

One of the most distinctive aircraft of World War II, the Dornier Do 335 had a unique configuration and great potential. Its main opponent would not be Allied aircraft but the relentless assault on German industry.

A high performance, twin-engine heavy fighter the Dornier Do 335 had an unconventional appearance with engines and propellers at both ends of the fuselage. This was the refinement of Dornier's work with inter-war flying boats where they had favoured the configuration due to the reduction in drag realized by placing the engines inline. In those installations, however, they were generally in engine pods separate from the fuselage. Work on what would become the Do 335 started in 1937 when Dornier commissioned a testbed for the rear engine drive shaft arrangement, the Goppingen Go 9. The success of this small wooden design with a pusher propeller mounted behind a cruciform tail proved the basic concept. However, it was not until 1942 that Dornier was allowed to submit a design for a requirement that could take advantage of the configuration, the RLM preferring they concentrate on flying boats and bombers. Originally asking for designs for a fast unarmed intruder aircraft the German Air Ministry redefined the requirement to encompass the day- and night-fighter, fighter-bomber, and reconnaissance roles. Consequently, due to the redesign work it wasn't until late 1943 that the Do 335 was ready for its first flight.

A Dornier Do 335A-0. Below was the seventh of 10-pre-production aircraft, most of which went to Eprobungskommando (EK) 335 for evaluation.

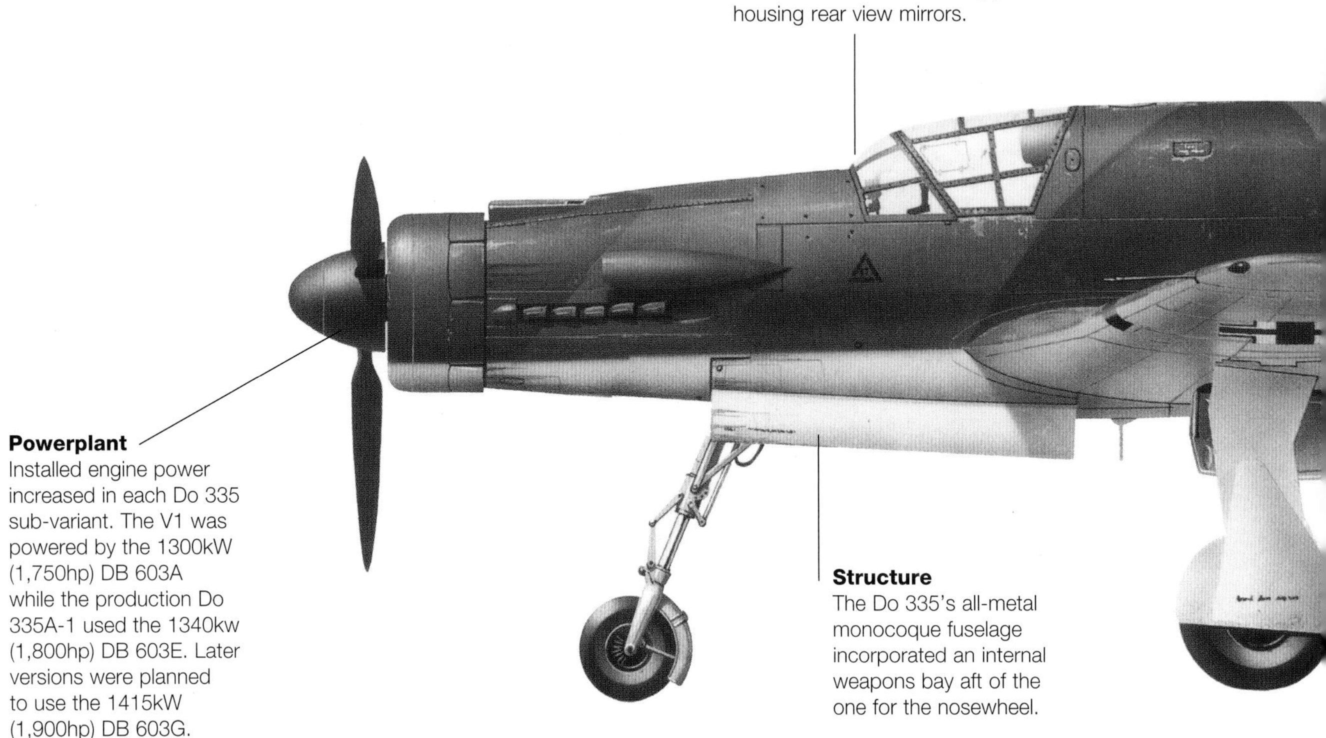

A low wing monoplane with tricycle undercarriage the Pfeil or Arrow was powered by a pair of Daimler-Benz DB 603s. The forward engine used a circular radiator mounted between the engine and the propeller, giving it the appearance of a radial engine. The rear engine meanwhile sat between the wing and the tailplane with an under-fuselage scoop feeding its radiator exhausting the air either side of the lower vertical stabiliser. A hollow shaft mounted on three bearings carried the drive to the propeller behind the rear control surfaces. The central fuselage contained a bay with capacity for a 500kg bomb with a large fuel tank above it. This location would ensure minimal changes in trim as fuel and weapons were expended. Although the first four aircraft featured no forward-firing armament subsequent airframes would feature two MG 151 15mm (0.59in) cannon above the front engine and a MK 103 30mm (1.18in) cannon firing through the propeller hub.

With its two DB 603s and minimal frontal area the Do 335 was fast with a high rate of climb, and favourable

Dornier Do 335A-0 VG+PH was completed at Dornier's Mengen plant in September 1944 before entering the test programme. It was captured by the Allies at Oberpfaffenhofen on the 29 April 1945, and transported to the USA for testing.

Ejection
In the event of the pilot activating the ejection system, the upper fin and rear propeller were separated by explosive bolts.

Intake
The underslung airscoop below the aft fuselage provided air to the rear engine's radiator.

Evaluation by EK 335 would be as close to operational service as the Do 335 would get. Even if the type had reached frontline units, it would have done little to stem the Allied advance.

Specifications: Do 335A-0

Type:	Twin engine fighter
Dimensions:	Length: 13.85m (45ft 5.25in); Wingspan: 13.80m (45ft 3.5in); Height: 5.0m (16ft 4.9in)
Weight:	9,510kg (20,966lb)
Powerplant:	2 x 1290kW (1750hp) Daimler-Benz DB603A inverted V-12 piston engine
Maximum speed:	732km/h (455mph)
Range:	1600km
Service ceiling:	10,700m
Crew:	1
Armament:	2 x 15mm (0.59in) MG 151/15 Cannon and 1 x 30mm (1.18in) MK 103 Cannon; 1 x 500kg bomb in weapons bay

handling for a heavy fighter. The lower fin required a shallow take-off and landing angle to avoid damage, although it was recommended to use its integrated bump stop after the main wheels had contacted the runway. To assist in reducing the landing run, by around 25%, the forward propeller's pitch could be reversed allowing it to be used for braking. Meanwhile, in the event of undercarriage failure, the lower fin could be jettisoned by explosive bolts to improve the chance of a successful recovery.

Service trials and evaluation of the Do 335 began in September of 1944 using pre-production aircraft, while the first production Do 335A-1s powered by DB 603E-1 engines were delivered shortly afterwards. These include underwing hardpoints allowing two fuel tanks or a pair of 250kg (550lb) bombs to be carried. A similar unarmed reconnaissance aircraft, the Do 335A-4 was constructed and equipped with two Rb 50/18 cameras in the weapons bay and extra fuel. Although more powerful engines were planned for later examples only the one example was built before the factory was overrun by the American Army.

Also cancelled by the changing circumstances of the war was the Do 335A-6 night-fighter. A considerable redesign this required a second crew member to operate the FuG 217J Neptun radar. This sacrificed volume from the main fuel tank, which was compensated for by deleting the weapons bay and placing additional fuel there. Design work for this had already been carried out for the two-seat trainer Do 335A-10 and 12, which were produced in limited

The ninth prototype, CP+UI, was built to pre-production standards and had improved undercarriage, DB603A-2 engines, and armament of two MG 151 and one MK 103 cannon.

numbers. The night-fighter also required antenna for the radar system mounted forwards of the wing and flame-dampers over the exhausts, the additional weight and drag being estimated to reduce performance by around 10%. The 335A-6 was supposed to have been produced in Vienna by Heinkel but ultimately only the V10 prototype flew without the Neptun radar fitted. Two prototype Zerstörer heavy fighters were also completed, the Do 335B-1 replaced the weapons bay with a fuel tank and the MG 151s were upgraded to 20mm (0.79in) versions. The B-2 meanwhile also added a Mk 103 30mm (1.18in) cannon to each wing. Both B models featured a revised armoured windscreen with V profile and a pillar on the centreline as well as a larger nosewheel.

When the Dornier factory at Oberpfaffenhofen was overrun 37 Pfeils had been completed with around 70 awaiting completion. A variety of models were on the drawing board including high altitude aircraft with increased wingspan, night-fighters with the armament of the B-1, and even a mixed power plant Do 535, which replaced the rear DB 603 with a jet engine. All of this was too little too late, and although the Do 335 was a sound design with great potential it had emerged at a time when German industry was reeling from the effects of the Allied bombing campaign and was unable to devote the resources necessary to even enter series production of the basic model.

Although it is not believed the Do 335 ever entered combat, it was encountered by Allied aircrew over Europe. Pierre Closterman, author of *The Big Show,* recalled coming across one when flying Typhoons with the RAF's 3 Squadron, the Pfeil's great speed ensured it was in no danger of being caught.

Ejector seat

To deal with the extra danger presented to an escaping pilot by the rear propeller and vertical tail, the Do 335 was one of the first aircraft to feature an ejection seat. This was more complex in operation than a modern example, which requires the pilot to pull a single handle to initiate the entire sequence. First a button to fire explosive charges in the rear propeller needed to be pressed, followed by one to do the same for the vertical tail, and then a third to prime the seat. The canopy then needed to be jettisoned by way of two levers at its leading edge before the ejection itself could be initiated via a trigger on the seat arm rest.

There are no records of the seat being successfully used; however, two pilots were found in the wreckage of their aircraft with their arms missing. The accepted wisdom was that these had been carried away with the canopy when the jettison levers were pulled.

Messerschmitt Bf-109K (1944)

The final version of the Bf 109 to enter mass production, the K-4 attempted to integrate all the lessons learnt during 10 years of operations. Entering service when the war was already lost, it was too little too late.

With multiple detail differences between later Bf 109G variants, even of the same sub-type, there was a desperate need to rationalize the design to ease the logistics and maintenance burden. This would also be an opportunity to incorporate the best of the design elements introduced on previous models.

Messerschmitt Bf 109K-4 of III./JG 27, a Gruppe that was fully equipped with the K-4 model by the end of 1944 and engaged in the fierce fighting over the Reich in the closing stages of the war.

Power was provided by the DB 605D series engine, providing 1,324kW (1,800hp) or 1,451kW (1,973hp) if the methanol-water injection system was fitted and working. The new wider engine also required a small bulge to the cowling just forward of and below the exhausts. The K-4, which was the only mass-produced K variant, featured the tall wooden fin introduced on the G-6 together with a wooden horizontal stabilizer, the Erla-Haube canopy, which improved the pilots' field of view. There was also a retracting long tailwheel to improve ground handling, a deeper oil cooler, wide-bladed propellers and a distinctive

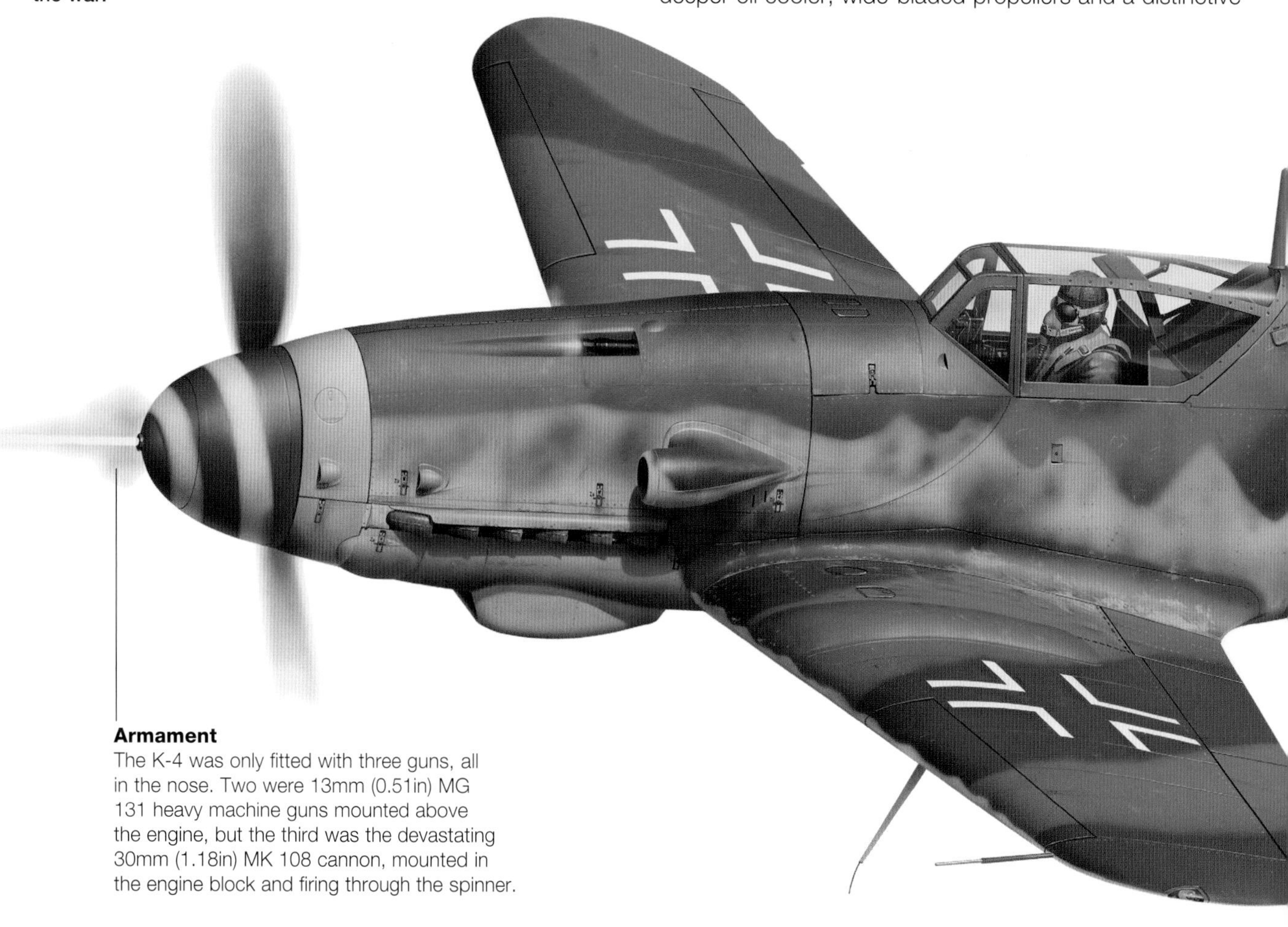

Armament
The K-4 was only fitted with three guns, all in the nose. Two were 13mm (0.51in) MG 131 heavy machine guns mounted above the engine, but the third was the devastating 30mm (1.18in) MK 108 cannon, mounted in the engine block and firing through the spinner.

forward-sloping antenna under the left wing for the FuG 16ZY radio. The wheel wells were squared off to allow doors to be fitted to the outer portion, although these were regularly removed in service, the 8km/h (5mph) speed advantage presumably not being considered worth the maintenance burden.

The K-4 also differed from earlier Bf 109s in having servo tabs for the ailerons to reduce the control forces. However, these were generally locked in place as the elevators did not feature them, which led to problems with control harmony due to the significant difference in the forces required to move the two control surfaces.

Home defence

Armament was the MK 108 30mm (1.18in) cannon firing through the propeller hub and a pair of MG 131 13mm (0.51in) machine guns mounted above the engines. Although theoretically an improvement on the 20mm (0.79in) cannon used in earlier aircraft, the MK 108 tended to jam if fired when manoeuvring – a problem that was never fully rectified. The Bf 109K-4/R1 also gained the R1 centreline bomb rack modification, while the R3 modification allowed the use of a 300-litre (79-US gallon) drop tank.

III. Gruppe of Jagdgeschwader 53 was the first unit to receive K-4s in October 1944 for home defence although as with most units it would continue to operate later-model Bf 109Gs alongside them. These would all be involved in the desperate final days of the Reich with 109s defending against massed bomber attacks and low-level air-to-air combat on the Eastern Front and providing ground support as the Allied armies advanced.

In an attempt to stem that advance, a large-scale ground and air offensive was planned for the Western Front. The air element was Operation Bodenplatte, launched on 1 January 1945. Of the 12 BF 109-equipped Gruppen that took part, eight were wholly or partly equipped with K-4s, the remainder using the earlier G-10 or G-14 models. Launching at first light, the fighters were led by Ju 88s

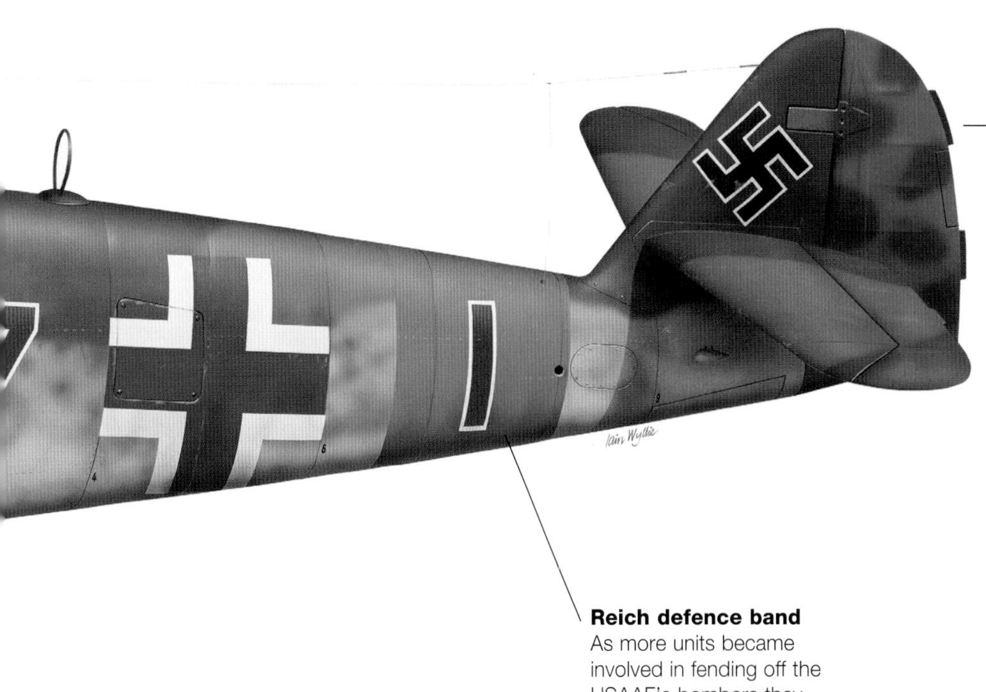

Rudder
The Bf 109 had a fabric covered rudder throughout its many variants and was fitted with a geared trim tab.

Reich defence band
As more units became involved in fending off the USAAF's bombers they painted distinctive coloured bands around their aircraft to enable them to reform more rapidly after an attack before commencing the next one.

to handle the navigation of each formation and achieved almost complete surprise at 27 airfields in France, Belgium and the Netherlands. The K-4-equipped IV. Gruppe of JG 53 took part in the attack on Metz-Frescaty Air Base in France just south of the border with Luxembourg. Here, 22 P-47s were destroyed on the ground while IV./JG 53 also shot down an unlucky Auster AOP aircraft on the way to the target. In exchange, the two Gruppen lost 17 aircraft and pilots – a pattern that was to be repeated across the operation with more than 150 Luftwaffe pilots lost in exchange for 450 Allied aircraft. The latter could soon be replaced, while the former had already been in short supply before Bodenplatte.

With the superior North American P-51D Mustang and Spitfire XIV roaming the skies over Germany, only the weather could provide refuge for the Luftwaffe. Both types were faster than the Messerschmitt, could climb at least as well, and were more manoeuvrable. When the weather did clear in late February, JG 53 lost 20 Bf 109Gs and 109Ks in only two days and would soon start to disband Staffeln and Gruppen as supplies of aircraft and pilots dwindled. The situation was no better to the East as the Soviet armies advanced and where the majority of combat took place at low level. The newer Soviet types, such as the Yak-3 and LaGG-7, had closed the performance gap and in some areas, such as manoeuvrability, outperformed the German aircraft.

Specifications: Bf 109K-4

Type:	Fighter
Dimensions:	Length: 9.02m (29ft 7in); Wingspan: 9.92m (32ft 6.5in); Height: 3.37m (11ft 1in)
Weight:	3,362kg (7,412lb)
Powerplant:	1 x 1,342kW (1,800hp) Daimler-Benz DB 605D inverted V-12 piston engine
Maximum speed:	680km/h (423mph)
Range:	650km (404 miles)
Service ceiling:	13,500m (44,290ft)
Crew:	1
Armament:	2 x 13mm (0.51in) MG 131 machine guns; 1 x 30mm (1.18in) MK 108 cannon

Protecting the Me 262

Messerschmitt's jet fighter, although almost untouchable at speed, was vulnerable to marauding Allied fighters while on approach to land. To address this, Bf 109 units would provide cover over Me 262 bases. On 2 March 1945, III./ JG 27 based at Hesepe was protecting the 262s operating from Rheine when Spitfire XIVs from 130 and 350 Squadrons based near Eindhoven appeared overhead conducting a fighter sweep.

Spotting the German fighters below them, the Spitfires dived into action. Blue 3 of 350 Squadron engaged in a turning fight with a Bf 109K-4, which then dived to the deck before pulling up in a steep turning climb. Despite this, the Spitfire was soon positioned to engage the Messerschmitt and opened fire. A further two K-4s fell to 350 Squadron for no losses. Overall, the day went badly for the defenders with five more Bf 109s lost, whereas only two Spitfires of 130 Squadron were lost, their pilots becoming prisoners of war.

Ground-attack role

As the end approached, Bf 109K-4s were being used in attempts to slow down the Allied advance on the ground. JG 53's III. and IV. Gruppen attacked bridges and advancing US troops along the Danube with limited effect. As fuel and supplies dwindled, the Gruppen were disbanded and their aircraft destroyed or left to rot almost as they landed from these missions in the final weeks of April 1945.

Around 1,700 Bf 109K-4s were produced, primarily by Messerschmitt at its Regensburg factory – an impressive achievement in light of the difficulties German manufacturing faced at the time. However, this inevitably took its toll with shortages of engines and guns meaning earlier G models were being produced alongside it. The difficulties obtaining high-octane fuel also meant the DB 605 could rarely provide anything like its rated power while reports of poor workmanship and potential sabotage by the forced labourers building the Bf 109 were widespread. These factors combined to prevent the Bf 109K-4 from regularly achieving its full potential.

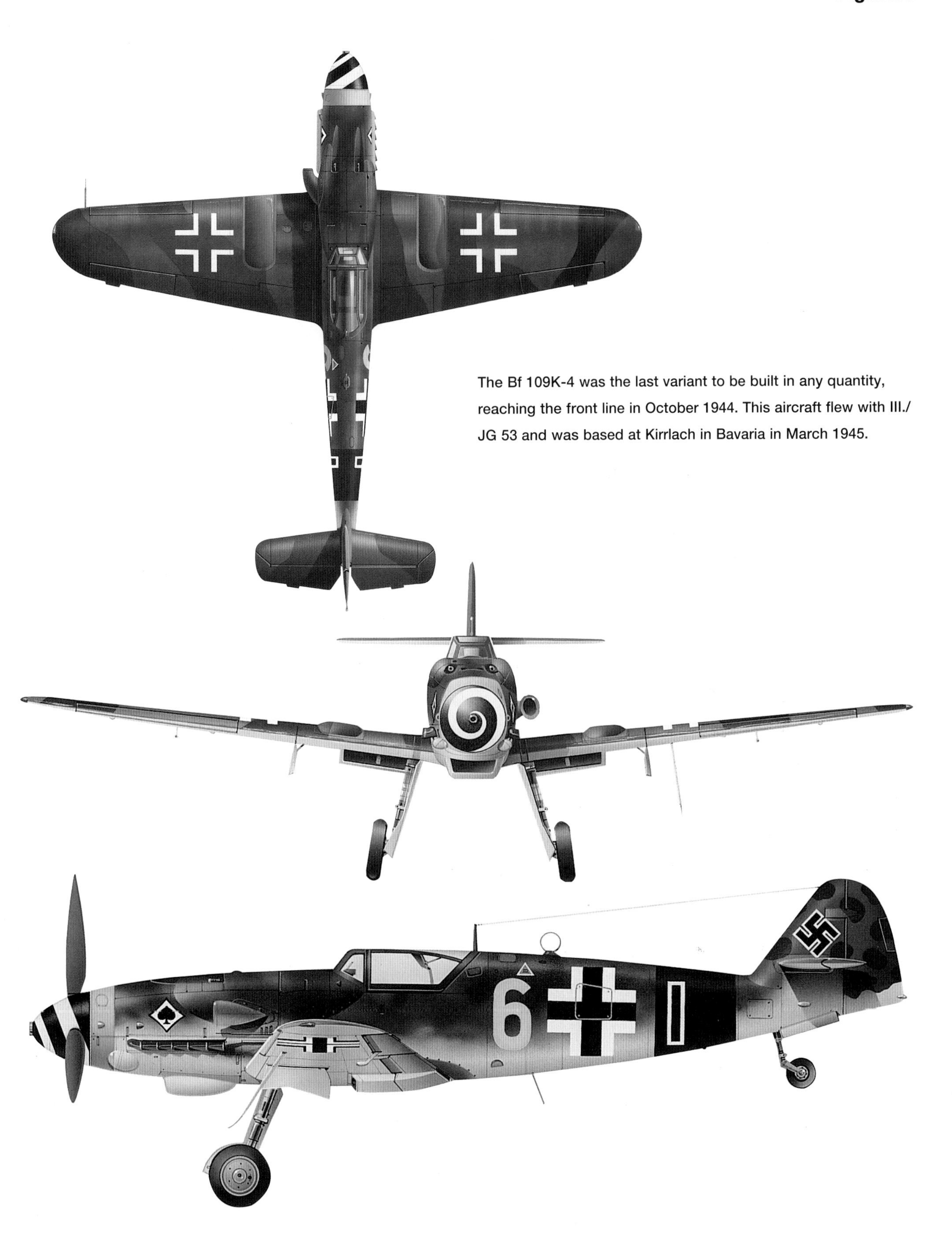

The Bf 109K-4 was the last variant to be built in any quantity, reaching the front line in October 1944. This aircraft flew with III./ JG 53 and was based at Kirrlach in Bavaria in March 1945.

Bombers

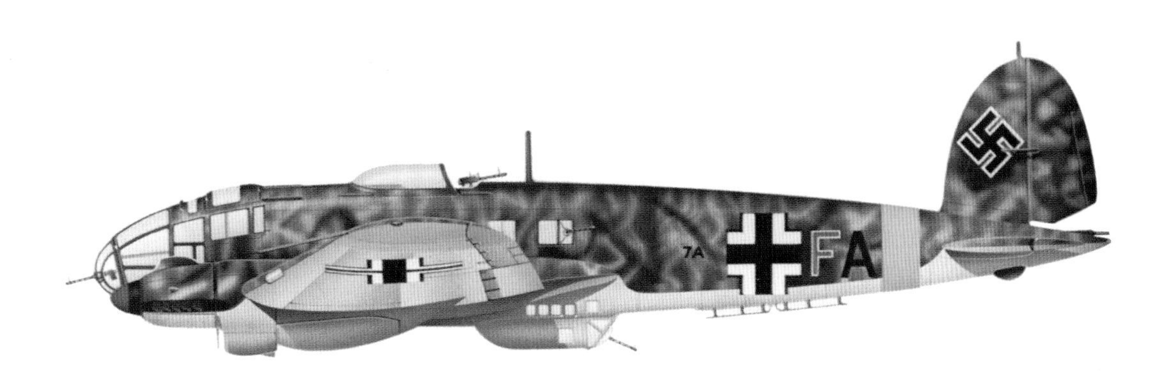

The Luftwaffe developed its bomber doctrine during the Spanish Civil War, and at Guernica had shown the world the horrors that it could inflict on the civilian population in unrestricted warfare. World War II would see Germany attempt to recreate this on a larger scale with its fleet of Do 17s, He 111s and Ju 88s – all of which were designed as high-speed tactical bombers. While attempts at developing a larger strategic bomber were disappointing, the Luftwaffe would pioneer the use of specialist close air support aircraft, foreshadowing the Cold War A-10. By 1944, operation of all of Germany's bomber aircraft would be greatly curtailed due to fuel shortages, with whole Staffeln left to sit on the ground as the Allies advanced across Europe.

Opposite: The Ju 88A-17 and A-4/Torp were dedicated torpedo-carriers, with a pair of bomb racks under the wings.

Dornier Do 17 & Do 215 (1934)

Conceived as a high-speed airliner, the Do 17 would have no success in that role and instead became one of the Luftwaffe's principal bombers in the opening stages of the war before being relegated to second-line roles.

First flying in 1934, the Dornier Do 17 – also known as the 'Flying Pencil' – started life as a high-speed mail carrier, with the option of carrying two passengers in a cabin ahead of the wing and a further four behind it. The narrowness of the fuselage rendered even this minimal accommodation cramped and Lufthansa were sufficiently unimpressed that the three prototypes were returned to Dornier by 1935. Fortunately for them, one of the test pilots suggested to the Reichsluftfahrtministerium that it could make the basis of a medium bomber. This led to an additional prototype with a bomb bay and the H-configuration tailplane that would come to be synonymous with the type. Further prototypes would refine the configuration, introducing a glazed nose position for the bombardier and a rear-facing gun position with a hand-operated 7.92mm (0.31in) MG 15 machine gun at the aft of the cockpit.

Assigned to an unidentified training unit in 1939, this is an example of the Do 17E-1 initial series-production version. Note the combination of four-digit code SA+HN and individual aircraft number '26', plus three-tone, upper surface splinter scheme.

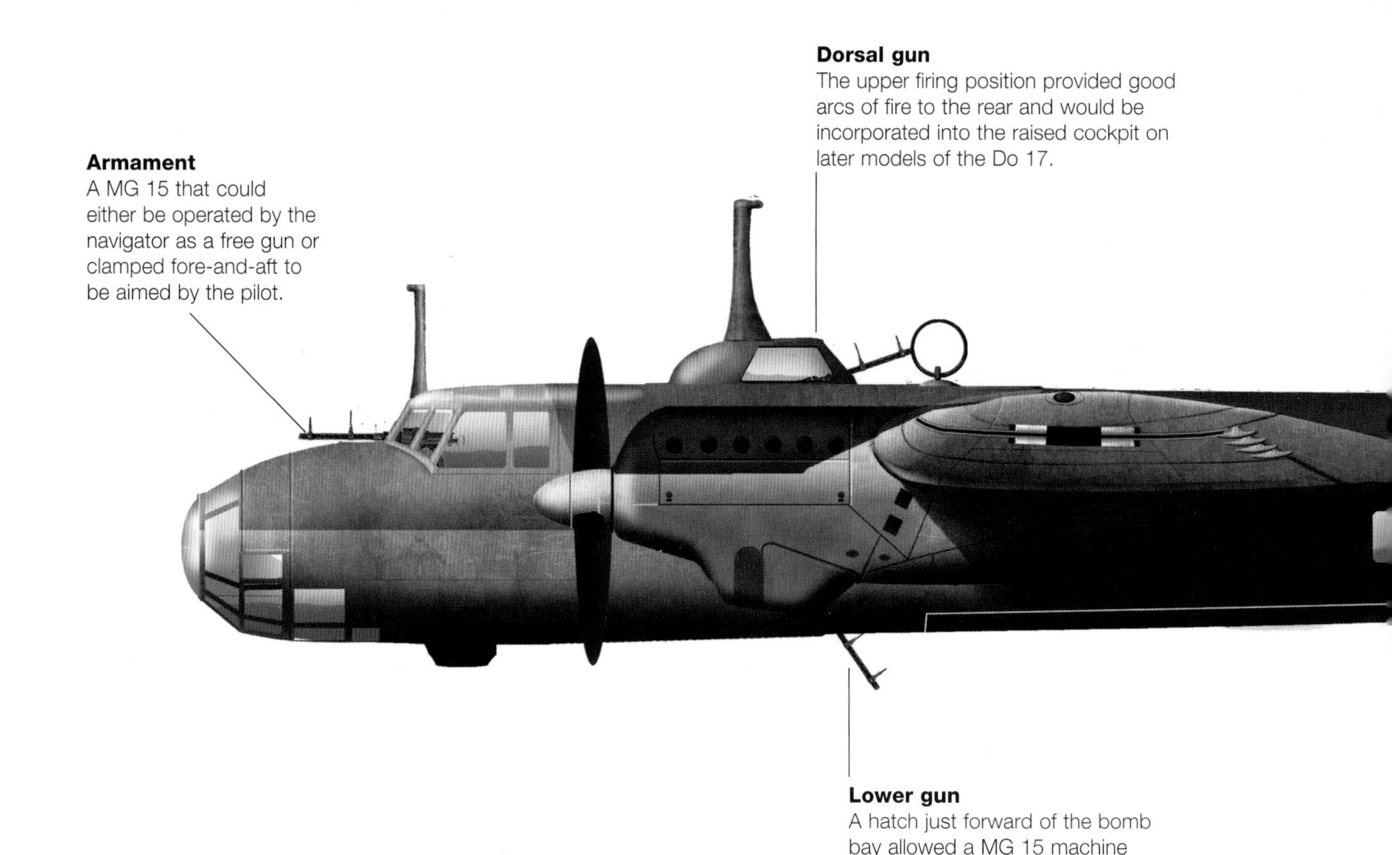

The Do 17E-1 would be the first production bomber. This could carry up to 750kg (1,654lb) of bombs and gained a downwards-firing machine gun just forward of the bomb bay. The F-1, meanwhile, was a reconnaissance version with cameras and an auxiliary fuel tank fitted in the bomb bay. Power for both versions was provided by the 562kW (754hp) BMW VI 7.3 liquid-cooled V-12 engines. Four bomber Geschwader and one reconnaissance Gruppe were soon formed and elements of each were sent to Spain in 1937 with the Condor Legion on the side of the Nationalists in the civil war. Both versions proved capable of evading the Republican fighters due to their superior speed.

The year 1937 also saw the introduction of the Do 17M and 17P. These were originally intended to be powered by the Daimler-Benz DB 600 but limited supplies of this engine led to them respectively using the 675kW (905hp) Bramo 323A and 645kW (865hp) BMW 132N air-cooled radial engines. The latter engine gave the P model greater

A trio of Do 17s overfly Athens in May 1941, shortly after German victory in the Battle of Greece.

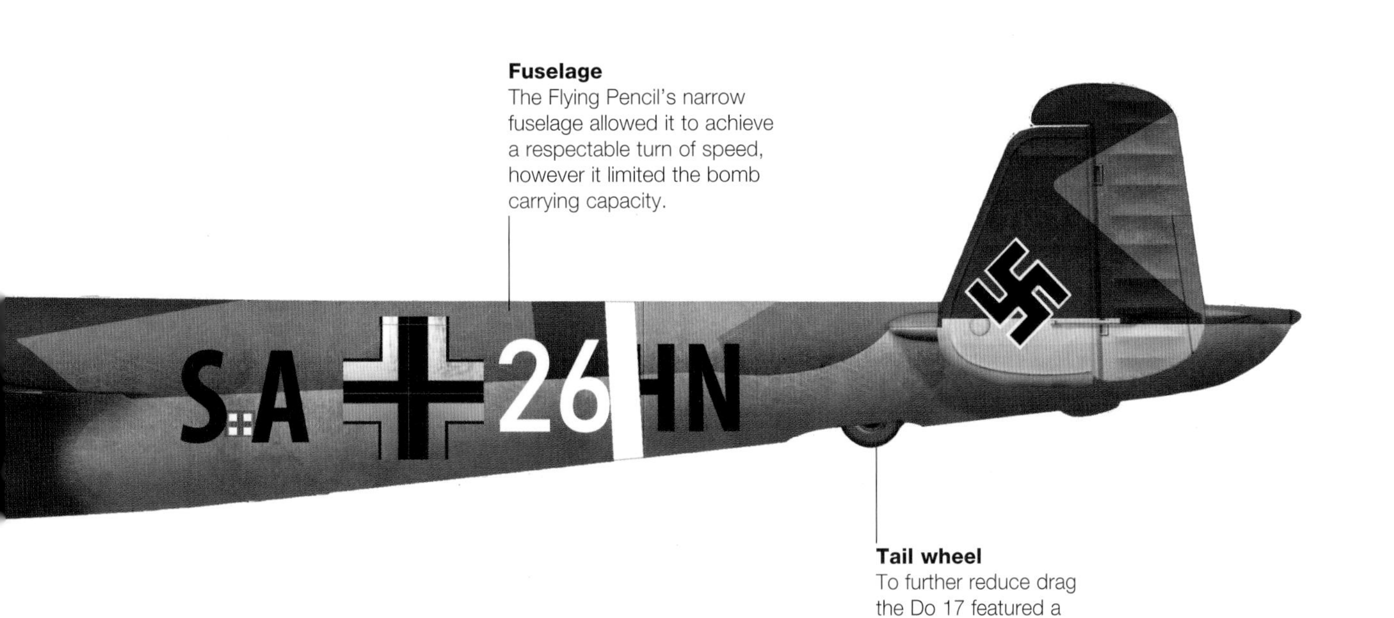

Fuselage
The Flying Pencil's narrow fuselage allowed it to achieve a respectable turn of speed, however it limited the bomb carrying capacity.

Tail wheel
To further reduce drag the Do 17 featured a semi-retracting tail wheel.

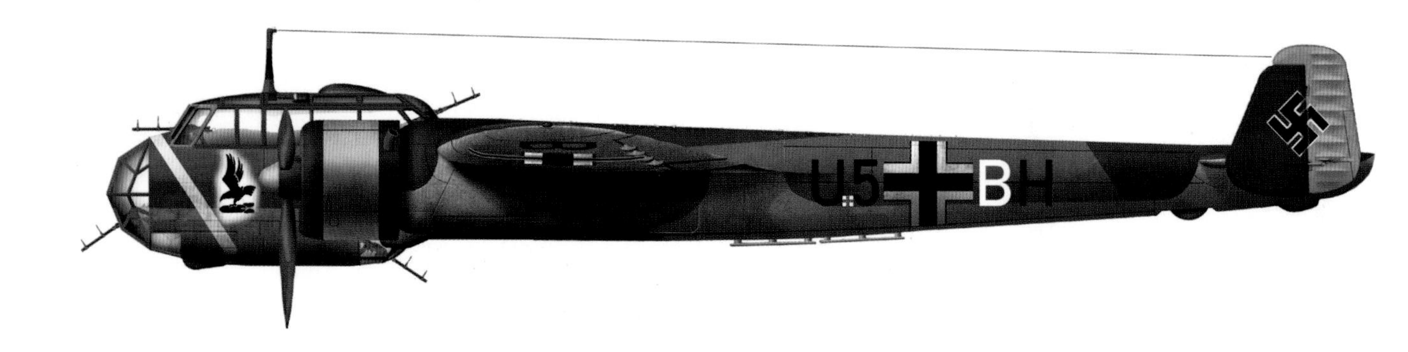

range for the reconnaissance role while both types had an expanded bomb bay that could carry up to 1,000kg (2,205lb) of bombs. They also gained a forward-firing MG 15 machine gun that could be operated by the bombardier or locked in position to be aimed by the pilot.

The next major production version was the Do 17Z. This was essentially a Do 17M but with a new cockpit section. This applied lessons learnt in the Spanish Civil War, when the type had proved vulnerable to attacks from below. A downwards-firing machine-gun position was placed at the rear of the new cockpit section just in front of the bomb bay, creating a distinctive step. At the same time, the cockpit rook was raised and fully glazed while the nose glazing was increased in area. The

A Dornier Do 17Z-2 of 1./Kampfgeschwader 2 'Holzhammer', based at Tatoi in Greece in May 1941 during Operation Marita, the German invasion of Greece.

Specifications: Do 17Z-2

Type:	Medium bomber
Dimensions:	Length: 15.79m (51ft 10in); Wingspan: 18.00m (59ft); Height: 4.56m (14ft 12in)
Weight:	8,840kg (19,489lb) maximum take-off
Powerplant:	2 x 746kW (1,000hp) Bramo 323P nine-cylinder radial engines
Maximum speed:	410km/h (255mph)
Range:	1,160km (721 miles)
Service ceiling:	7,000m (22,966ft)
Crew:	4
Armament:	2 x 7.92mm (0.31in) MG 15s firing forwards; 2 x MG 15s firing from side windows; 1 x MG 15 firing from upper and lower aft positions; a bombload of up to 1,000kg (2,205lb)

Do 17Z-1 used the Bramo 323A engines of the 17M. However, these left the type underpowered when carrying a full bombload, consequently the Z-2 gained the 746kW (1,000hp) Bramo 323P with two-speed superchargers. In this configuration, the full bombload of 1,000kg (2,205lb) could be carried although the radius of action was then reduced to 330km (205 miles).

Early war service

At the outbreak of the war, around 200 of the 370 Do 17s in operational service were Z-1 and Z-2 models with the remainder being made up of Do 17E-1s and 17M-1s. III./KG 2 were one of the first units involved, their Do 17Z-2s bombing the railway bridge at Dirschau in the opening hours of the invasion of Poland. Despite lacking the outright speed of the Ju 88s, the Do 17s were still able to outpace the Polish air force's fighters and losses were relatively light. This would change as German forces advanced westwards in 1940 and faced more capable French and Royal Air Force aircraft and losses began to mount.

The opening of the Battle of Britain in 1940 saw the Do 17Z-2s of KGs 2, 3, 76 and 77 in the vanguard of the attack on Channel shipping, targeting the coastal convoys that carried a significant proportion of the UK's internal trade. As the battle moved overland, the limitations of the Do 17's defensive armament led to various field modifications to add more machine guns. At the same time, the Kampfgeschwader favoured low-level operations in an attempt to avoid being intercepted by the RAF's Spitfires and Hurricanes. Despite this, the Dornier's lack

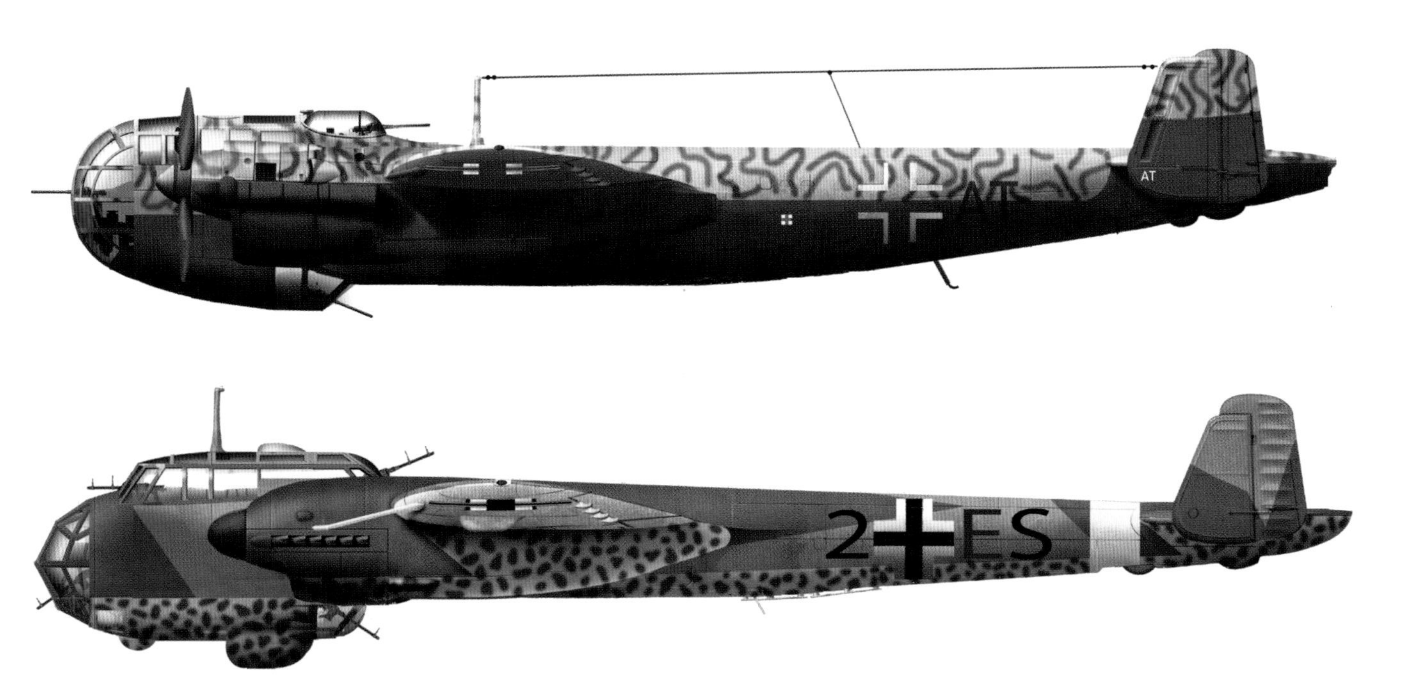

Above upper: A Dornier Do 217M-1 from Kampfgeschwader 2. This example was assigned to the 9. Staffel (9./KG 2) and stationed in France in 1944.

Above lower: A Dornier Do 215B-4 from 3./Aufklärungsgruppe Ob.d.L., this aircraft was operational in Ukraine in August 1941.

Glider tug

Although withdrawn from frontline service as a bomber by 1942, the Do 17 found a new lease of life as a glider tug for the DFS 230 transport gliders of the Air Landing Wings. The gliders themselves could carry up to nine men or 1,200kg (2,646lb) of cargo. The Do 17s and DFS 230s of the First Air Landing Group, I./LLG 1, arrived on the Eastern Front in 1942. Although too late to participate in the Battle of Stalingrad in 1942–43, they assisted in the resupply and evacuation of the 17th Army at the Kuban bridgehead on the Black Sea, losing five aircraft in the process. I./LLG 1 would again see action after the invasion of Normandy when the French Resistance staged an uprising on the Vercors Massif in South-west France. Two squadrons of DFS 230s were used to transport troops on to the plateau to crush the uprising, again towed by Do 17s. Although I./LLG 1 would be disbanded shortly afterwards, Do 17s would continue as target tugs until the last days of the Reich, including the futile attempts to resupply German forces in Budapest in early 1945.

of armour led to a high attrition rate, around 150 being lost during the Battle of Britain, and by the autumn of 1940 the Luftwaffe was starting to withdraw the type from frontline service, production having ended earlier that year. By the launch of Operation Barbarossa in June 1941, only KG 2 was still operating the type as a bomber.

The Do 215 was initially proposed for export to Yugoslavia and was essentially a Do 17Z powered by the Gnome-Rhône 14 radial engine. An 865kW (1,160hp) DB 601A-powered version was then offered to the Swedish Air Force, who placed an order for 18 aircraft. Their delivery was embargoed in August 1939 as war loomed in Europe and they were taken on charge by the Luftwaffe for the reconnaissance role as the Do 215B-0. Further aircraft were then ordered, leading to a total production run of 105 aircraft in addition to the 2,139 Do 17s produced.

The Do 17 served with the Finnish and Spanish Air Force while Yugoslavia licence-built the Do 17M as the Do 17K from 1939. These aircraft were either destroyed or captured when Germany then invaded the country in 1941, some being passed on to the Croatian Air Force.

Henschel Hs 123 (1935)

Intended as an interim type while the Ju 87 was developed, the Hs 123 would barely have looked out of place in World War I. Despite this, it would remain in service until 1944, when only a lack of airframes led to its retirement.

About a year before the existence of the Luftwaffe was publicly acknowledged on 26 February 1935, it issued a requirement for a dive-bomber. With the primary aim of the German air force being close support of ground forces and with the leadership's experience from World War I, it was thought this would be a core element of future aerial warfare. Consequently, the requirement covered both a state-of-the-art aircraft, which would become the Ju 87, and a more basic design that could be in service within a couple of years. Henschel and Fieseler both developed designs powered by the BMW 132A radial engine, and both flew in early 1935, the Henschel design soon proving superior and being ordered into production.

The He 123 was a sesquiplane design, a biplane where one wing is much smaller than the other, with an open cockpit for the pilot. The BMW 132Dc engine in the production aircraft produced 649kW (870hp), which allowed it to carry up to 450kg (992lb) of bombs, two 50kg (110lb) bombs under each wing and a 250kg (551lb) bomb on the centreline, which swung down on a crutch to clear the propeller when dropped in a dive. The aircraft also carried two 7.92mm (0.31in) machine guns, which extended through the engine cowling and were

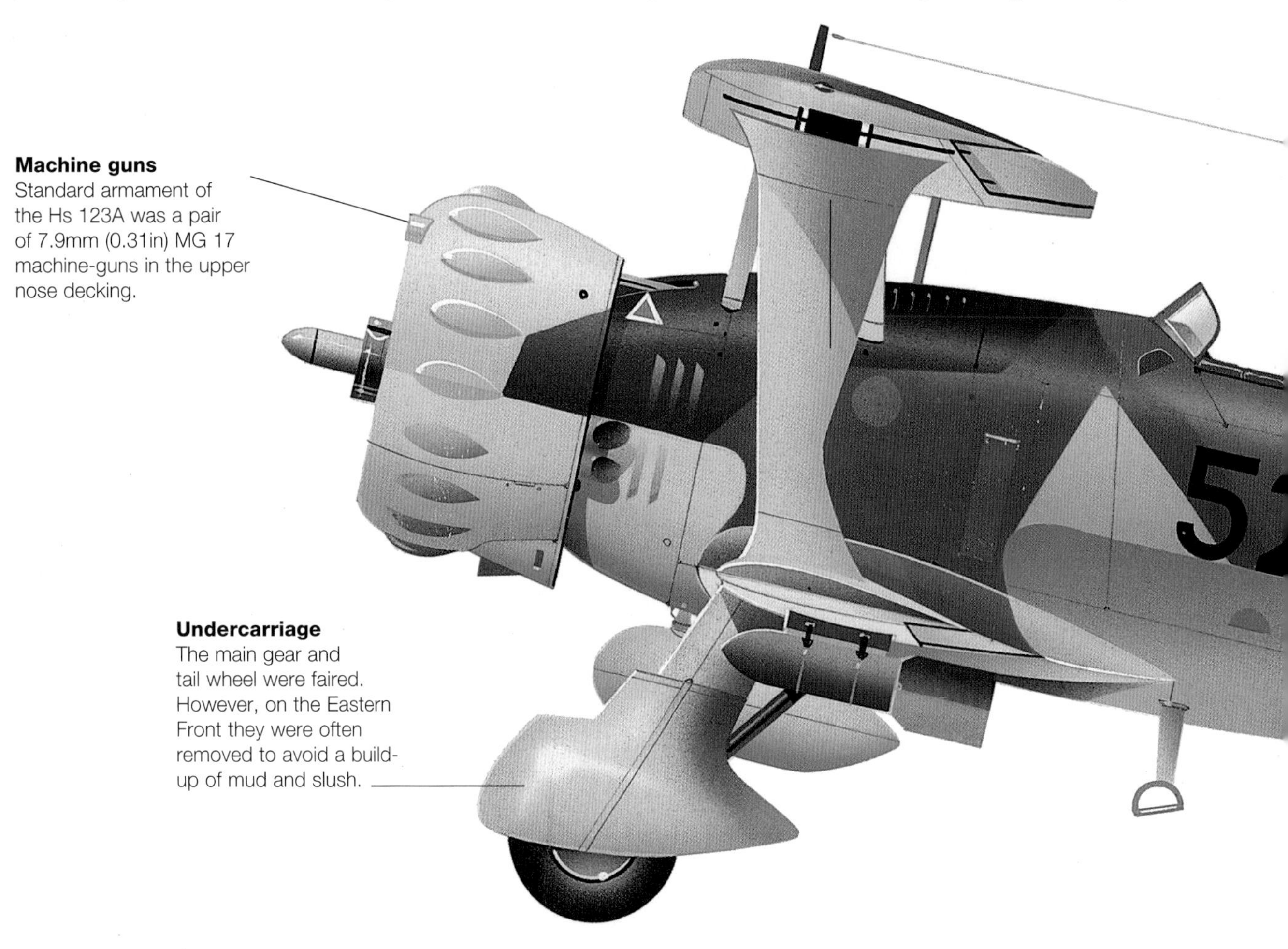

synchronized to fire through the propeller. Stukagruppe I./162 was the first unit to receive Hs 123s in mid-1936 while in early 1937, five aircraft were sent to join the Condor Legion, which was supporting the Nationalists in the Spanish Civil War. Here, they were so successful in the close air support role that the Nationalists purchased the aircrafts and ordered another 11.

In Germany, meanwhile, by 1937 the Ju 87 was already entering service and replacing the Hs 123 in the Stukagruppen. With the Stuka also gaining the close air support role, the Hs 123's days appeared numbered and by late 1938 only one Gruppe, II./LG 2, remained equipped with the type. Defying the odds, II./LG 2 was in the vanguard of the attack on Poland when World War II opened in Europe. Here, it proved effective operating at low level over the Polish troops. Predominantly using their machine guns and 50kg (110lb) bombs for attacks on the enemy troops, the pilots also had a terror weapon in their toolbox: in certain RPM ranges they could use the sharp noise of the engine to simulate machine-gun fire to disperse troops and terrify horses. The Henschel's rugged simplicity also meant the type was easy to maintain and proved exceptionally reliable while operating from field strips near the front line.

In 1941, with the end of the 'Phoney War' and blitzkrieg in the West, the Hs 123 was again called upon to support the 6th Army sweeping through Belgium, Luxembourg and into France. With the Stuka being employed in the tactical bomber role, the Henschel was the only aircraft available to provide close air support to the troops. Being forward based, II./LG 2 regularly flew more sorties than any other unit as the Nazi war machine advanced across Europe. II./LG 2 had reached Cambrai in north-eastern France when the Armistice with France was signed. With no German army on the far side of the English Channel there was no call for the Hs 123's abilities in the Battle for Britain and the unit was withdrawn to Germany to re-equip with the Bf 109E.

Amazingly, the remaining Hs 123s would return to service with 10./LG 2 to see service in the Balkans before taking part in Operation Barbarossa, the invasion of the Soviet Union. Primarily operating on the central and northern fronts,

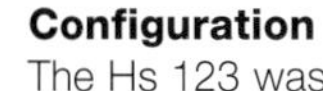

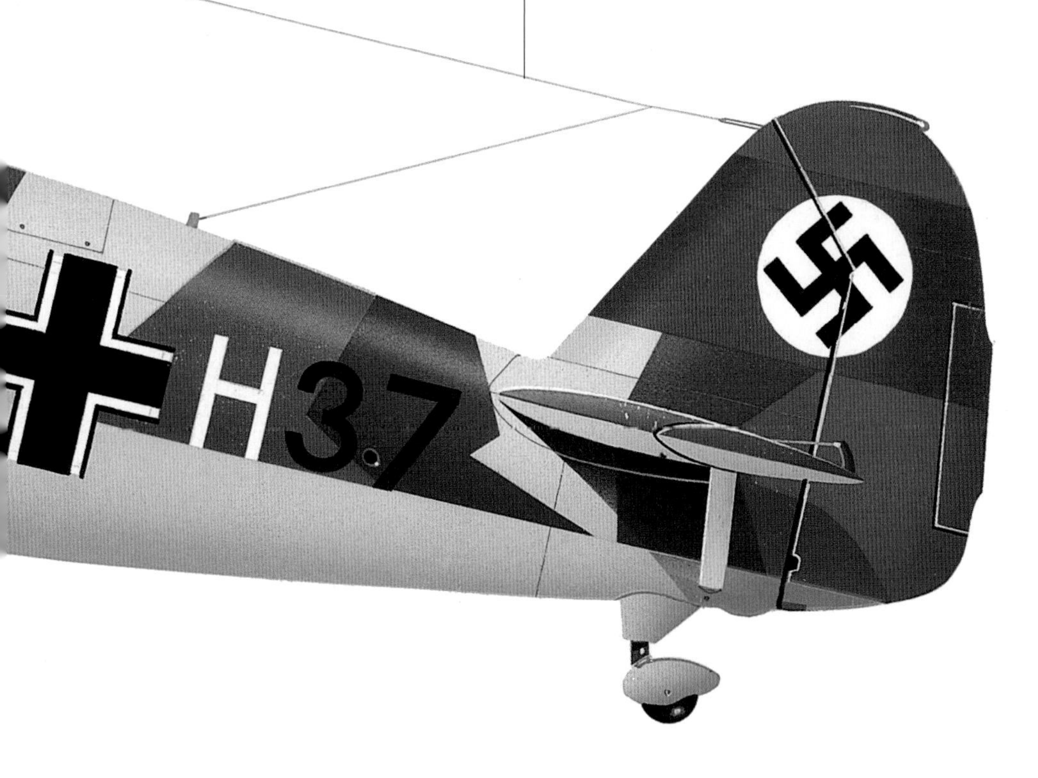

This Henschel Hs 123A-1 was based at Fürstenfeldbruck with 7./Stukageschwader 165 'Immelmann' during October 1937.

This Hs 123A-1 of Schlachtgeschwader 1 sports the Infanterie-Sturmabzeichen emblem of the close support units on its forward fuselage. Four SC 50 bombs are carried under its wings.

Spanish Civil War

German support to the Nationalist side in the Spanish Civil War began in August 1936 with men and equipment of what would become the Condor Legion being dispatched to Tablada airfield near Seville. The Henschel Hs 123 joined them in late 1936 and was originally intended to operate as a dive-bomber. However, it was found to be less than ideal for the role. Unable to maintain a steady dive, it couldn't achieve the accuracy required for the method of attack. The decision to move to close air support, essentially operating as flying artillery, leveraged the Henschel's ability to accurately place its small bombload at low level when operating near friendly troops. It was also able to absorb a significant level of damage and continue flying, which enabled it to survive while operating at low level over the front line. Although the small force was slowly whittled down to two aircraft by the summer of 1937, this didn't dissuade the Spanish Air Force from acquiring a dozen of their own, which they referred to as *Angelito* – 'Little Angel'.

10./LG 2 demonstrated the superiority of the Hs 123 over the Bf 109E in the close air support role, the Messerschmitt suffering undercarriage and engine problems the more rugged Henschel was immune to.

By winter, the Henschel pilots were supporting the army in the Battle of Moscow despite the extreme cold experienced while flying an open-cockpit aircraft. The only major modification to the aircraft for the conditions was the removal of the wheel spats to ease ground handling in the mud and snow. Around this time in January 1942, the Henschel Staffel was again redesignated, this time becoming 7 Staffel of Schlachtgeschwader 1 – the first Geschwader to officially be dedicated to ground attack. 1942 saw the unit operating in Crimea and along the Southern Front, taking part in the Second Battle of Kharkov and the Battle of Stalingrad, this constant action whittling down the number of available aircraft. To address this, airframes were taken from training schools and parts salvaged from unserviceable airframes, production having finished in 1938. In fact, so popular was the Hs 123 as a ground support aircraft that at the beginning of 1943 the commander-in-chief of Luftflotte 4, Generaloberst Wolfram von Richthofen, asked if production could be restarted. Alas, with the jigs having been dismantled in 1940 after no further orders had been forthcoming, this proved impossible. Consequently, the little Henschel's last major action would be the Battle of Kursk before 7./SG 1 finally transitioned to the Ju 87 in early 1944.

Approximately 255 Hs 123s were produced, including aircraft for Spain and the Republic of China, which used them in 1937 against Japanese warships along the Yangtze during the Battle of Shanghai.

Specifications: Henschel Hs 123A-1

Type:	Single-seat close support aircraft
Dimensions:	Length: 8.33m (27ft 4in); Wingspan: 10.50m (34ft 5in); Height: 3.22m (10ft 7in)
Weight:	2,217kg (4,888lb)
Powerplant:	1 x 656kW (880hp) BMW 132Dc air-cooled radial engine
Maximum speed:	341km/h (212mph)
Range:	860km (534 miles)
Service ceiling:	9,000m (29,528ft)
Crew:	1
Armament:	2 x 7.92mm (0.59in) MG 17 machine guns; racks for 4 x 50kg (110lb) bombs under the wings or 2 x 20mm (0.79in) MF FF cannon pods; 1 x 250kg (551lb) bomb under the fuselage

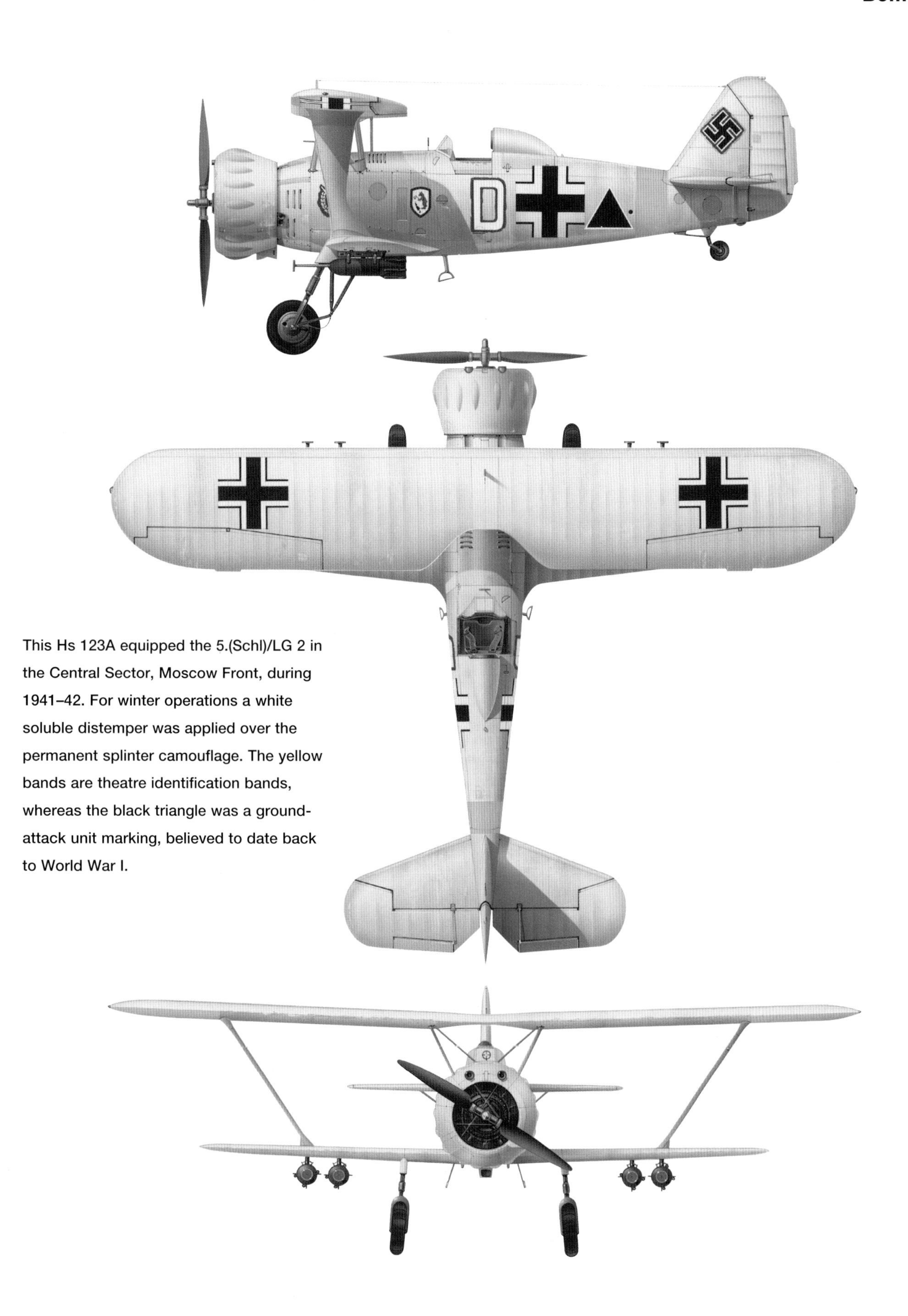

This Hs 123A equipped the 5.(Schl)/LG 2 in the Central Sector, Moscow Front, during 1941–42. For winter operations a white soluble distemper was applied over the permanent splinter camouflage. The yellow bands are theatre identification bands, whereas the black triangle was a ground-attack unit marking, believed to date back to World War I.

Junkers Ju 87 Stuka (1935)

The Stuka is remembered as a symbol of the success of Nazi Germany's blitzkrieg operations in 1939 and 1940. Although rapidly outclassed in its original dive-bomber role, the Ju 87 saw service with the Luftwaffe until the end of the war.

The Sturzkampfflugzeug (dive-bomber, abbreviated to Stuka) concept was tailored to support Nazi Germany's blitzkrieg combined-arms doctrine. A form of flying artillery support, work on three prototypes commenced in 1934 with the design team headed by Hermann Pohlmann, and the Ju 87 was first flown in September 1935. The earliest prototypes featured twin tailfins and a Rolls-Royce Kestrel engine. However, this aircraft was lost in dive tests when the tail unit collapsed. The second prototype employed a single fin and a Junkers V-12 Jumo 210A inline engine. The

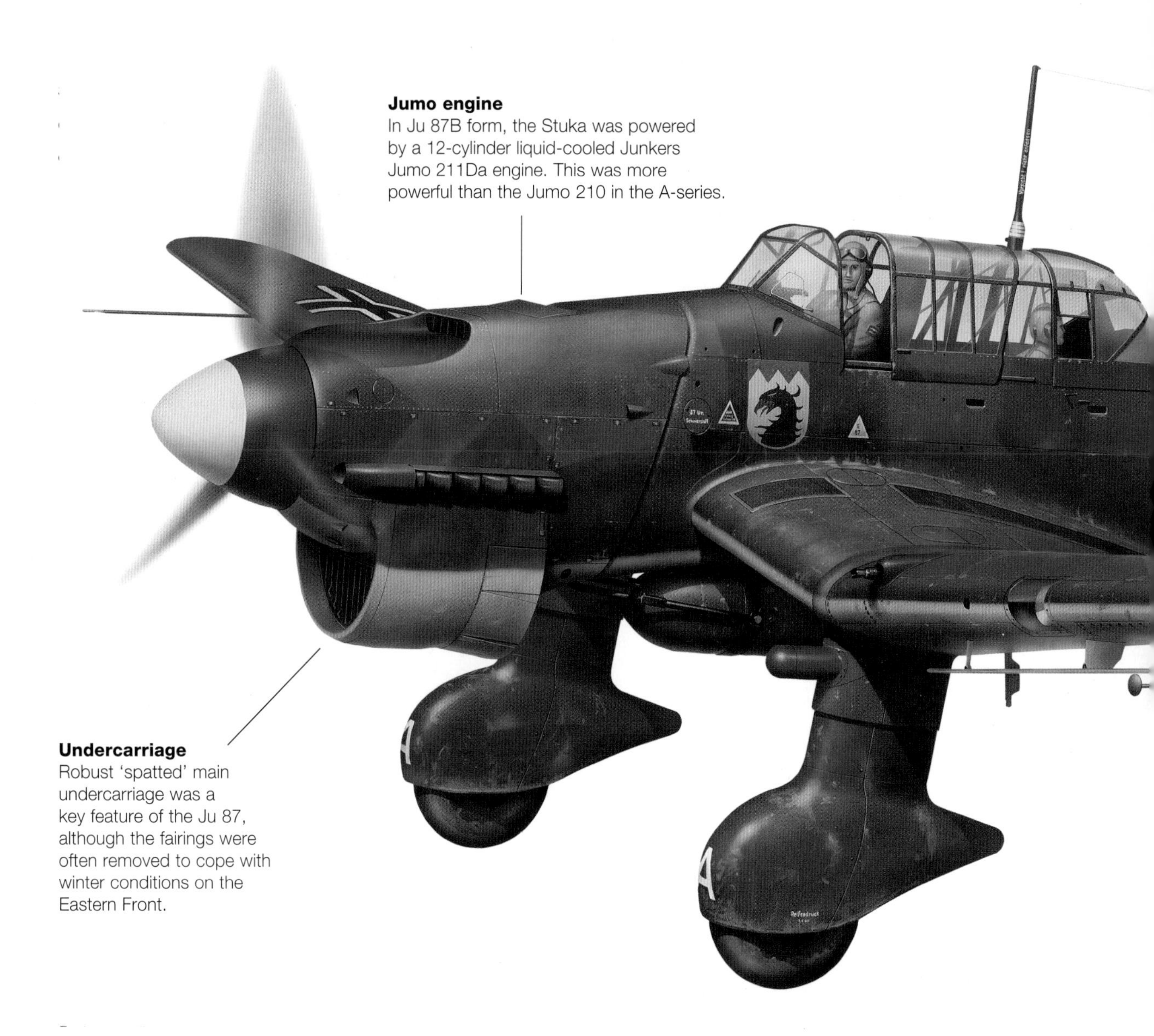

prototypes were followed by a pre-production batch of Ju 87A-0 aircraft, now with the Jumo 210Ca engine.

When the initial Ju 87A-1 production version entered service in spring 1937, it soon began to replace the Henschel Hs 123 biplane in the close support units of the fledgling Luftwaffe. Small numbers of Ju 87A-1s and B-1s were deployed by the Condor Legion for testing under operational conditions in the Spanish Civil War during 1938–39. Here, they met little in the way of fighter opposition and did much to convince Luftwaffe leaders of the viability of the Stuka concept while primarily operating as a tactical bomber. By the time of the invasion of Poland in September 1939, the Luftwaffe included nine Stukagruppen, all equipped with Ju 87s. The first combat mission of World War II was arguably flown by three Stukas as they attacked the Dirschau Bridge over the Vistula 11 minutes before Germany officially declared war on Poland. It was here that the dive-bomber would develop a notorious reputation as a terror weapon, striking lines of communication, bridges, rail targets and airfields.

On occasion, the Luftwaffe further exploited the psychological effect of the Stuka attacks by fitting the aircraft with sirens, so-called 'Trumpets of Jericho'. However, the Ju 87 relied upon air superiority in order to deliver its tactical blows with the required accuracy, and in the course of the Battle of France, it revealed itself

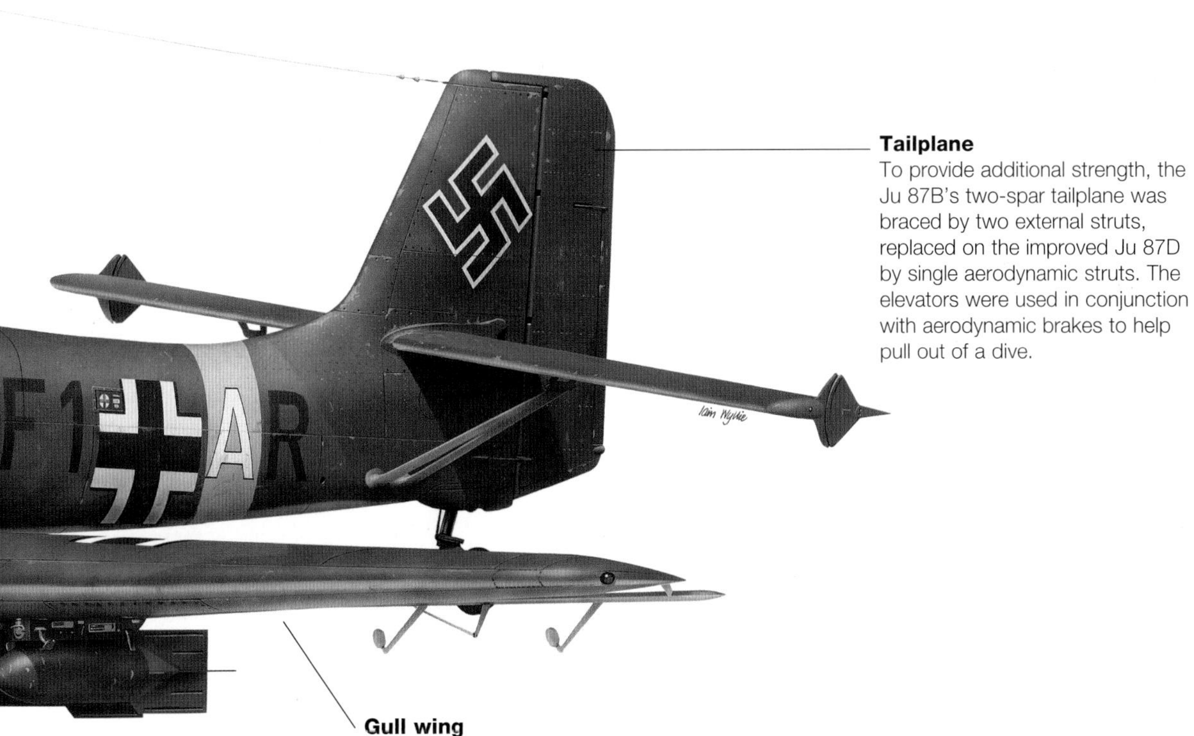

This Ju 87B-2 was on strength with Sturzkampfgeschwader 77, which flew the type, in different versions, throughout the war, including the campaigns in Poland, the Battle of Britain, Greece and the Balkans, and finally on the Eastern Front until 1943.

Ju 87-Ds wear temporary winter camouflage for operations on the Eastern Front in 1942. The aircraft are armed with AB 500 cluster bomb containers and centreline SC 250 bombs.

vulnerable to fighter interception. The Stuka's attack was, in theory at least, simplicity itself. Approaching the target at medium altitude, after carrying out his checks the pilot could either roll on to the target as it disappeared behind the wing or push the nose into a bunt. With the dive brakes open, the correct dive angle, usually 90 per cent, was achieved with reference to marks on the canopy. The target was then manoeuvred into the bombing site by use of the ailerons as the speed built up to around 600km/h (373mph) in 1,300m (4,265ft) of descent. A contact altimeter on the instrument panel was preset to show a warning light at the correct drop height. When this illuminated, the pilot pressed the bomb release button. The bomb swung out on a crutch to clear the propeller before dropping away. At the same time, an automatic pull-out was initiated, which retracted the dive brakes and pulled out of the dive at 6g, aiming to clear the ground by 450m (1,476ft). It was not uncommon for the crew of two to at least grey out slightly during the automatic recovery as the blood drained from their heads.

Specifications: Ju 87D-1

Type:	Dive-bomber and assault aircraft
Dimensions:	Length: 11.50m (37ft 9in); Wingspan: 13.80m (45ft 3in); Height: 3.90m (12ft 9.5in)
Weight:	6,600kg (14,551lb) maximum take-off
Powerplant:	1 x 1,044kW (1,400hp) Junkers Jumo 211J-1 inverted-Vee piston engine
Maximum speed:	410km/h (255mph)
Range:	1,535km (954 miles)
Service ceiling:	6,100m (20,013ft)
Crew:	2
Armament:	3 x 7.92mm (0.31in) machine guns; a bombload of up to 1,800kg (3,968lb)

Norwegian campaign

During the Norwegian campaign, when the Stukas enjoyed success in attacks on British shipping, efforts were made to extend the range of the Ju 87, resulting in the Ju 87R, equipped with underwing fuel tanks. The next stage in the conquest of Europe was the invasion of France, which began on 10 May 1940. As German infantry crossed the Meuse River on the Franco–Belgian border, Stukas hammered the French artillery while in a co-ordinated effort the Luftwaffe's fighters kept the skies clear for the vulnerable dive-bombers. To maintain momentum and

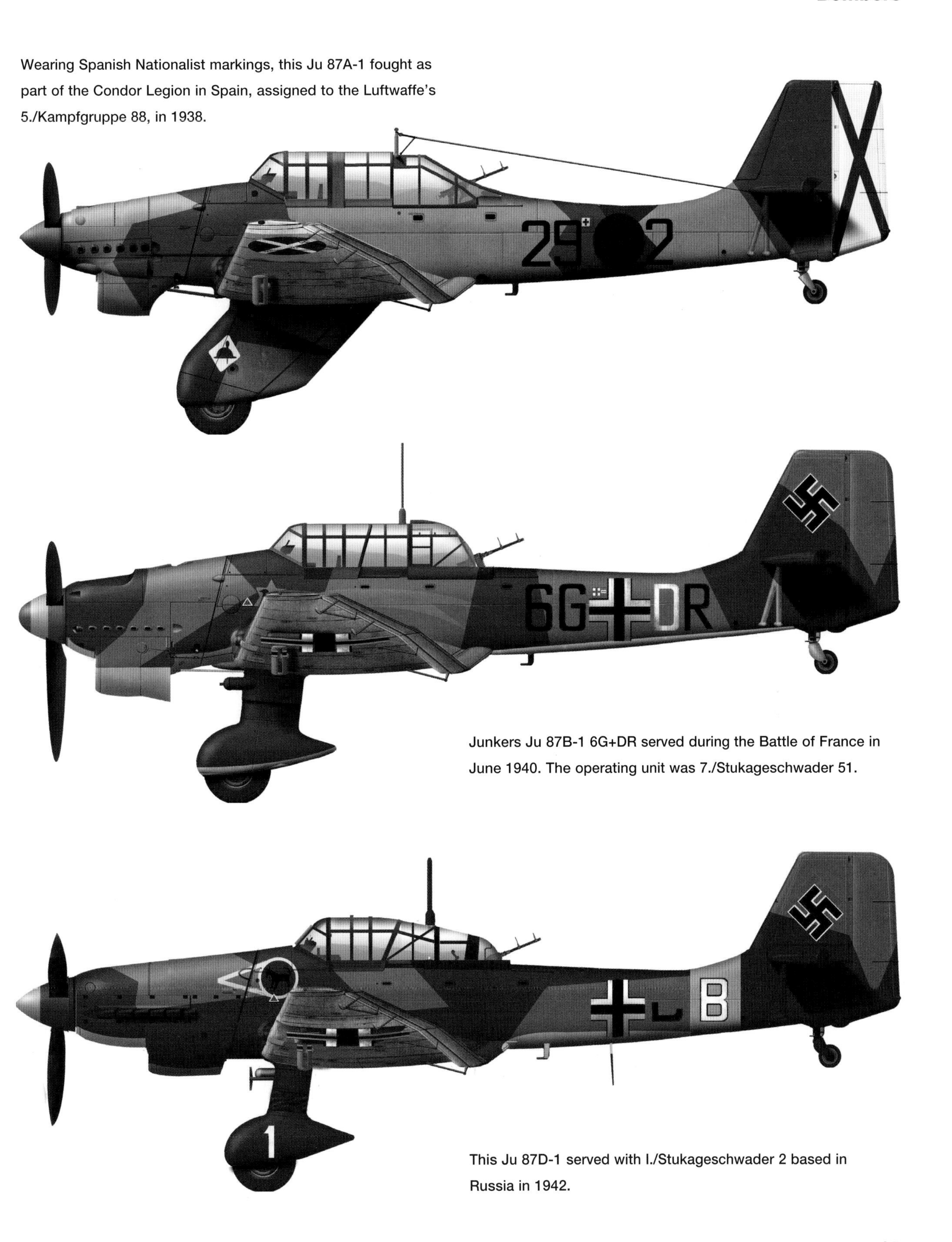

Wearing Spanish Nationalist markings, this Ju 87A-1 fought as part of the Condor Legion in Spain, assigned to the Luftwaffe's 5./Kampfgruppe 88, in 1938.

Junkers Ju 87B-1 6G+DR served during the Battle of France in June 1940. The operating unit was 7./Stukageschwader 51.

This Ju 87D-1 served with I./Stukageschwader 2 based in Russia in 1942.

Hans-Ulrich Rudel

On the Eastern Front, the greatest exponent of the tank-busting Stuka was undoubtedly Hans-Ulrich Rudel. Originally an observer/gunner, Rudel had retrained as a pilot by the time the Nazis launched their invasion of the USSR. On the first day of the campaign alone, he flew four combat missions. In spring 1943, Rudel began flying the anti-tank Ju 87G, the development of which he had contributed to. In the first day of operations at the Battle of Kursk, Rudel used his Ju 87G to destroy 12 T-34 tanks. By the time the war came to an end, Rudel has posted claims for a battleship, cruiser and a destroyer sunk, in addition to 519 tanks destroyed, in the course of around 2,530 combat missions. He was himself shot down on 30 occasions. Rudel was described as Hitler's favourite pilot, a fact attested to by his being the sole recipient of the Golden Oak Leaves to his Knight's Cross.

break through to the open countryside beyond, the Ju 87s were constantly rearmed and relaunched, some crews flying nine sorties in a day.

The Battle of Britain saw the Luftwaffe fighter arm stretched to the limit in defence of the bomber raids, while the Ju 87s were pressed into a strategic bombing role for which they were ill-suited. Tasked against the RAF's 'Chain Home' radar network, it was hoped the accuracy of the Stuka would cripple the vital network. However, the system proved far more resilient than thought and although some damage was caused it was never completely offline. During a 10-day period that summer, the Stuka arm lost 66 dive-bombers and crews, including 17 aircraft from Sturzkampfgeschwader 77 that fell in a single day of operations. With a top speed of 315km/h (196mph) to a Hurricane's 540km/h (336mph), the Ju 87 was highly vulnerable to British fighters even when escorted by Bf 109s. Nevertheless, before they were withdrawn from the theatre, the Stukas left their mark in heavy raids launched against British airfields along the south coast.

A change of fortunes was presented by the war in the Mediterranean where the Stukas of X. Fliegerkorps were ordered to find and sink the Royal Navy carrier HMS *Illustrious* that had been responsible for the attack on the Italian fleet in November 1940. Estimating that four hits would sink her, the Ju 87 crews trained against floating mock-ups and on 10 January 1941 their opportunity came. Within range of the Stuka base in Sicily, over the course of six minutes *Illustrious* was hit by six bombs, one of which pierced her armoured flight deck. Despite this, she was able to reach Malta, where over two weeks she was repaired sufficiently to depart for a year-long refit. The Stuka units involved in the attack would next turn their attention to North Africa. Here, they were involved in operations against the Free French forces at Bir Hakeim and, crucially, in the Second Battle of Tobruk, where the Germans advanced to the Nile Delta.

Battle of Crete

At this time, German Ju 87 units were also involved in the fighting in the Balkans, supporting Italian forces. Perhaps their greatest contribution was during the Battle of Crete, where with limited air opposition they were again able to roam unhindered and acted as long-range artillery in support of the German *Fallschirmjäger* or paratroopers. Using a new 50kg (110lb) air burst fragmentation bomb, they spent several days softening up the British defences before the first airborne assaults landed and took crucial points on the island. Once this was done, further men and materiel could be flown in.

A refined version of the crank-winged Stuka appeared in late 1941, in the form of the Ju 87D, with a more powerful 1,044kW (1,400hp) Jumo 211 powerplant. The overall design was also improved to reduce drag with a new engine cowling and canopy. To cope with the increased payload, the Ju87D-5 onwards had extended wingtips to provide greater lift. The Ju87D models saw battle during the invasion of the Soviet Union and were introduced to the campaign in North Africa in 1941.

It was on the Eastern Front that the Stuka enjoyed its greatest success, albeit in a close support role rather than as the dive-bomber originally envisaged. With the Soviets initially having no air cover of their own, the Ju 87 was again able to operate freely. Before the winter of 1941, the Stuka had helped the Germans advance to within 40km (25 miles) of Moscow. Winter, however, would bring new challenges: hydraulic systems froze, engines needed to be warmed by heaters before they could be started and instruments proved unreliable. Worse was to come with

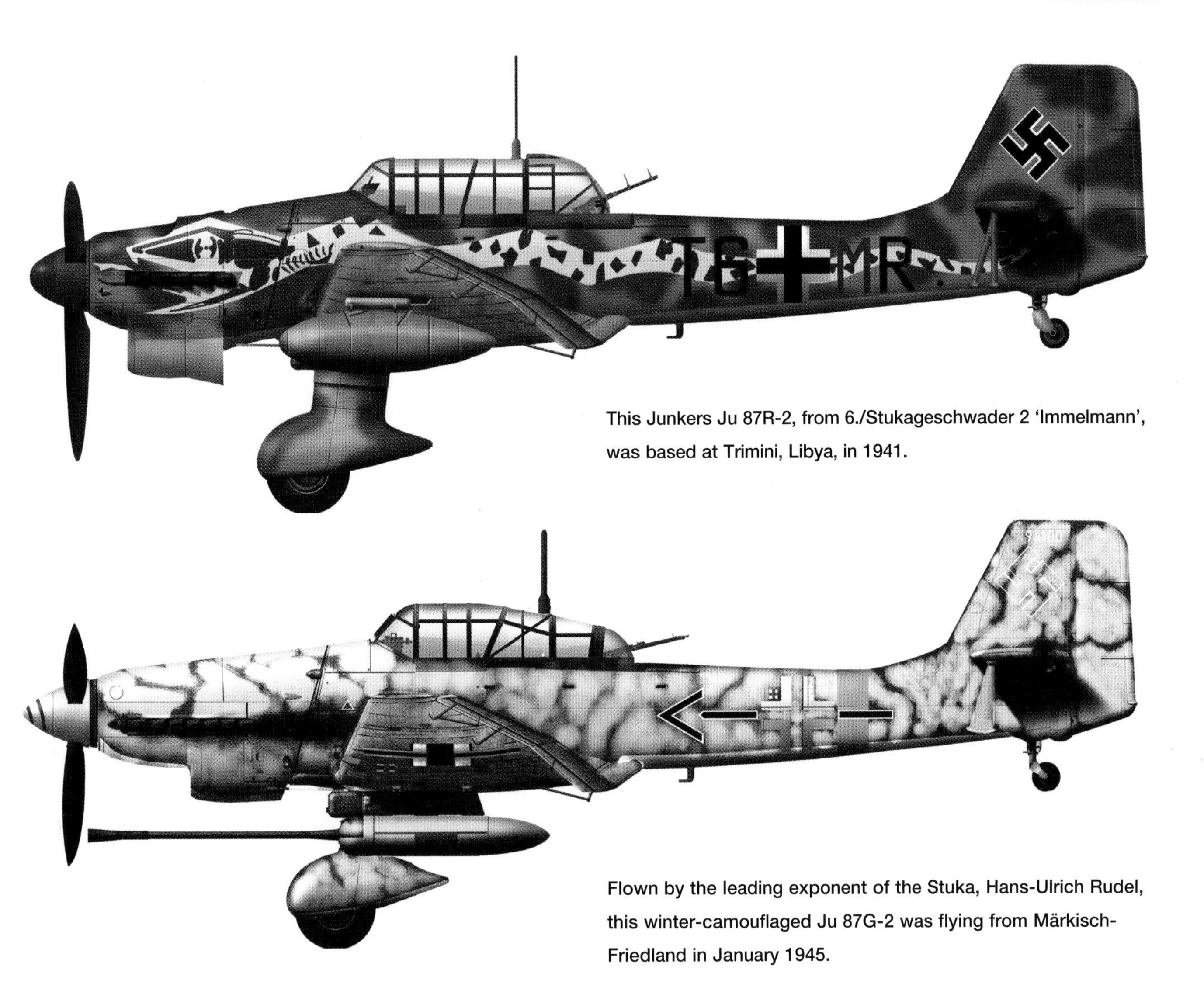

This Junkers Ju 87R-2, from 6./Stukageschwader 2 'Immelmann', was based at Trimini, Libya, in 1941.

Flown by the leading exponent of the Stuka, Hans-Ulrich Rudel, this winter-camouflaged Ju 87G-2 was flying from Märkisch-Friedland in January 1945.

the thaw of 1942 as the Soviet Air Forces at last began to field aircraft that could counter the Luftwaffe. With dive-bombing becoming untenable, the Sturzkampfgeschwader switched to operating at low level armed with armour-piercing bombs and gun pods housing 20mm (0.79in) cannon. From the Ju 87D-3 onwards, greater armour protection was also provided in recognition of the increased vulnerability to ground fire. However, as the tide of the war turned against the Luftwaffe, the Ju 87 was found increasingly vulnerable, and by 1943 losses were such that the aircraft was switched to the night-assault role. The Ju 87D-7 featuring extended exhaust pipes over the wings specifically for the role as well as a further increase in power with the Jumo 211P engine.

The Ju 87G represented a specialist anti-tank aircraft, armed with a pair of 37mm (1.5in) cannon under the wings. This, the final operational version of the Stuka, served with seven Staffeln. Converted from Ju 87D-5 models, the Ju 87G had the dive brakes deleted as surplus to requirements. Originally designed for anti-aircraft use and relatively heavy, the weapons had a muzzle velocity of 850m (2,789ft) per second, making them suitable for penetrating the toughest of armoured targets. As an alternative to the guns, the Ju 87G could carry bomb armament.

Total Ju 87 production amounted to 5,709 aircraft. When the conflict came to an end, a total of 125 Stukas were still on strength with the Luftwaffe, albeit now replaced in their primary close support role by the Focke-Wulf Fw 190 fighter. As well as service in German hands, the Ju 87 was also employed during the war by Bulgaria, Hungary, Italy and Romania, while Japan had received a sole Ju 87A-1 in 1938 for evaluation.

Heinkel He-111 (1935)

Ostensibly designed as a fast passenger and mail carrier, the He 111 would become one of the most well-known bombers of World War II. Seeing service in the Spanish Civil War, the lessons learned would pave the way for the definitive glazed-nose variants that would serve from the invasion of Poland onwards.

The Heinkel He 111 was originally designed in response to a 1934 contract from Deutsche Lufthansa for a high-speed mail and passenger aircraft. Even at this stage, one eye was kept on developing it into a bomber for the secretly re-formed Luftwaffe. The prototype first took flight in February 1935 and shared a family resemblance with the He 70 single-engined fast passenger plane. Powered by a pair of 562kW (754hp) BMW VI engines, the He 111 had an elliptical wing and a clean fuselage that tapered to the tail, where the horizontal and vertical stabilizers were also elliptical. The main difference to the wartime aircraft was the conventional nose and stepped windscreen for the cockpit. Even with the relatively low-power engines, the prototype He 111 could reach 344km/h (214mph) and there was immediate interest from the Luftwaffe after its first flight on 24 February 1935, two days before its own existence was officially announced. In fact, Lufthansa would only use a few aircraft as mail carriers and no

Crew
The standard crew was five. The pilot sat back in the glazed section, offset to port. The navigator/bombardier sat alongside for take-off, but for operations moved forwards to the extreme nose. In the rear was the radio operator/dorsal gunner. Two further gunners were carried, to operate the weapons in the beams and ventral gondola, which was known to the crew as the 'Stertebett' (death bed).

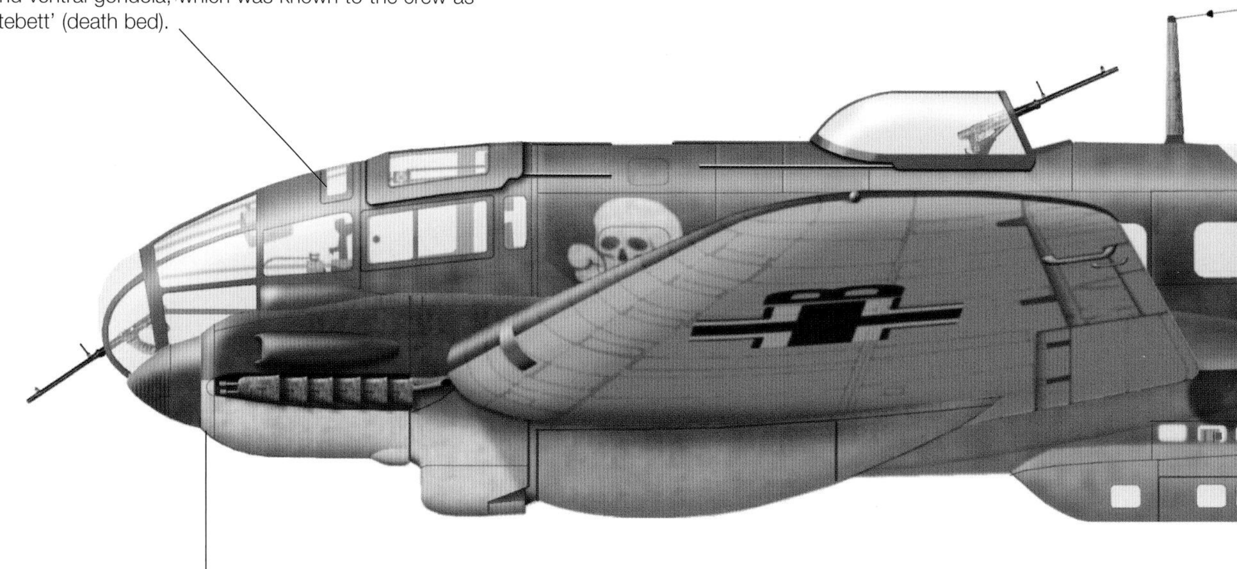

Powerplant
The He 111P powered by the 809kW (1,085hp) DB 601A-1. Given demand for that engine the otherwise identical He 111H-0/H-1 aircraft had the Jumo 211A-1 rated at 753 kW (1,010 hp).

commercial passengers would ever fly on one, although four examples would be converted for clandestine aerial photography in Lufthansa markings.

Although 10 He 111A-0 bombers were built, the performance was considered inadequate with the power available when carrying a bombload and they were soon sold to China. The He 111B would therefore be the first bomber version to see frontline service, in late 1936. These featured the Daimler-Benz DB 600CG with 709kW (951hp) giving a top speed of 370km/h (230mph), which was faster than many contemporary fighters. Defensive armament comprised three 7.92mm (0.31in) machine guns on flexible mounts in the nose, a hatch on the top of the fuselage and a retractable dustbin underneath it. Up to 1,500kg (3,307lb) of bombs could be carried internally in a slightly unusual arrangement: due to the low mounting of the wing, they had to pass through the wing spars. With the size of the opening fixed, the bombs were hung vertically in two racks either side of the centreline. These could carry four bombs, each of which was suspended nose up, causing it to topple through 180 degrees as it fell from the aircraft. Due to the size of the gap in the wing spar, larger bombs would have to be carried under the wings on later models.

Condor Legion

By 1937, the Condor Legion supporting the Fascist forces in the Spanish Civil War was finding the Ju 52 transports it was using as makeshift bombers were too vulnerable to the fighters the Republican forces fielded. Consequently, in February of that year the first three He 111Bs arrived to operate alongside a similar number of Do 17Es in

A Heinkel He 111P of 2./KG54 based in France during 1940. KG 54 were known as the *Totenkopf*, literally 'Dead person's head', hence the skull and crossbones marking on the front fuselage.

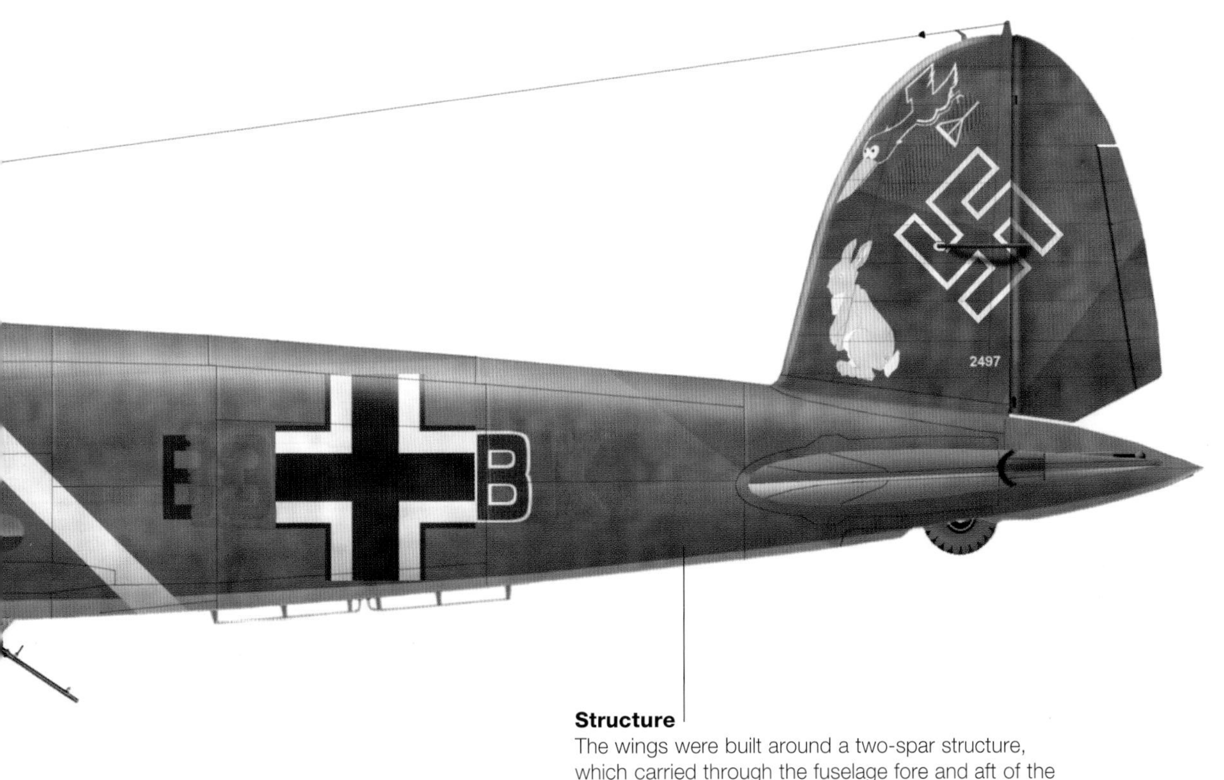

Structure
The wings were built around a two-spar structure, which carried through the fuselage fore and aft of the bomb bays. Interspar fuel tanks were situated inboard and outboard of the engine nacelles. The rear fuselage was largely empty, providing stowage space for the master compass and emergency dinghy.

Kampfgruppe 88. Although both carried a maximum bombload of 1,500kg (3,307lb), the He 111 had a 60km/h (37mph) speed advantage over the Dornier. Able to outpace the I-15 and I-16 that had become the Republican Air Forces' primary fighters, the He 111 crews began to develop a sense of invulnerability that would not endure the coming world war. Meanwhile, the B models were being replaced by the more powerful He 111E with Jumo 211A-1 engines producing 746kW (1,000hp) and able to carry 2,000kg (4,409lb) of bombs. At the civil war's conclusion in March 1939, just under 100 He 111Bs and Es had served with the Condor Legion, with 22 lost to enemy action in two years of heavy fighting. The remaining aircraft were left for the Spanish Air Force, the Luftwaffe being in the process of re-equipping with the He 111P.

He 111P variant

The DB 601-powered P variant would introduce the glazed streamlined nose that would distinguish the type for the rest of its production. Designed to address pilot complaints regarding visibility downwards and while taxiing, the whole cockpit section was comprised of glazed panels save for the bombardier's position at the right lower area where his site was positioned. The pilot, meanwhile, sat on the left to give him a clear view ahead, the instrument panel being on the cockpit ceiling above. To avoid distortion from rain on the windscreen making landing in poor weather challenging, the pilot's seat could be raised, allowing his head to pass through a sliding panel in the roof. A retractable windscreen then provided protection from the elements. Meanwhile, protection for the gun positions was improved with a cupola for the dorsal gunner and a gondola under the fuselage provided a more streamlined position for the lower gunner. The wing had been simplified to ease production with straight leading and trailing edges. Originally used on the He 111F and G models, the P was the first major production variant to feature it. Despite being 25 per cent heavier than the B, the reduced drag from the redesigned nose gave the P an 8km/h (5mph) speed advantage.

The He 111P equipped around half of the operational He 111 units at the start of the war in Europe in 1939. With operations in Poland and Norway seeing little in the way of modern fighter opposition, the pattern of the Spanish Civil War was repeated, with the Kampfgeschwader seeing relatively few losses. Experience in Poland did, however, lead to the He 111P-4, which gained extra armour protection, a forward-firing gun in the gondola and on either beam, and the ability to carry bombs larger than 250kg (551lb) on external racks. At the same time, the internal bomb racks could be replaced

Rugged Heinkel

Although the He 111's defensive armament was considered weak by its crews, it was able to absorb a lot of punishment. Learning lessons from the Spanish Civil War, the Luftwaffe developed self-sealing fuel tanks. The tank itself was surrounded by three layers, the first 3mm (0.118in) chrome leather, the next 3mm unvulcanized rubber, and then a 0.5mm (1/100in) covering of lightly vulcanized rubber. When a bullet or metal fragment pierced the tank, the leaking fuel would react with the outer layer of rubber causing it to swell and form a seal.

From late 1939, it would be fitted as standard to Luftwaffe bombers and was highly effective against the 7.7mm (0.3in) rounds then in use by the RAF. During the invasion of Norway, one He 111 was intercepted by Hurricanes and hit 233 times. Despite this, it managed a four-hour return flight to Trondheim at 160km/h (100mph), the undercarriage having locked down when the hydraulics failed, creating a high amount of drag.

Specifications: He 111P-4

Type:	Medium bomber
Dimensions:	Length: 16.4m (53ft 10in); Wingspan: 22.6m (74ft 2in); Height: 4m (13ft 2in)
Weight:	13,300kg (29,322lb) maximum take-off
Powerplant:	2 x 820kW (1,100hp) DB 601A-1 V-12 piston engines
Maximum speed:	395km/h (245mph)
Range:	2,400km (1,491 miles)
Service ceiling:	8,000m (26,247ft)
Crew:	5
Armament:	5 x 7.92mm (0.31in) MG 15 machine guns; a bombload of up to 2,000kg (4,409lb)

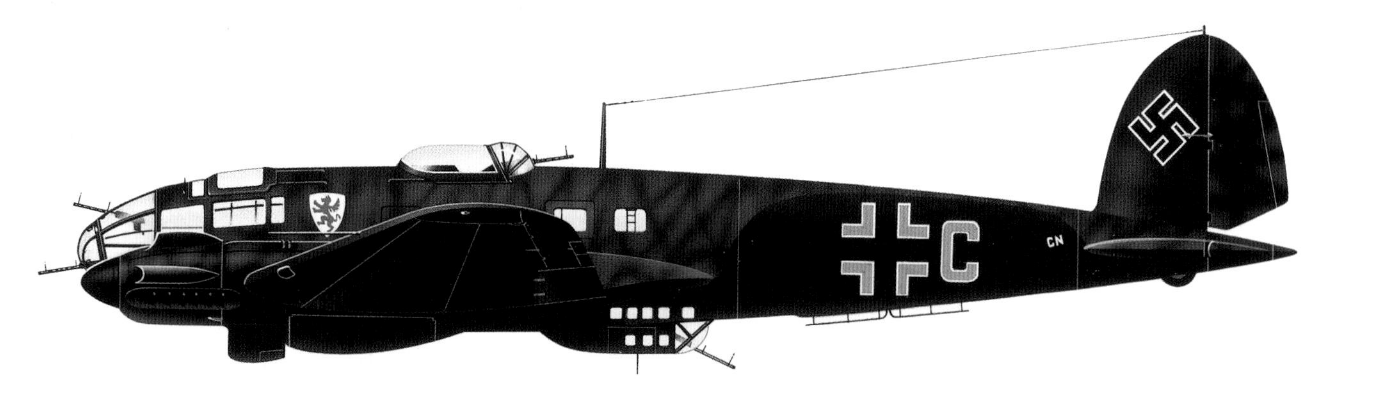

by an extra fuel tank. KG 54 would be one of the longest users of the He 111P, operating the type during the offensive in the West, including the devastating attack on Rotterdam on 14 May 1940, which saw almost 91 tonnes (100 US tons) of bombs dropped on the heart of the city.

Kampfgeschwader 55 operated this He 111P-2 on the 'Night Blitz' nocturnal raids against the United Kingdom in late 1940 and early 1941, flying from Dreux and Villacoublay in France.

Battle of Britain

He 111Ps were still active during the Battle of Britain, generally flown by staff crews, with the commander of KG 55 being shot down over RAF Middle Wallop in one. KG 55 were still using He 111P-2s during the night raids over Britain in late 1940 while a handful were still in active service when the attack on the Soviet Union began in June 1941. However, with the DB 601 engine being prioritized for the Bf 109 and 110, they were rapidly superseded by the Jumo 211-powered He 111H in frontline service. Consequently, only around 400 P models were built before production ended in early 1940.

With the central aircraft still bearing its factory radio code of NO+GO, these He 111Ps are awaiting delivery to their operational squadrons.

Junkers Ju 88 (1936)

One of the most versatile warplanes to see service in World War II, the Ju 88 excelled in roles as diverse as medium bomber, anti-shipping strike and close air-support and was a mainstay of the Luftwaffe throughout the conflict.

The Junkers Ju 88 was first conceived as a high-speed medium bomber and in this form the initial Ju 88 V1 prototype was first flown in December 1936. Of all-metal construction, the aircraft was initially powered by a pair of Daimler-Benz DB 600A V-12 engines. After a total of nine prototypes had been completed, in 1939, construction switched to a pre-production batch of 10 Ju 88A-0 aircraft. By now, the basic design had been revised to include a four-man cockpit with a distinctive multifaceted glazed nose. Since the Ju 88 was also expected to undertake diving attacks, dive brakes were added under the outer wings and the inner wings were strengthened for the carriage of bombs. Conventional bombing was carried out using a sight in a ventral gondola that also held a rearward-firing MG 15 machine gun for self-defence. Dive-bombing attacks, meanwhile, were carried out at an angle of 60 degrees with the pilot using a swing-down sight.

By the outbreak of the war, the Luftwaffe was receiving series-production Ju 88A-1 bombers. Around 60 of these had been completed by the end of 1939 although production was slow to build up from an initial one aircraft a week. After service with a test unit that summer, the first

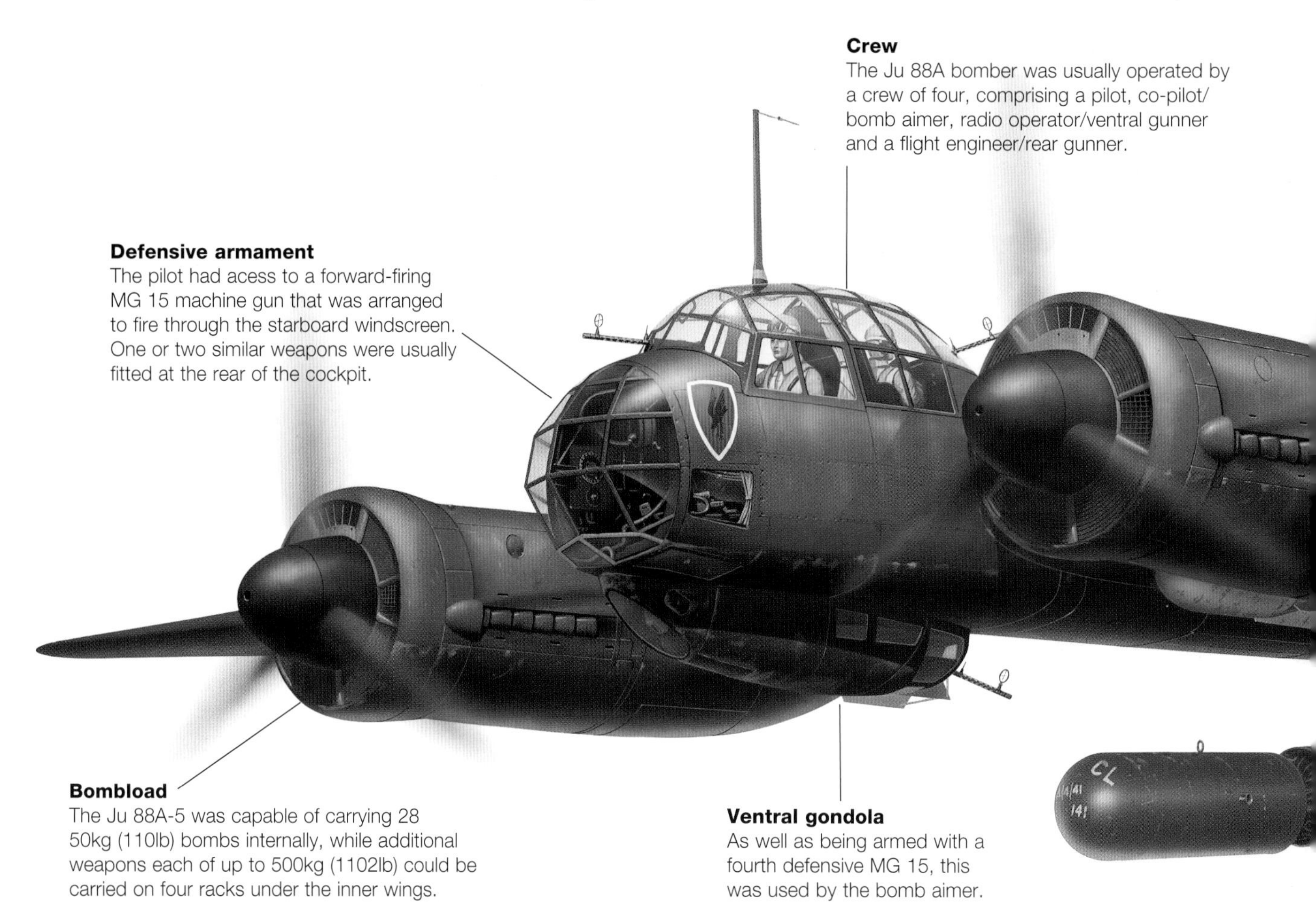

Crew
The Ju 88A bomber was usually operated by a crew of four, comprising a pilot, co-pilot/bomb aimer, radio operator/ventral gunner and a flight engineer/rear gunner.

Defensive armament
The pilot had acess to a forward-firing MG 15 machine gun that was arranged to fire through the starboard windscreen. One or two similar weapons were usually fitted at the rear of the cockpit.

Bombload
The Ju 88A-5 was capable of carrying 28 50kg (110lb) bombs internally, while additional weapons each of up to 500kg (1102lb) could be carried on four racks under the inner wings.

Ventral gondola
As well as being armed with a fourth defensive MG 15, this was used by the bomb aimer.

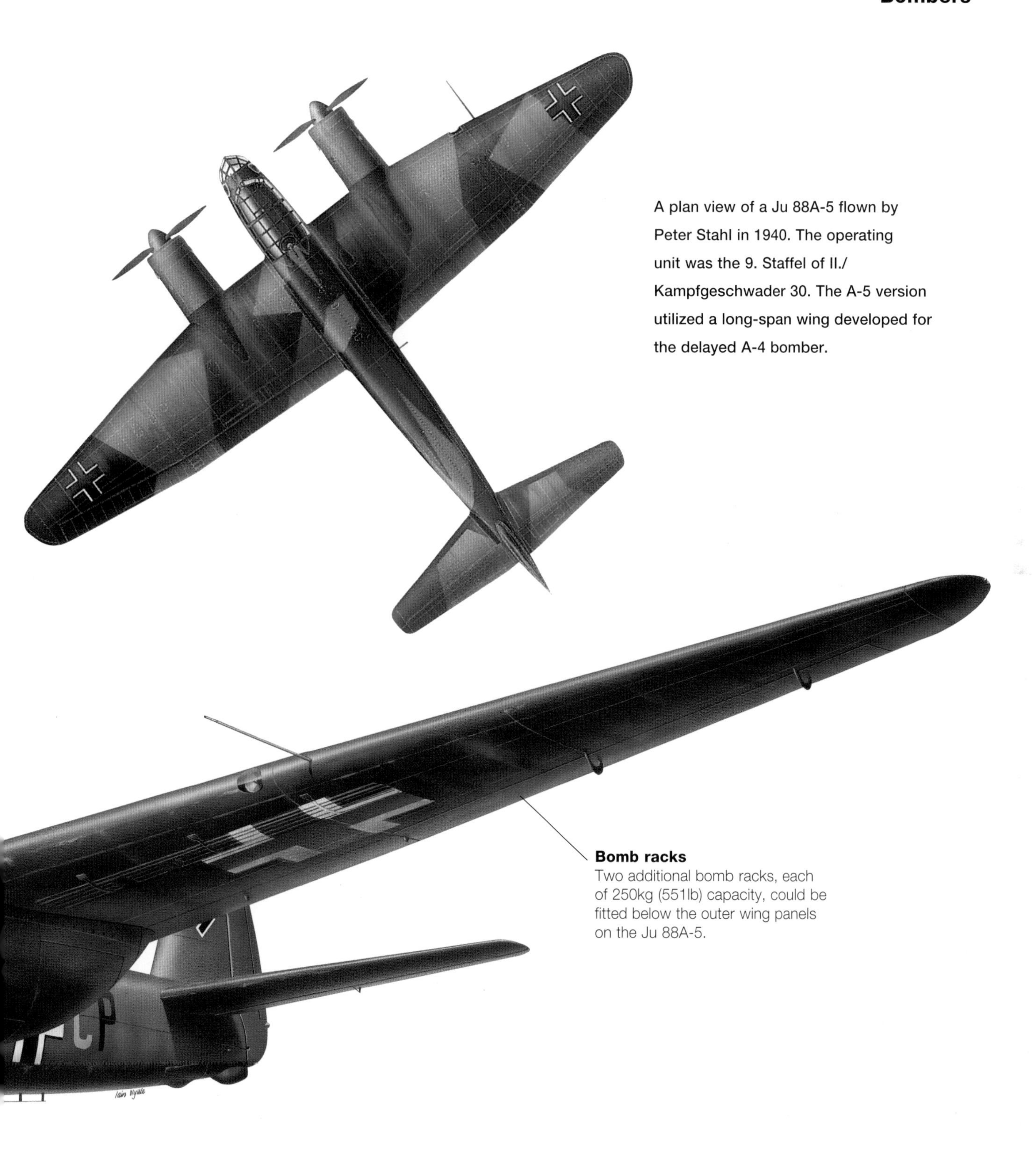

A plan view of a Ju 88A-5 flown by Peter Stahl in 1940. The operating unit was the 9. Staffel of II./ Kampfgeschwader 30. The A-5 version utilized a long-span wing developed for the delayed A-4 bomber.

Another KG 30 Ju 88A-5. Based at Westerland-Sylt, the I. Gruppe of this wing was the first to use the Ju 88 in combat, its initial mission involving an anti-shipping attack on British warships in the Firth of Forth on 26 September 1939.

A profile view of the Junkers Ju 88A-5 flown by Peter Stahl during the Night Blitz over Britain in 1940.

combat missions were carried out by I./Kampfgeschwader 30 in September that year against the British Home Fleet at Scapa Flow. A few days later, however, on 9 October, Ju 88As of 1./KG 30 were the first Luftwaffe aircraft to be shot down over the mainland UK, and the first to be shot down by Spitfires of Fighter Command.

A total of seven Gruppen had converted to the Junkers bomber by the time Germany invaded Norway, and production had increased to 300 aircraft a month. 1940 also saw the appearance of the Ju 88A-2 version, with provision for rocket-assisted take-off gear. This was followed by the A-4 and similar A-5 with increased wingspan and strengthened landing gear, the A-5 being something of a compromise, making do with the Jumo 211B-1 while difficulties were resolved with the 211J used in the A-4.

The Ju 88 was heavily involved in the Battle of Britain with their first target again a naval base, this time at Portland on the south coast. Here, they carried out dive-bombing attacks on the fuel installations while the escorting fighters held off aircraft of Fighter Command, although ultimately five Ju 88As were lost in the raid. The following day, 12 August, saw KG 51 attacking yet another naval base, this time at Portsmouth, and again causing damage but losing 10 aircraft to the defenders. Much like the RAF at the beginning of the war, the Luftwaffe was finding daylight raids against well-defended targets rapidly took a toll on the attacking bombers. By 15 September, Battle of Britain day, the Ju 88 units were starting to conduct night attacks, although daylight raids were still being undertaken. Forty-six Ju 88As from LG 1, KG 51 and KG 54 bombed London over a four-hour period. This took advantage of the RAF's poor night air defence system at this early stage of the war and none of the Junkers were lost to enemy action. The Night Blitz, as it came to be known, would last until the May of 1941 when units started to be withdrawn in preparation for Operation Barbarossa.

At the same time, the autumn of 1940 saw I./KG 30 moving to Sicily to support Italian forces fighting in Albania

Mistel

In the final, desperate months of the war, the German war machine called upon ever more unconventional ideas in an effort to turn the tide of the fighting. One such was the use of redundant Ju 88s as unmanned missiles. In the Mistel (Mistletoe) composite aircraft weapon, the Ju 88 provided the explosive-filled lower component, which was guided to the vicinity of the target by a manned fighter mounted above the centre fuselage in a piggyback fashion. Once the bomber had been directed at the target, it was released, and the fighter returned to base.

The initial Mistel I combined the Ju 88A-4 and a Bf 109F fighter. Among the airframes diverted to the Mistel programme were the small batch of Ju 88G-10 ultra-long-range heavy fighters, with lengthened fuselages for additional fuel carriage. The Mistel combinations found some success in the last months of the conflict in Europe.

Specification: Ju 88A-5

Type:	Medium bomber
Dimensions:	Length: 14.4m (47ft 3in); Wingspan: 20.00m (65ft 7.5in); Height: 4.8m (15ft 9in)
Weight:	14,000kg (30,865lb) maximum take-off
Powerplant:	2 x 895kW (1,200hp) Junkers Jumo 211B-1 liquid-cooled piston engines
Maximum speed:	470km/h (292mph)
Range:	1,790km (1,112 miles)
Service ceiling:	8,200m (26,903ft)
Crew:	4
Armament:	5 x 7.92mm(0.31in) MG 15 or MG 81 machine guns; 500kg (1,102lb) of bombs in an internal bay, up to 3,000kg (6,614lb) externally

and Greece. One of the Ju 88's most significant acts of the war would occur on the night of 6 April 1941 when Operation Marita, the invasion of Greece, was launched. After laying mines in the port of Piraeus, an aircraft piloted by Hauptmann Hans-Joachim Herrmann bombed the freight Clan Frazer that was unloading at the jetty. With 227 tonnes (250 US tons) of munitions on board when the bombs hit, the explosion disintegrated the 10,886-tonne (12,000-ton) cargo ship, smashed windows 11km (7 miles) away, and effectively rendered British forces without a port through which they could land supplies. The one attack had significantly shortened the German campaign to take over the country.

Invasion of the Soviet Union

Eighteen Ju 88 Gruppen were committed to the attack on the Soviet Union when it commenced on 22 June 1941. Attacking in almost complete surprise, the bombers faced no air- or ground-based opposition despite the clear skies. Their first targets were airfields, which were attacked in shallow dives from 3,000m (9,843ft) with a mix of general-purpose bombs. Soviet aircraft were still lined up in ranks, making the German attack all too easy. Even when fighters appeared, the Ju 88s were able to outrun the Russian Polikarpov I-15s and I-16s. As the war progressed, the Ju 88s on the Eastern Front would also be pressed into the close support role, although not without significant losses, the type being more vulnerable than the Do 17 or He 111.

The Ju 88A remained the principal medium bomber of the Luftwaffe until the end of the war, with a variety of sub-variants. The A-6 was essentially the same as a Ju 88A-5 but with a heavy framework running from wingtip to wingtip ahead of the propellers and fuselage. This was intended to cut the mooring lines for barrage balloons, however in service, the extra weight was found to reduce the speed significantly. Most aircraft were subsequently reworked as A-6/U, losing the framework and gaining a FuG 200 Hohentwiel radar for the anti-shipping role.

For operating in desert regions, the A-9 was a modified A-1, while the A-10 and A-11 were similarly modified A-4s

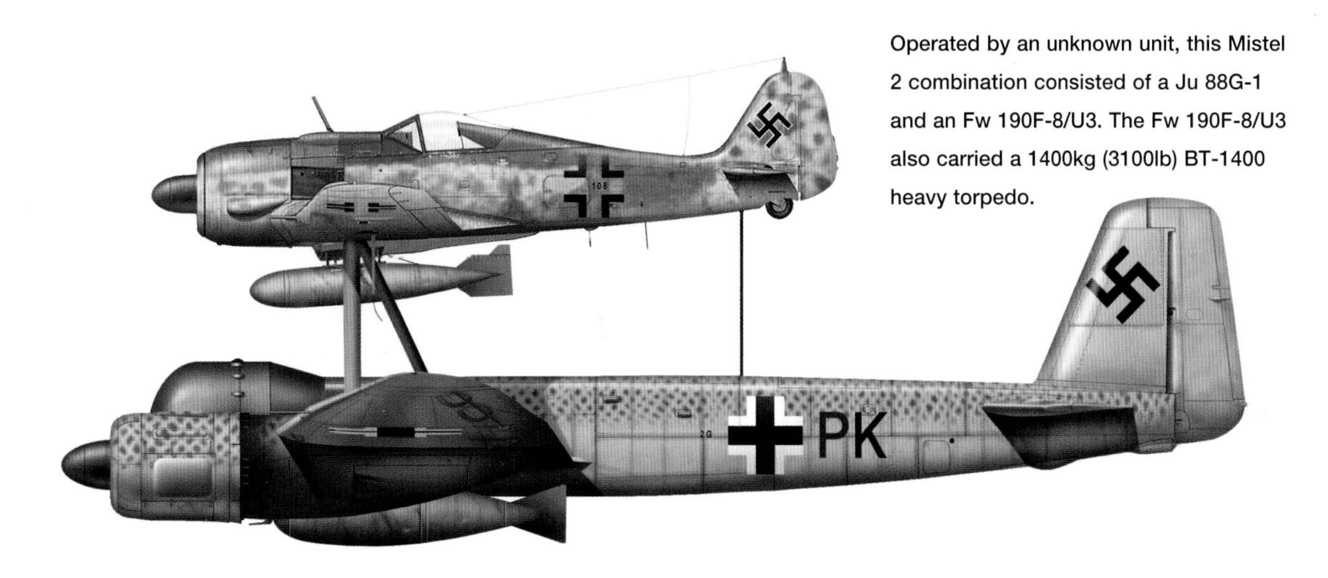

Operated by an unknown unit, this Mistel 2 combination consisted of a Ju 88G-1 and an Fw 190F-8/U3. The Fw 190F-8/U3 also carried a 1400kg (3100lb) BT-1400 heavy torpedo.

and A-5s. These featured sand filters for the engines, sun blinds and a desert survival kit among other modifications and were used principally to support the Afrika Korps. The A-13 and A-14 were modified for the ground-attack role with extra armour and provision for gun pods to be mounted under the wings, or in the ventral gondola for the A-14. They were also employed in the anti-shipping role, the weight of firepower being devastating to smaller vessels. The A-15, meanwhile, featured an enlarged bomb bay allowing up to 3,000kg (6,614lb) of bombs to be carried internally.

A further anti-shipping variant was produced, based on the A-4: the A-17 was modified for the torpedo attack role being able to carry one under each wing. The maritime-attack Junkers saw service in the Balkans, the Mediterranean and the Atlantic. Perhaps the most successful missions were those flown against the Allied North Cape convoys in 1941–42, when Norway-based Ju 88As from III./ KG 26 and KG 30 sank around 27 merchantmen and seven naval vessels. Ju 88As were also involved in the attacks on Operation Pedestal, a relief convoy to Malta, as both dive- and torpedo-bombers, again helping to inflict heavy losses. Conversely, when the Allies landed at Normandy on D-Day, 125 Ju 88A-17s were positioned to attack the invasion fleet. Despite there being over 4,000 potential targets, including battleships, cruisers and landing craft, the attacks on the night of 6 June failed to hit a single one. A combination of Allied night-fighters, friendly fire from German flak units and fire from the fleet itself prevented any successful attacks from being carried out.

The Versuchskommando fur Panzerbekampfung (Anti-tank Test Command) employed this Ju-88P-1 for trials work. The unit was formed at Rechlin in July 1942 and sent elements to the Eastern Front for combat trials.

A Junker Ju-88A-17. Converted from the Ju 88A-4, the A-17 was designed for the anti-shipping role and could carry two LT F5b torpedoes. This example was operated by 3./KG 26 from Bardufoss, Norway, in February 1945.

Reconnaissance version

The Ju 88Ds were dedicated reconnaissance versions. Again using Jumo 211 engines, these had the bomb bays repurposed to house fuel tanks and two or three cameras were placed in the rear fuselage. The D-3 and D-4 were tropicalized versions of the D-1 and D-2 respectively while the D-5 standardized on a three-camera installation. The H-1, meanwhile, was a maritime reconnaissance version intended to locate convoys in the vastness of the Atlantic Ocean. For this role, the fuselage was lengthened by 2.29m (7ft 6in) to allow extra fuel to be carried and the Hohentwiel radar was fitted to the nose.

A handful of specialized anti-tank Ju 88Ps were built and tested in combat. The P-1 carried a 75mm (3in) PaK 40 cannon, based on an army anti-tank weapon. Two 37mm (1.5in) cannon were carried in a ventral fairing on the P-2, which was trialled against US heavy bombers as well as tanks. Despite the promise of the weaponry, its weight and that of the additional armour for the low-level environment meant the aircraft were found to be too unwieldy to be introduced into service.

Ju 88S

The final bomber version to be produced, although in small numbers, was the Ju 88S. To improve performance, the airframe was cleaned up and the nose glazing streamlined. Power was from the BMW 801 series of engines, giving the Ju 88S-1 a top speed of 610km/h (379mph).

As the Allied armies pushed towards Germany, their increased focus on the oil industry made it progressively harder for the Luftwaffe to conduct operations. Through the second half of 1944, Ju 88 units were withdrawn and then disbanded as it became untenable to operate them. Some airframes went to the Mistel programme while others sat on the ground never to fly again, no Ju 88 bombers serving until the end of the war.

In total, 14,500 Ju 88s were built. As well as with the Luftwaffe, the Ju 88A-4 also served with the Finnish, Hungarian, Italian and Romanian air forces during the war. The French Air Force also operated them from around September 1944 using commandeered examples. Refurbished by SNCASE, the A-4s were used against German positions in the Gironde estuary and would remain in service into the 1950s.

The *Totenkopf* badge identified KG 54, which operated this Ju 88A-5 on standard bomber duties. After action in the Battle of Britain, KG 54 served in Russia, before returning to the West in 1944 for nocturnal raids on Britain.

Heinkel He 177 Greif (1939)

Conceived as the Luftwaffe's first strategic bomber, the Greif, or 'Griffin', would have a troubled development that would delay its entry to service to the point that it was surplus to requirements.

Although development started in 1936, the first He 177 would not fly until November 1939 due to the Luftwaffe prioritizing tactical medium bombers, such as the He 111 and Do 17. This was exacerbated by their main proponent of strategic bombing, General Walther Wever, dying in a plane crash on the day the 'Bomber A' specification for what would become the 177 was issued. This called for an aircraft with a top speed of 540km/h (336mph) able to carry 2,000kg (4,409lb) of bombs over a radius of 1,600 km (1,000 miles), or half the payload over 2,900km (1,802 miles).

To achieve this, Heinkel designed a streamlined aircraft with remotely controlled defensive guns, coupled engines and an evaporative steam-cooling system that dispensed with the need for radiators, instead using the surface of the wings. However, although practical for smaller engines, it soon became apparent that the cooling system would not work for the 1,940kW (2,602hp) Daimler-Benz DB 606 engines, being unable to dissipate enough heat, so a

This He 177A-5/R2 hailed from the 4. Staffel, II Gruppe of Kampfgeschwader 100, during the time the unit was based at Bordeaux-Mérignac, France, in 1944.

Weapons
Weapons carried by the He 177A-3 included the Hs 293 radio-controlled missile and, in the A-3/R7 and all A-5 versions, a range of anti-ship torpedoes, including the LT 50 glider torpedo.

Engine
The He 177A-5 standardised with the DB 610 engine (coupled DB 603s) as opposed to the DB 606 (coupled DB 601s) of the earlier variants. Despite tests with one aircraft that had identified and fixed 56 potential causes for engine fires, these problems continued. It was felt that to incorporate the modifications would have severely disrupted the production lines.

conventional higher-drag radiator installation was designed. The remote-controlled gun barbettes, meanwhile, were too advanced for the Luftwaffe, who insisted on conventional manned turrets, which again created more drag. While these issues were being resolved, the Luftwaffe mandated that the He 177 be capable of dive-bombing, bombsights at the time not providing an equivalent level of accuracy for conventional bombing. This required a much stronger, and therefore heavier, structure to withstand the resulting loads on the airframe. The most obvious outcome of the redesign work was the addition of an extra undercarriage leg on each side of the aircraft to support the additional weight. Despite all this, the Luftwaffe were sufficiently impressed with the design to order six prototypes in November 1938.

Test flight

A year later, the first He 177 took to the air for a short test flight. This revealed that the rear control surfaces were slightly undersized, together with a slight vibration and heating issues with the engines. The control surface issue was easily fixed but the last would never be fully resolved. Moreover, the prototype could only achieve a maximum speed of 460km/h (286mph) – well below the original specification – while the range was also similarly affected. Worse was to come as three of the prototypes were lost in accidents, along with their crews. Despite this, 35 pre-production He 177A-0s were ordered for development work, their main contribution being to demonstrate that the aircraft was too big to be a dive-bomber. Test dives would regularly require lengthy repairs to the aircraft on completion and with the development of more accurate bombsights, the requirement was dropped.

The summer of 1942 saw the first operational He 177A-S being received by I./KG 40 at Bordeaux-Mérignac to take over from the ageing Fw 200 in the maritime patrol role. The A-1s were found to be so inadequate they were almost immediately withdrawn and were not replaced until the following year, by the improved He 177A-3. This had

Tail gun
The He 177 included a 20mm (0.79in) MG 151 in the tail, plus seven other machine guns as defensive armament.

Camouflage
The He 177s of KG 40 and KG 100 wore a variety of overwater camouflage, this example being standard for most of KG 100's aircraft.

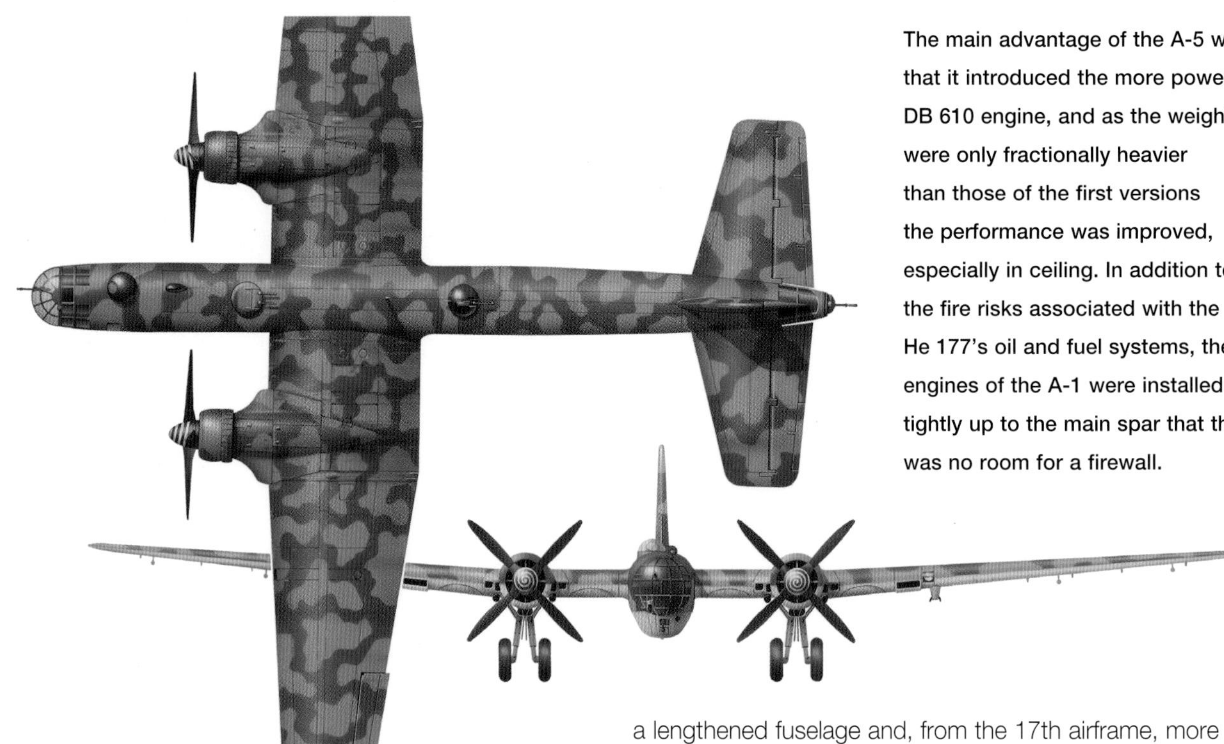

The main advantage of the A-5 was that it introduced the more powerful DB 610 engine, and as the weights were only fractionally heavier than those of the first versions the performance was improved, especially in ceiling. In addition to the fire risks associated with the He 177's oil and fuel systems, the engines of the A-1 were installed so tightly up to the main spar that there was no room for a firewall.

a lengthened fuselage and, from the 17th airframe, more powerful DB 610 engines producing 2,200kW (2,950hp). This would be followed by the similarly powered A5, which also had a strengthened wing to carry heavier loads outboard of the engines. As well as long-range patrols, KG 40 would undertake massed attacks on Allied convoys with the Hs 293 remote-controlled glide bomb. Each Greif could carry one under each wing, or a single example under the forward bomb bay. Boosted to flying speed by a 10-second rocket burn, the bomb had a range of up to 8km (5 miles) when launched from altitude. An operator in the aircraft would track the bomb by flares on its rear and guide it towards a suitable target. Although a number of ships were sunk in the Atlantic and Mediterranean, the bombers were vulnerable to Allied fighters while carrying the bombs, and when trying to control them. On one occasion, a Liberator patrol aircraft managed to disrupt the Heinkels sufficiently to prevent any bombs from hitting their targets. At the same time, the Greif's reliability and poor weather were causing an unsustainable loss rate: the first two Hs 293 attacks had sunk two ships between them while accounting for 12 He 177s and eight crews. KG 40 would soon move to night attacks until, after suffering even worse losses over the D-Day armada, they were withdrawn to Norway.

The He 177 was also used in the conventional bombing role, where it had more success during the Operation

Specifications: He 177A-5/R2

Type:	Long-range heavy bomber
Dimensions:	Length: 22.0m (72ft 2in); Wingspan: 31.45m (103ft 2in); Height: 6.4m (21ft)
Weight:	31,000kg (68,343lb) maximum take-off
Powerplant:	2 x 2,200kW (2,950hp) Daimler-Benz DB 610 coupled V-12 engines
Maximum speed:	360km/h (224mph)
Range:	5,500km (3,418 miles)
Service ceiling:	8,000m (26,247ft)
Crew:	6
Armament:	2 x 20mm (0.79in) MG 151 cannon in the nose and tail; 1 x 7.92mm (0.31in) MG 81 in the nose; 2 x MG 81s in the lower rear cabin; 2 x 13mm (0.51in) MG 131s in the remote control barbette; 1 x MG 131 in the upper turret; a bombload of up to 5,600kg (12,346lb)

Steinbock raids on the UK in early 1944. On these raids, they could carry as much as 5,600kg (12,346lb) of bombs from the Rhine to London, however they often represented only 5 per cent of the 120–200 aircraft in a raid. On the Eastern Front, the situation was even worse since they were operating from poorly equipped airfields. Although some massed raids were carried out, the lack of bulk fuel storage would often result in sorties going unflown as the expected tankers failed to turn up in time, or at all. Inexplicably, when the Soviet offensive began in June 1944, none other than Reichsmarschall Göring ordered KG 1 to take its He 177s and operate them in the low-level anti-tank role. Predictably, they suffered huge losses, losing a quarter of the aircraft involved.

Fuel shortages

When the Greif finally started to enter service in numbers in the summer of 1944 it faced the cruellest of blows when Allied bombing attacks on German fuel installations finally started to bite. With aviation fuel output falling to under 10 per cent of what it had been at the start of the year, it was now simply impractical to operate heavy bombers. A sortie by 80 bombers could use as much as 454 tonnes (500 US) tons of fuel – a day's output for what remained of the fuel industry. The survivors of the production run of 1,094 aircraft would see out the war while sitting at airfields around occupied Europe, where they were strafed by Allied pilots.

The weapon most associated with the Heinkel He 177 in the anti-shipping role was the Henschel Hs 293A missile. These could be carried under the wing or, as here, on a special pylon fitted to the blanked-off forward bomb bay.

Engine trouble

To minimize the frontal area of the engine installation, Heinkel chose the DB 606, and later the DB 610 engines. These were two DB 601, or 605, liquid-cooled V-12s attached to a common gearbox to drive a single propeller. The gearbox also allowed the engines to be individually disconnected from the propeller, permitting them to be operated as individual engines for starting and shutdown. Unfortunately, the engines had an Achilles heel due to the closeness of the two inner banks, which led to the inner exhaust manifold becoming extremely hot and igniting any oil or fuel spills. This was not helped by the tendency for the fuel injectors to leak and the omission of a firewall to save weight. These problems contributed to the appalling loss rate for the early He 177, 19 of the 130 He 177A-0s being lost without even seeing combat. Although a number of recommendations were made to resolve the problems, 56 different causes being identified, the delays in implementing them made it untenable to do so and the nickname *Reichsfeuerzeug*, 'Reich's lighter', would endure.

Henschel Hs 129 (1939)

Essentially the only aircraft of World War II designed specifically for the anti-tank role, the Hs 129 was compromised as an aircraft but highly effective in its intended role. Despite this, it failed to have a major impact on the war due to a failure by the Luftwaffe to prioritize its production.

In a potentially far-sighted move based on its early experience in the Spanish Civil War, in 1937 the Luftwaffe issued a requirement for a close air support aircraft. This would require extensive armour to survive at low level over the battlefield while carrying an armament of at least two 20mm (0.79in) cannon, while being small enough to be propelled by two low-powered engines. Henschel proposed a single-seat low-winged monoplane with fuselage-mounted weapons. To minimize the amount of armour protection required, the cockpit was painfully small, with engine instruments mounted on the inside of the nacelles to save space. Indeed, it was so cramped that the prototype required large control forces due to the short control column. Despite this, the Hs 129 was chosen in preference to the competing Focke-Wulf design, which was based on the Fw

8./Schlachtgeschwader 1's fleet included this Hs 129B-1, based at El Aluin, Tunisia, in February 1943.

Cockpit
The triangular section of the fuselage made for a very cramped cockpit, with little room for instruments. These were located on the inner surfaces of the engine nacelles.

Armour
To protect it and its pilot from the intense small arms fire encountered in the low-level attack mission, the Hs 129 incorporated much armour. The entire nose section formed an armoured 'bath' for the pilot, built from 12mm (0.47in) plate on the undersides and 6mm (0.23in) to the sides. Cockpit glass was 75mm (3in) thick.

Internal armament
This comprised a pair of MG 151/20 20mm (0.79in) cannon and a pair of 7.9mm MG 17 machine-guns mounted either side of the fuselage. This aircraft has the R2 field conversion which added a jettisonable 30mm MK 101 cannon in fairing under the fuselage.

189, primarily on cost grounds – the Henschel being around a third cheaper.

Eight pre-production aircraft were ordered as the Hs 129A-0, which were broadly the same as the prototype. The most distinguishing feature was the triangular cross-section fuselage, with a broad base. This housed the cramped cockpit, which was surrounded by 6mm (0.25in) and 12mm (0.5in) armour plates totalling some 1,080kg (2,381lb) in weight and almost comically small windows made of 75mm (3in)-thick armoured glass. As well as the engine instruments, the weapons sight was also externally mounted in front of the V-shaped windscreen. The weapons themselves consisted of two 20mm (0.79in) MG FF cannon in the fuselage just aft of the cockpit and a 7.92mm (0.31in) MG 17 machine gun in each wing root with distinctive channels recessed into the sides for the passage of the rounds. Power was provided by two 342kW (459hp) Argus As 410A-1 air-cooled V-12 engines with a distinctive spinner ahead of the propeller that drove the pitch control mechanism. The nacelle for each engine was armour-plated and also housed the main landing gear, which remained partially exposed when retracted to reduce damage in the event of a wheels-up landing.

Revised cockpit

Initial trials were not promising, re-emphasizing the shortcomings of the prototype with the lack of power and impeded visibility making it a poor choice for a frontline airplane. Although Henschel had plans for a much-revised larger aircraft, for the sake of expediency, the Hs 129B would only feature a revised cockpit and more powerful engines. The cockpit gained larger windows to significantly improve lookout, although engine gauges and the Revi C 12/C weapon site would remain external, and a less angular nose profile. The engines, meanwhile, were swapped for 522kW (700hp) Gnome-Rhône 14M38

Undercarriage
The single-strut main undercarriage retracted hydraulically into the rear of the engine nacelle, where a portion of the wheel was left exposed to minimise damage in a wheels-up landing.

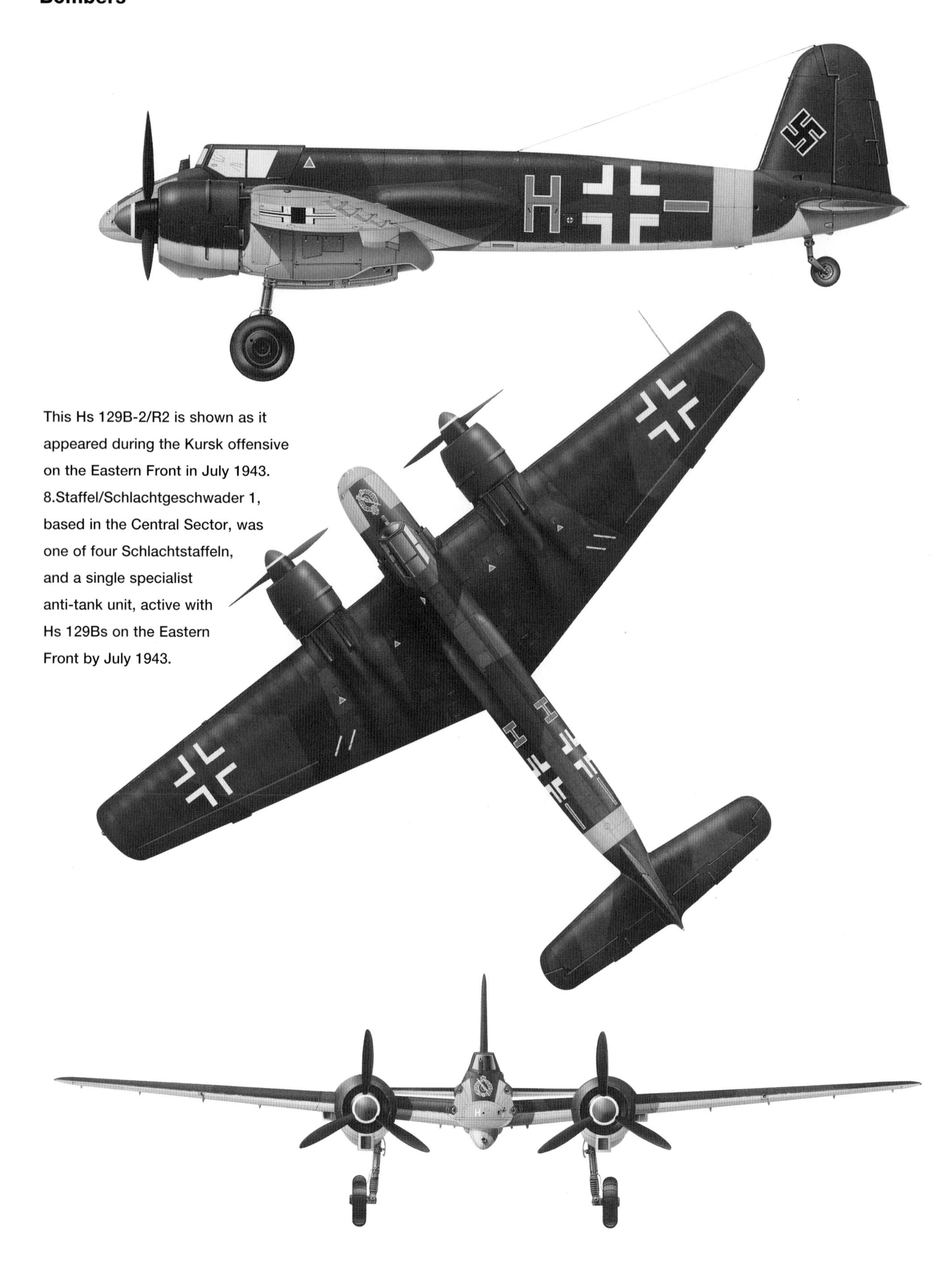

This Hs 129B-2/R2 is shown as it appeared during the Kursk offensive on the Eastern Front in July 1943. 8.Staffel/Schlachtgeschwader 1, based in the Central Sector, was one of four Schlachtstaffeln, and a single specialist anti-tank unit, active with Hs 129Bs on the Eastern Front by July 1943.

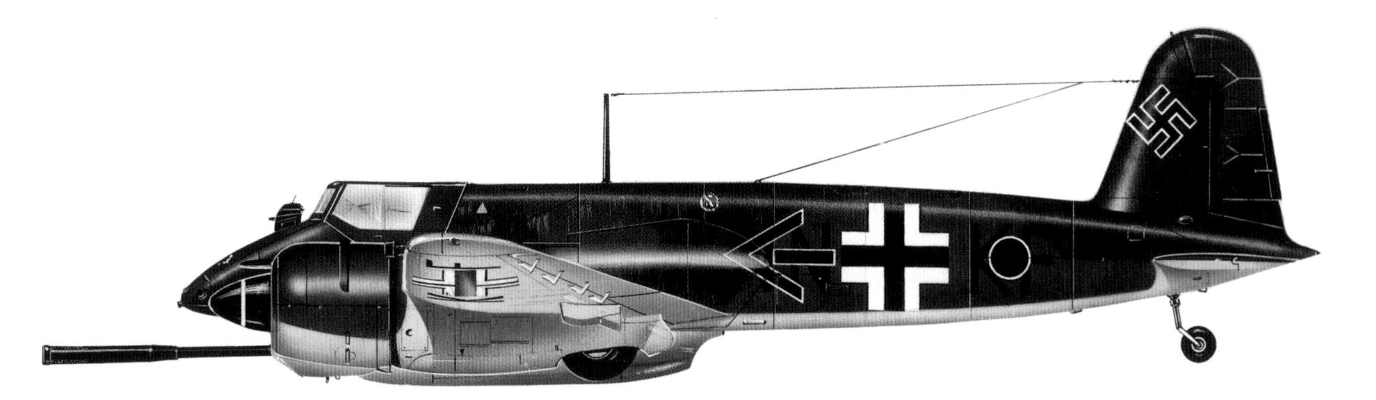

This Hs 129B-3 was flown by 14.(Panzerjäger)/ Schlachtgeschwader 9, one of only two operational units to fly the B-3 with the BK 7,5 cannon carried in a jettisonable fairing.

air-cooled radials acquired from Occupied France. To combat the high control forces, an electric trim system was provided, which could be used to fly the aircraft. Other improvements included replacing the MG FF cannon with the MG 151 of the same calibre but with a higher muzzle velocity and provision for external carriage of a variety of stores. Although still not a pilot's aeroplane, the Hs 129B was much improved on the A model and well suited to the close air support role for which it was intended.

Close support role

It was put into immediate high-priority production due to the invasion of the Soviet Union and experience with the earlier Hs 123s in the close air support role. However, build-up of the planned Schlachtgeschwader force was delayed by a stream of modifications and fixes to the basic design of the Hs 129B, which delayed series production reaching the planned 40 aircraft per month until the middle of 1943. An area that would require continual attention were the Gnome-Rhône engines, which although more powerful than the Argus that they replaced, consistently ran hot and would suffer from the effects of sand and dust in North Africa and the Soviet Union despite multiple attempts to rectify both issues.

At the same time, options to increase the available power were explored. First, the 14M38 version of the Gnome-Rhône was trialled, providing 611kW (819hp), but this still suffered the overheating problems the earlier engine had, and the plan wasn't taken any further. The other serious option was the Isotta Fraschini Delta air-cooled V-12, which produced 570kW (764hp). However, with the Italians surrendering to the Allies in September 1943, this would become impossible and the Hs 129 would soldier on with the 14M38 for the rest of the war.

The first operational flying was done by SchG 1 on the southern part of the Eastern Front against the Soviet Union in mid-1942. Results were impressive with the Hs 129B demonstrating its abilities against Soviet armour. This was enhanced by the addition of the Mk 101 30mm (1.18in) cannon, which was carried under the fuselage on some aircraft and had tungsten-carbide armour-piercing rounds. SchG 1 was particularly successful in the battles around Kharkov and Reichsmarschall Göring was keen that every fighter Geschwader should have a Staffel attached, although

Specifications: Hs 129B-2

Type:	Close air support aircraft
Dimensions:	Length: 9.75m (31ft 12in); Wingspan: 14.2m (46ft 7in); Height: 3.25m (10ft 8in)
Weight:	5,250kg (11,574lb) maximum take-off
Powerplant:	2 x 522kW (700hp) Gnome-Rhône 14M38 air-cooled radials
Maximum speed:	407km/h (253mph)
Range:	680km (423 miles)
Service ceiling:	9,000m (29,528ft)
Crew:	1
Armament:	Armament: 2 x 7.92mm (0.31in) MG 17 machine guns in the wing roots; 2 x 20mm (0.79in) MG 151 cannon in the fuselage sides; 1 x 30mm (1.18in) MK 103 cannon in the ventral pod or 250kg (551lb) of external ordnance

this was only actually achieved by JG 51. Late 1942 saw the Hs 129B operating on a second front in response to the Allied success at El-Alamein. 4./SchG 2 beginning operations from El Adem in November again achieved great success against Allied armour using the MK 101 30mm (1.18in) cannon. 5./SchG 1, meanwhile, began operating from El Aouina in Tunisia shortly afterwards. Both units would, however, suffer from the effects of dust on the Gnome-Rhône engines, whose intake filters were a continual weak point. Both units were evacuated back to Europe in the first half of 1943, leaving a number of aircraft behind for the Allies to examine.

New cannon

Continual attempts to improve the lethality of the Henschel saw a broad range of weapons fitted to the diminutive aircraft. The MK 103 30mm (1.18in) cannon replaced the earlier MK 101 in the summer of 1943, the higher firing rate increased the effectiveness of the round against armoured targets, and was widely fitted to the Hs 129B-2. Alternatively, some aircraft were fitted with a 37mm (1.5in) BK 3,7 cannon based on the Flak 18 anti-aircraft gun. Again carried under the fuselage, this weapon required the wing root guns to be removed to free up space for the ammunition. With a much higher muzzle velocity than the MK 103, the BK 3,7 was deadly against the turret and engine area of Allied tanks although the extra weight and low rate of fire made the Hs 129 more vulnerable to enemy aircraft when operating with it.

Taking this concept to the extreme was the Hs 129B-3/Wa, which was fitted with the 74mm (2.9in) PaK 40 anti-tank cannon. This fired a 3.2kg (7lb) round at just under 1,000m/s (3,281ft/s), which could destroy a tank with a single shot even against the thickest section of armour. However, only 26 rounds were carried and the 1,500kg (3,307lb) mass of the cannon, nearly half the empty weight of 3,800kg (8,378lb), made the Hs 129 extremely difficult to handle. Consequently, as a precaution, the weapon could be jettisoned in an emergency. Even so, only 20 Hs 129B-3 were completed due to the performance shortcomings.

Other weaponry carried by the Henschel included a pod bearing four MG 17 machine guns and 50kg (110lb) bombs, either one under each wing or four

A wrecked Hs 129B of 5.(Pz)/Schlachtgeschwader 1 at El Aouina in Tunis being examined by US troops in May 1943 after the Axis forces in Africa had surrendered.

on the centreline hardpoint replacing the external cannon. Alternatively, 24 SD 2 anti-personnel bomblets could be carried in place of a 50kg (110lb) bomb. It was also possible to carry a Rb 20/30 reconnaissance camera internally when the Rustatz 5 modification was incorporated.

Operation Citadel

In July 1943, four of the five Hs 129 Staffeln on the Eastern Front were involved in Operation Citadel, the attack on the Kursk salient. On the 8th, a surprise Soviet attack north-west of Charkow saw the 16 aircraft units attacked in relay, one attacking the tanks, two more approaching the target area, and the fourth returning to refuel and rearm. Attacking the tanks from the sides and rear where armour was thinnest, the Soviets were completely repulsed. The whole operation itself saw the Luftwaffe destroy 1,100 tanks and 1,300 vehicles. Despite this, the lack of German reserves and the need to respond to the invasion of Sicily meant that the encounter was a strategic victory for the Soviets.

Although demonstrably a battle-winning weapon, Hs 129 operations were overseen by the General der Jagdflieger, whose primary concern was fighter operations. It wasn't until late 1943 that a General der Schlachtflieger was established, by which point it was essentially too late to reverse the situation. Henschel continued production into mid-1944 but this was affected by Allied bombing raids and was stopped completely when the focus turned to the production of emergency fighters. With no new aircraft being delivered, the Schlachtgeschwader slowly dwindled in number until only a handful of aircraft were being operated by 10./SG 9 at the end of the war.

In total, 870 Hs 129 were produced, which was far less than would have been needed to change the course of the war, despite its unique capabilities. As well as the Luftwaffe, the Romanian Air Force also operated three squadrons of Hs 129B-2s, initially against the Soviet Union and then after August 1944 against Germany, the country having switched sides. Hungary also evacuated four aircraft in 1943, which remained in German markings.

Downward-firing weapons

As well as various large-calibre cannon, a novel system was trialled on three Hs 129B-0s that was not unlike a downwards-pointing version of the Schräge Musik anti-bomber system. The Sondergerärte 113 A (SG 113A) Förstersonde rocket mortar comprised a funnel-like fairing in the aft fuselage and housed six downwards-pointing 75mm (3in) smooth-bore tubes. The tubes housed a 45mm (1.8in) armour-piercing shell contained in a sabot pointing downwards, a steel cylinder of equivalent weight, and between them an explosive charge. When triggered, the charge would fire the shell towards the ground while the steel cylinder would fly out of the top of the tube, counteracting the recoil. Firing was automatic with the pilot flying the aircraft over armoured vehicles and a photocell in the nose triggering the firing mechanism. However, this would prove to be the Achilles' heel of the system as trials at Tarnewitz Waffenprüfplatz demonstrated that the photocell would regularly fail to pick out the target vehicles. Consequently, the armour-destroying potential of the SG 113A was never used operationally.

British Royal Air Force technicians of No1426 (Enemy Aircraft) Flight RAF – known as the 'Rafwaffe' – examine a captured Henschel Hs 129B, originally of Schlachtgeschwader 1, on 2 March 1944, at RAF Collyweston in Northamptonshire, United Kingdom.

Heinkel He 111H (1939)

Successor to the He 111P, the He 111H would become the most widely produced variant of the He 111. As well as operating as a conventional bomber, it would serve as a torpedo-bomber and cruise missile launcher and be pressed into service as a transport before the war was over.

While the He 111P introduced the glazed nose to the 111, it used the same DB 601 engines as the Bf 109 and 110 fighters. To avoid a production bottleneck, the decision was made to use the 783kW (1,050hp) Jumo 211 engine, which was in lower demand. The resulting He 111H entered service in 1939 and by the eve of the invasion of Poland made up over half the operational strength of 705 He 111s, which themselves made up 75 per cent of the Luftwaffe's medium bomber force. The invasion saw the He 111s along with Do 17s tasked with attacking airfields, logistics

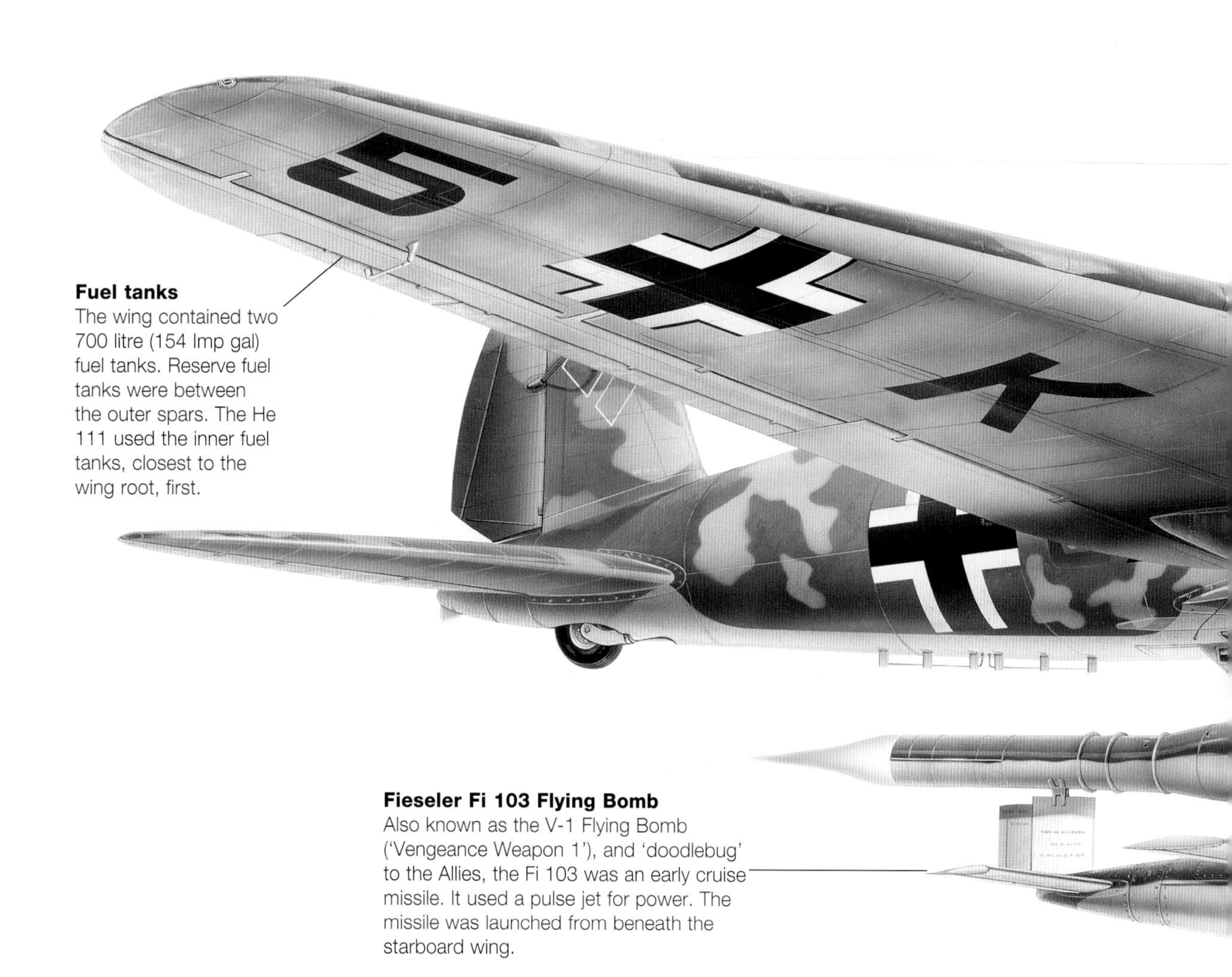

hubs and major cities. The initial attacks were conducted en masse, typically two Gruppen operating together, and saw little opposition with relatively few losses. It wasn't until the sixth day of action that the Polish Air Force managed to press home an attack when IV./KG 26 lost four of its number to an attack by around 20 Polish fighters.

This, however, would prove to be the exception rather than the rule and the crews of many of the aircraft that were shot down survived. With the invasion of Poland complete, the He 111 Geschwader were relocated to the West in anticipation of action against France and the United Kingdom. Forbidden from attacking land targets in Britain during the Phoney War period, they concentrated on the ports and sea lanes. The He 111s were also tasked with reconnaissance missions over the North Sea, fitted with cameras on the inner wing bomb pylons. During one such anti-shipping mission, He 111Hs of KG 26 would be involved in the sinking of two German destroyers near the Dogger Bank due to a breakdown in communication

A Heinkel He 111H-22 (with Feiseler Fi 103). Following experiments at Peenemünde, the German secret weapons establishment, in 1943, several Heinkel 111s were modified to carry a Feiseler Fi 103 (V-1) missile. The type was assigned to the newly formed III Gruppe, Kampfgeschwader (KG) 3, which became operational at Venlo and Gilze-Rijn in the Netherlands in July 1944. By August 1944, III/KG 3 had launched 300 V-1s against the British capital, London.

Dorsal gun
The He 111H-22 included a single 13mm (0.51in) MG 131 in the dorsal gun turret, as a defence against fighter attacks from above.

Crew
The standard crew for the He 111 was five: a pilot, who sat in the glazed section; a navigator/bombardier, who sat in the nose; a radio operator/dorsal gunner; and two further gunners, who operated machine guns in the beams and ventral gondola, which was known to the crew as the *Stertebett* ('death bed').

A Fi 103 installed under the port wing of an He 111H-22. The white cable attached to the engine was used to start the pulse jet engine prior to the missile's launch.

between the Luftwaffe and Kriegsmarine. By late 1939, the He 111H-3 had entered production. This gained a forward-firing 20mm (0.79in) MG FF cannon in the front of the gondola for use in the anti-shipping role and an additional crew member to operate it. Armour protection was also increased and the more powerful Jumo 211D-1 engine with 895kW (1,200hp) installed to compensate for the extra weight. The H-4 followed in early 1940. This had a reduced internal bomb load of 1,000kg (2,205lb) but had external pylons to carry weapons up to 1,800kg (3,968lb) that would not fit internally. Fuel capacity was also increased, using space in the bomb bay, while later aircraft gained the more powerful Jumo 211F-1 with 1,040kW (1,395hp). The He 111H-1 to H-4 would comprise the majority of the type involved in the Battles of France and Britain.

The blitzkreig through the Low Countries into France followed a similar pattern to Poland. The He 111-equipped KG concentrating on airfields and railheads, occasionally also being called to attack towns and cities. On one occasion, I./KG 4 were also called upon to resupply paratroopers, the Ju 52 units being unable to spare sufficient airframes. The He 111 kept the troops supplied with ammunition for three days until ground forces reached them. Unlike the Polish campaign, the Luftwaffe faced at least some modern aerial opposition and when the Dutch Fokker G-1A and Belgian Hurricanes got airborne they were able to successfully engage the He 111s. However, without an integrated radar command and control system the resistance was uncoordinated and unable to seriously impact the course of the war.

Specifications: He 111H-16

Type:	Medium bomber
Dimensions:	Length: 16.4m (53ft 10in); Wingspan: 22.6m (74ft 2in); Height: 4m (13ft 2in)
Weight:	14,000kg (30,865lb) maximum take-off
Powerplant:	2 x 1,000kW (1,341hp) Jumo 211F-2 V-12 piston engines
Maximum speed:	430km/h (267mph)
Range:	1,950km (1,212 miles)
Service ceiling:	8,500m (27,887t)
Crew:	5
Armament:	1 x 20mm (0.79in) MG FF cannon; 1 x 13mm (0.51in) MG 131; up to 7 x 7.92mm (0.31in) MG 15 machine guns; a bombload of up to 2,000kg (4,409lb) internally (8 x 250kg/551lb bombs), or 1 x 2,000kg (4,409lb) bomb externally

Defensive shortcomings

With the invasion of France starting on 10 May 1940, the same broad pattern would be repeated with the defenders being reactive in the face of the onslaught. However, the He 111 Kampfgeschwader would begin to see the shortcomings in their own aircraft. Unlike in Spain, where the Luftwaffe had gained its pre-war experience, and the attack on Poland that had opened the war, they now faced modern fighters that could keep pace with their bombers. The He 111's defensive armament did not allow a significant weight of fire to be directed against an attacking fighter, the mounts giving restrictive fields of fire to each gun with little overlap. At the same time, where the rifle calibre guns used by the British Hurricanes would require multiple passes to inflict significant damage, the single 20mm (0.79in) cannon on the French Air Force's Morane-

Saulnier MS 406 and Dewoitine D.520 were a different proposition, the larger shells knocking holes in the fuel tanks that the self-sealing system could not close, leading to fires and structural failures the crews were unable to recover from. In order to address the problem at least partially, He 111s would gain extra defensive weaponry, including extra guns in the nose and twin MG 81 machine guns in the beam positions.

The Luftwaffe had 15 Gruppen of He 111s assigned to the forces for what would become the Battle of Britain, of which around 90 per cent were Hs. This represented just under a half of the medium bomber force, the rest being equipped with Ju 88s or Do 17s. Although a formidable force, it now faced a fully integrated command and control system equipped with a comprehensive radar network. Initially concentrating on ports and coastal convoys, the He 111 KGs were regularly met by multiple squadrons of

An He 111H-2. The aircraft depicted here, Wk Nr 3340, 'Yellow B' of 9./KG 53 , is shown with the wing bars carried (for fighter identification and station-keeping) during the big Luftwaffe daylight raids on London during Sunday 15 September 1940 – the climax of the Battle of Britain. This aircraft was in fact damaged in action on that day and force landed at Armentiers with two wounded crew members; recent computerised research suggests that it was probably attacked by Spitfires of No. 66 (Fighter) Sqn.

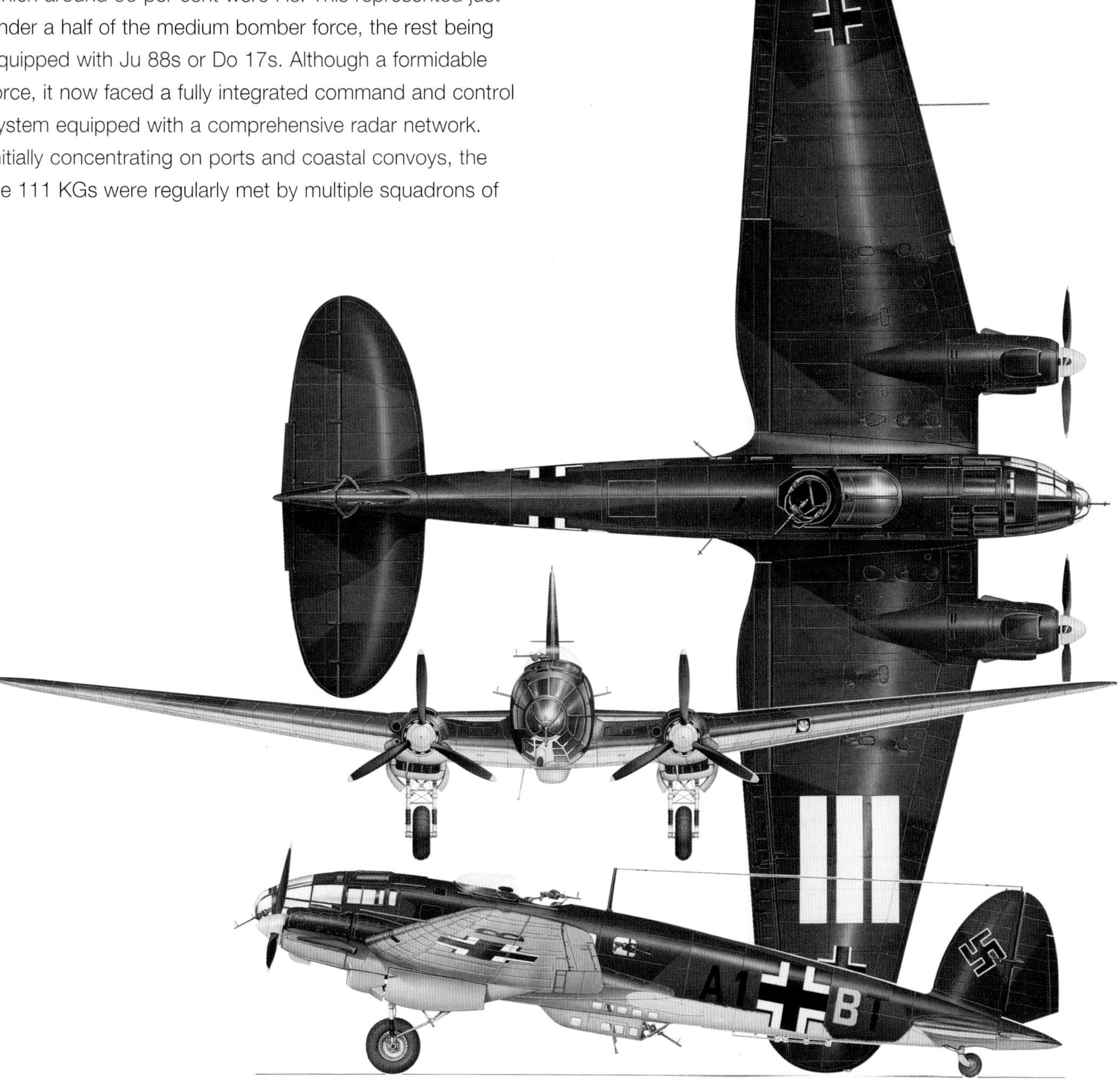

Hurricanes and Spitfires. In their first raid on 11 July, I./KG 55 were met by 87, 145 and 601 squadrons over the naval base at Portsmouth and although successful in hitting the target they lost three He 111s as well as three of the escorting Bf 110s. This pattern would be repeated throughout the bombing campaign until the decision was made to switch to night bombing in September. This gave the Kampfgeschwader a degree of invulnerability for the next six months as the RAF's night-fighter capabilities took some time to develop, and saw devastating raids carried out on Britain's industrial cities, such as Coventry.

Night bomber

To aid in the night bombing campaign, the Luftwaffe initially devised the Knickebein, a bent-leg radio navigation aid. This used the standard airfield approach receiver to guide the bombers to targets over the UK using transmitters based on the European mainland. A steady tone indicated to the pilot that he was on course, while a series of dots indicated that he was right of track and a series of dashes that he was to the left. When a second beam was intercepted, the bomber was over the target area and the bombs could be released. The next stage was the more accurate X-Gerät (X-Gadget) system. This required specialist equipment, and its installation resulted in the He 111H-14, which was primarily used by KG 100 in the pathfinder role. A further development was Y-Gerät, which was mainly used by III./KG 26. This only used one beam and an early form of distance-measuring equipment to indicate when the aircraft was over the target. In time, the RAF would develop countermeasures to all three systems as the electronic warfare battle evolved.

The He 111H-5 entered service in 1941. This used the Jumo 211F-1 engine as per the later H-4s while also gaining the ability to carry two 1,000kg (2,205lb) bombs, or one 1,800kg (3,968lb) bomb externally. These aircraft were extensively used in the Night Blitz against British cities. The He 111H-6 was broadly similar to the H-5 but could be armed with two LT F5b torpedoes for the anti-shipping role. When operating in this role, the H-6s would be guided to a convoy by a shadowing aircraft, making their approach at low level to maintain the element of surprise. Nearing the target, the formation would spread out, each crew selecting a ship to attack before launching both torpedoes in quick succession at a range of around 1,000m (3,281ft). The torpedo would take just under 20 seconds to reach its target, giving the ship almost no time to manoeuvre out of its way.

The H-6s would perform this mission to devastating effect in June 1942 against Convoy PQ 17 north of

This He 111H-6 of Gefechtsverband (Combat Formation) Kulmey flew from Immola airfield in Finland, July 1944.

Norway. He 111Hs would also operate in the anti-shipping role in the Mediterranean using torpedoes and bombs, including attempts to block the Suez Canal by sinking shipping in the main channel.

Operation Barbarossa

As preparations began for the invasion of the Soviet Union the majority of the He 111 force was withdrawn to central Europe to ready themselves for the upcoming assault. In the opening salvos, they struck at Moscow and other targets deep inside enemy territory. However, as the war progressed, they would be pressed into tactical support for the army and interdiction. The latter involved attacks on trains and was a particularly dangerous mission for the He 111H-6s of KG 55, who specialized in the role for a period. The lack of manoeuvrability made low-level attacks in unfamiliar terrain almost as dangerous as the flak encountered at higher levels.

February 1942 would see the He 111s again called on to carry out aerial resupply of an isolated force, this time a garrison of 3,500 men at Kholm in north-eastern Russia. While a return to strategic and tactical bombing would resume as German forces pushed towards the Caucasus oilfields, November would see them again employed as

Flown by 9./Kampfgeschwader 44, this He 111H-20 was based at Breslau in March 1945.

A young Luftwaffe pilot sits at the controls of a Heinkel He 111H. The He 111 typically had a crew of five: pilot, navigator/bombardier/nose gunner, ventral gunner, dorsal gunner/radio operator and side gunner.

transports. With the Soviets surrounding the German Sixth Army at Stalingrad, He 111s began flying supplies in and wounded troops out. In addition to internal loads, the He 111H-8/R2, H-11/R2 and H-14/R2 could also tow a Gotha Go 242 assault glider with additional supplies.

One type that didn't see service at Stalingrad was the He 111Z, which was a specialized glider tug made by joining two He 111H-6 fuselages. The wing outboard of the engine on the adjoining sides was replaced by a new section carrying a fifth engine and connecting the two aircraft together. With a total of 4,928kW (6,700hp), the He 111Z was able to tow the massive Me 321 transport glider, originally intended for the invasion of Britain, although when the 30.8 tonne (34 US ton) glider was fully loaded, two booster rockets were required on the Heinkel to enable the whole assembly to get airborne. Not used at Stalingrad due to the Soviet seizure of two airfields required to accept the unwieldy gliders, they were used for other operations on the Eastern Front from 1943.

Eastern Front raids

By 1943, the He 111 should have been replaced by the four-engined He 177 in the bomber role, but this was beset with delays and the earlier Heinkel was forced to soldier on. Operations were predominantly carried out on the Eastern Front, where the Luftwaffe could still maintain sufficient air superiority for the He 111 to conduct raids with some chance of success. Rail depots were a favoured target to interrupt the flow of materiel to Soviet forces, while the tank factory at Gorky was also subject to a bombing campaign in a vain attempt to shut down production.

A small victory would come the He 111 KGs' way in June 1944 with a unique opportunity to attack the Eighth Air Force. After bombing the Schwarzheide oil refinery in eastern Germany, the Eighth's B-17s landed at Soviet airfields in Ukraine, the intention being to carry out the operation in reverse a few days later. Detected by a Ju 88 reconnaissance aircraft, the He 111s of KGs 4, 53 and 55 launched an attack on the night of 21 June. Of 72 B-17s at the Poltava Air Base, 44 were wrecked and 26 suffered damage. A further 33 aircraft were damaged or destroyed and 1.8 million litres (396 million gallons) of fuel destroyed.

The return attack on Drohobycz in Hungary contained only half the B-17s that had set out from England five days earlier. This success would be the exception rather than the rule, though, and as the year progressed, He 111s would find themselves flying much-needed supplies into besieged areas almost as often as they conducted bombing raids.

Fading force

By the beginning of 1945, the shortage of fuel was being felt by all bomber units as fighters were prioritized for the defence of the Reich. By April, only around 20 He 111s were reported as being operational. World-beating at the start of the conflict, the He 111 should have long since been retired, but the inability of German industry to produce a replacement bomber left the Luftwaffe with

Cruise missile carrier

The first operational use of the Fieseler Fi 103 cruise missile (or Vergeltungswaffe Eins, 'Vengeance Weapon One' – the V-1) weapon was from land sites in the Pas-de-Calais area of France. These were sufficiently close to be able to target London, the first strike to do so happening on the night of 15 June 1944 in response to D-Day. From 9 July, specially modified He 111H-22s joined in the attack, each being able to launch one V-1 in flight from either the left or right wing. This expanded the potential target area for the new weapons and gave an alternative if Allied attacks on the distinctive land launch sites made it untenable to continue using them. The H-22s were converted from H-16, H-20 and H-21 airframes. In addition to the mounting points for the V-1 missiles, they were powered by Jumo 213E engines producing 1,508kW (2,022hp) when using methanol-water injection. This helped compensate for the extra drag from the 2,150kg (4,740lb) missile and enabled the Heinkel to accelerate to the speed required for the pulse jet engine to work.

Operating from bases in France and Holland, the H-22s would operate on nights with poor weather to maximize their chances of evading Allied fighters. Making their approach at low level over the sea to reduce the likelihood of being detected by radar, while approaching their launch point the aircraft would climb to a minimum of 500m (1,640ft). They would then accelerate to 320km/h (200mph), at which point the V-1's pulse jet engine would be started, the exhaust flame illuminating the darkness. Ten seconds later, the missile would be released, and the Heinkel would reverse course and descend back to low level. Initially operated by III./KG 3, in autumn 1944 it was redesignated I./JG 53, with II. and III./KG 53 converting to the role to provide a full Geschwader of missile carriers.

Despite this vote of success, the air-launched missile was if anything less accurate than the land-based version, which had only a 25 per cent success rate at hitting its targets. Up to half the missiles flew into the sea shortly after being launched. Of those that managed to stay airborne, few made it to the vicinity of their target. Of 50 fired at Manchester on the night of 23 December 1944, 30 were seen to cross the coast, of which 11 hit within 24km (15 miles) of the target, only one of which was actually within city limits. The attackers, meanwhile, lost one He 111 to RAF night-fighters.

In fact, as well as allowing attacks to be maintained against the UK after the launch sites in France were overrun, the main contribution to the war effort was in tying down the enemy's night-fighters. Whenever weather conditions were suitable for the He 111H-22s to operate, the RAF would mount Mosquito patrols over their bases and the likely launch areas. The glare from the rocket motor ensured the Heinkels would be detected at missile launch if there were fighters in the vicinity. KG 53 officially ceased operations on 14 January 1945 due to fuel shortages. It is estimated that the He 111H-22 force lost 77 aircraft during the six months it was operational, 16 to night-fighters, against a nominal strength of 90 for KG 53. In exchange for this they launched approximately 1,200 V-1 missiles, of which at best 20 per cent reached their intended target.

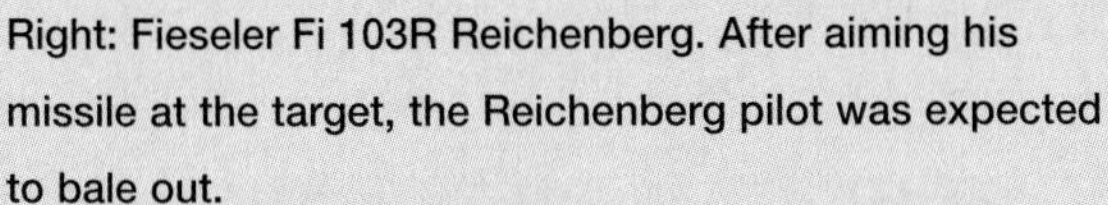

Right: Fieseler Fi 103R Reichenberg. After aiming his missile at the target, the Reichenberg pilot was expected to bale out.

no choice but to continue using it. Heinkel produced over 7,200 He 111s, more than 6,000 of which were H variants. A further 30 were constructed in Romania, while CASA in Spain produced 200 during and after the war, the later aircraft ironically being fitted with Rolls-Royce Merlin engines.

Junkers Ju 188/Ju-388 (1942)

An outstanding aircraft, the Ju 188 was an evolution of the Ju 88 bomber, which could carry more, farther and faster. However, it failed to fulfil its true potential due to delays caused by the Luftwaffe chasing even higher performance with the failed Bomber B programme.

The Ju 188 traced its origins to the original proposal for the Ju 88 in 1936, when Junkers had offered an alternative, Ju 88B, design featuring a streamlined, glazed crew compartment with multiple sections of curved glazing. Despite offering improvements in drag and visibility, this was rejected in favour of the more conservative faceted nose section seen on the Ju 88A. Junkers would persist, however, and in 1940 a Ju 88B was flown. Essentially mating the new cockpit on the front of a Ju 88A-1 airframe, power was provided by two 1,194kW (1,601hp) BMW 801 radial engines. A small batch of Ju 88B-0s were subsequently built based on the A-4 airframe with wider-span wings and additional bomb racks on the outboard wings. Although the Luftwaffe's tests were successful, the marginal performance gain over the baseline A-4 was not considered worth the disruption to the production line. The handful of B-0s were subsequently converted to reconnaissance aircraft with

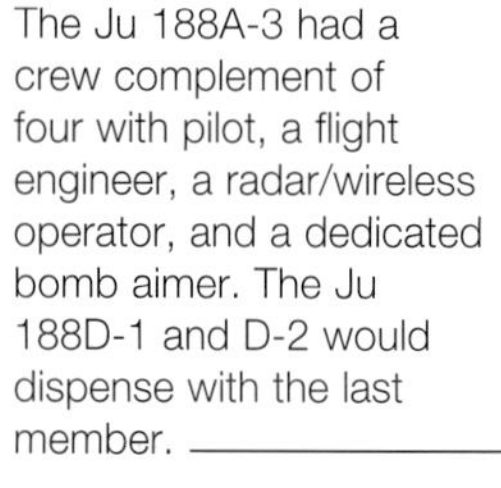

Crew
The Ju 188A-3 had a crew complement of four with pilot, a flight engineer, a radar/wireless operator, and a dedicated bomb aimer. The Ju 188D-1 and D-2 would dispense with the last member.

Wing
By comparison with the Ju 88B, the Ju 188 had a wing of greater span, with both wingtips and ailerons being extended outboard to give a distinctive pointed outline in plan view. The slotted dive-brakes of the V1 and V2 prototypes were omitted from production Ju 188 versions.

the removal of all bombing equipment and the installation of long-range fuel tanks.

By 1942, it was becoming clear that the contenders for the Bomber B replacement Schnellbomber (fast bomber) programme – the Do 317, Fw 191 and Ju 288 – were in no imminent danger of being delivered. With an urgent need to improve the capabilities of its Kampfgeschwader in the face of ever-improving Allied aircraft, the Luftwaffe looked to further develop an existing type. The Ju 88 was an obvious choice and Junkers had been quietly making improvements and development plans since the Ju 88B-0 had flown. Two prototype Ju 88s, V27 and V44, were co-opted to the programme with wider-span pointed wings for improved high-altitude performance. The second aircraft was further improved with a larger squared-off vertical stabilizer, which was shared with the Ju 88G series. The underwing bomb shackles, meanwhile, were moved inboard of the engines.

New design

The new design was chosen for production in late 1942 with the designation Ju 188 to give the impression the design was new. The Ju 188A-1 was fitted with the 1,290kW (1,730hp) Jumo 213 while the otherwise identical Ju 188E-1 used the 1,250kW (1,676hp) BMW 801. Deliveries of the former were delayed, leading to the Ju 188E entering service first in May 1943 with 1 Gruppe of KG 6 while the first operational raid was carried out by I./KG 66 in August against factories in Lincoln. Together with III./KG 2, both units would also use the Ju 188 during Operation Steinbock, the renewed bombing offensive against the Greater London area, also known as the 'Baby Blitz', which took place between January and May 1944.

Despite the Luftwaffe's original aim of having a force of fast bombers able to roam almost at will over enemy territory, no German manufacturer had produced anything that could survive in the face of equally modern fighters.

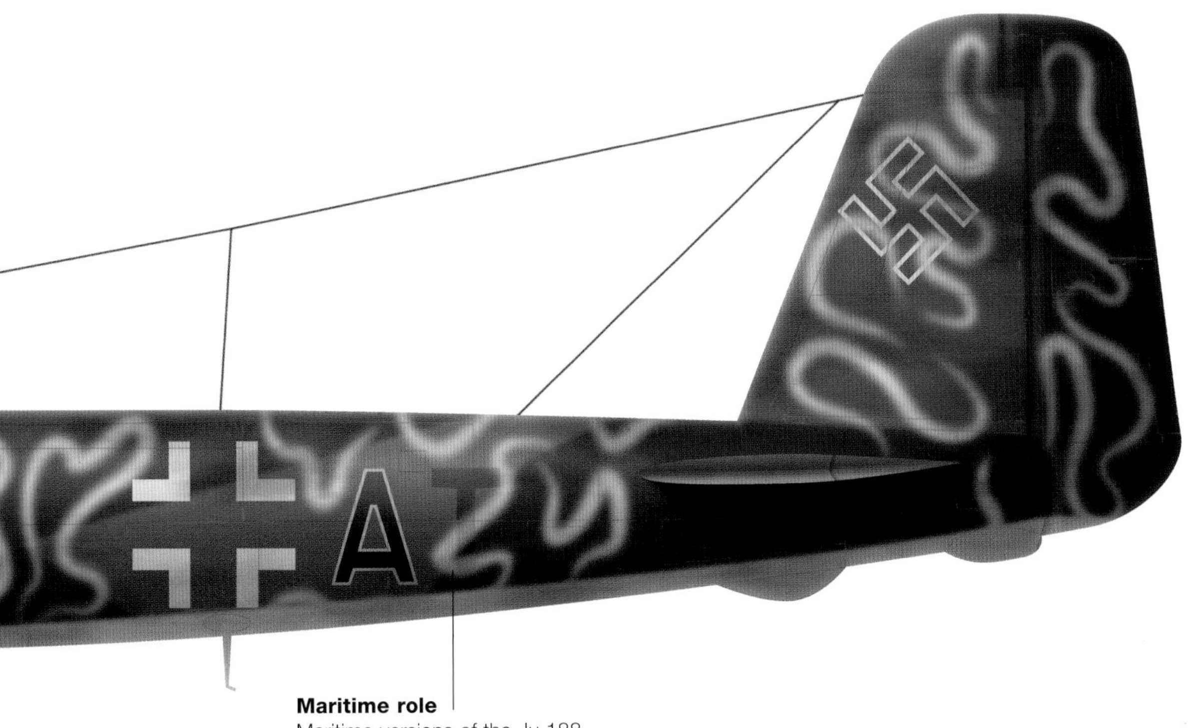

Maritime role
Maritime versions of the Ju 188 included the torpedo-armed Ju 188A-3 and Ju 188E-2, and the reconnaissance-configured Ju 188D-2 and Ju 188F-2.

Wearing the 'Wellenmuster' camouflage pattern typical for operations in the Mediterranean Sea, this Ju 188A-3 was flown by III./Kampfgeschwader 26 in April 1945.

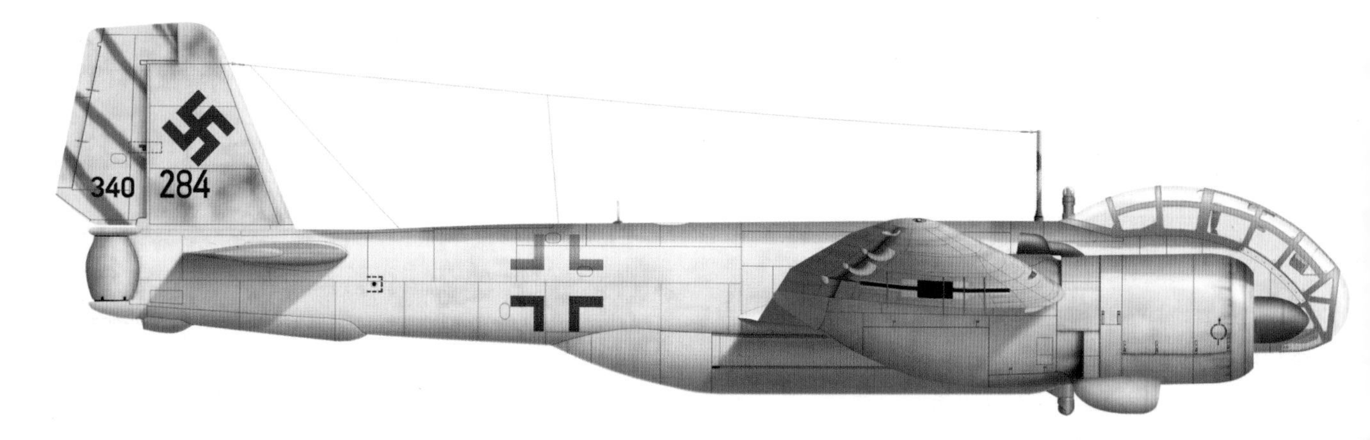

Consequently, unlike the RAF's Mosquito, defensive armament was always an important consideration and the Ju 188A and E featured a turret above the cockpit with either a 20mm (0.79in) MG 151 cannon or, on at least some E models, a 13mm (0.51in) MG 131. Further upper defence was provided by a second rearwards-firing 13mm (0.51in) MG 131 immediately beneath and behind the turret. Further rearwards-facing armament was provided by twin 7.92mm (0.31in) MG 81Z machine guns beneath the cockpit while an additional MG 151 was in the nose for protection from frontal attacks. Various attempts were also made at developing a tail turret, either the remotely controlled FA15 barbette, which featured two MG 131s, or when that proved inaccurate and prone to unreliability, a manned version. Perhaps fortunately for the potential rear gunners this was taken no further than a prototype as it was likely to be all but impossible to escape from in an emergency.

As a photo-reconnaissance aircraft the ventral pannier held cameras, while at the tail of this Junkers Ju 388L-1 was an FA 15 remotely controlled gun turret and a FuG217 Neptune warning radar. Only a handful were produced.

Specifications: Ju 188E-1

Type:	Medium bomber
Dimensions:	Length: 14.95m (49ft 0.6in); Wingspan: 22.0m (72ft 2in); Height: 4.44m (14ft 7in)
Weight:	15,195kg (33,500lb) maximum take-off
Powerplant:	2 x 1,300kW (1,743hp) BMW 801D-2 air-cooled radial engines
Maximum speed:	500km/h (311mph)
Range:	2,500km (1,553 miles)
Service ceiling:	1,000m (3,281ft)
Crew:	4
Armament:	3 x 13mm (0.51in) MG 131 machine guns; 1 x 20mm (0.79in) MG 151 cannon; a bombload of up to 3,000kg (6,614lb)

Although the new cockpit design improved the forward view for the pilot considerably compared to the Ju 88, the amount of framing required for the large number of separate panels was found to be obstructive by some Allied test pilots when evaluating captured examples. Rather than a conventional instrument panel there was a cluster of primary flight instruments attached to the central canopy pillar. A console on the left-hand fuselage wall held the engine controls and instruments with further switches on the wall itself. Pilots transferring from the older Junkers found the new arrangement to be much more logically laid out and ergonomic. The Ju 188's handling, meanwhile, was at least the equal of its famous forebears, being stable and responding crisply to the controls, the more powerful engines and increased wing area making it faster and able to operate at higher altitudes. All in all, it was a marked improvement over the Ju 88 so far as its operators were concerned.

With an urgent need for reconnaissance aircraft, Ju 188As were frequently completed as Ju 188D-1s or D-2s. These lost the MG 151 mount in the nose and the bombardier, leaving a crew of pilot, flight-engineer and radar/radio operator. They gained various combinations of Rb 50/30, 70/30, NRb 40/25 and 50/25 cameras and additional fuel to increase the range to 3,395km (2,110 miles). The BMW-engined equivalents were the F-1 and F-2. Both the D-2 and the F-2 carried the FuG 200 Hohentwiel maritime radar, recognizable by the array of antennae mounted either side of and below the nose. Able to detect ships at around 70km (43.5 miles) and land at 150km (93 miles), these aircraft were involved in searches for Allied shipping in the North Sea and around Norway's North Cape, often working with the A-3 and E-2 torpedo-bomber variants. These latter aircraft also carried the FuG 200 system as well as having a bulge on the right-hand side of the fuselage, which contained equipment to adjust their two torpedoes' steering gear prior to launch. Although excellent aircraft, their contribution to the war would be minimal since the overland reconnaissance D-1 and F-1 failed to penetrate the UK's airspace sufficiently to observe the build-up of forces for D-Day. Meanwhile, at sea, the convoy battle was already all but won, while the presence of Mosquitos over the North Sea made it an increasingly dangerous hunting ground for the Ju 188s.

High-altitude variant

Late 1943 saw Junkers putting considerable effort into high-altitude variants of the Ju 188. Some of these would enter service as the Ju 388; however, the S and T would become intruder and reconnaissance versions of the

Aéronavale Ju 188s

After the Armistice in June 1940, with the agreement of the French authorities, German aviation manufacturers set up overhaul and repair facilities in France. Junkers established a plant at Villacoublay that primarily worked on the Ju 88. However, in early 1944 it was subcontracted to the Ju 188 programme – a move that also allowed the French Resistance to smuggle details of the aircraft to the Allies. After the war, the Aéronavale quickly adopted the Ju 188 as its main shore-based patrol aircraft and had around 30 Ju 188Es and Fs refurbished for operations. In a further twist, SNCASE were contracted to deliver a further 12 new-build Ju 188Es using parts from French and German factories. At least some of the Aéronavale aircraft used the 1,600kW (2,146hp) Arsenal 12H development of the Jumo 213 also built at Villacoublay. Although relatively short-lived in frontline service, the French Ju 188s were also used by 10S Flotilla in the test and development programmes for guided missiles, piston and jet engines, themselves often based on German technology.

A Junkers Ju 288V-103 protoype, eventually developed into the Ju 288C-1. Only 22 Ju 288s were built before the programme was discontinued.

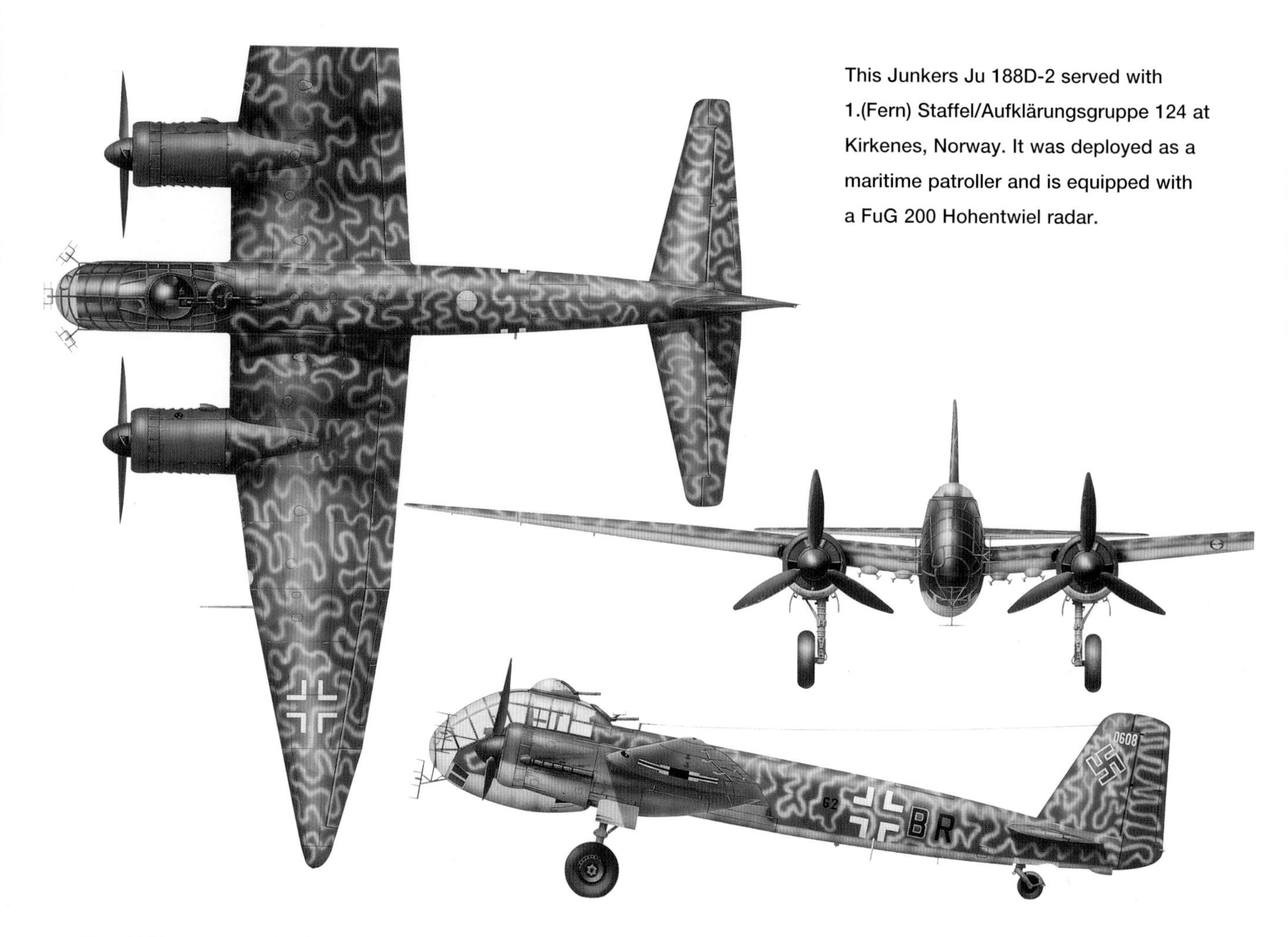

This Junkers Ju 188D-2 served with 1.(Fern) Staffel/Aufklärungsgruppe 124 at Kirkenes, Norway. It was deployed as a maritime patroller and is equipped with a FuG 200 Hohentwiel radar.

188. With a pressurized crew compartment and Jumo 213E-1 engines producing 1,260kW (1,690hp) when boosted with nitrous oxide, they were capable of reaching 11,500m (37,730ft). With an internal bomb load of 800kg (1,764lb), the Ju 188S-1 could reach 685km/h (426mph) at that altitude, while the T-1 equipped with two large Rb cameras was even faster at 700km/h (435mph). At those speeds and altitudes, the Ju 188S and T would have proved very difficult to intercept. However, ever-changing priorities saw the S stripped of its pressurization system, equipped with a 50mm (1.97in) BK 5 cannon, and delivered as low-level ground attack aircraft as the Ju 188S-1/U. The Ts meanwhile appear to have been moved to Junkers' Merseburg factory and reworked as Ju 388s. The Ju 388 was intended to fulfil the role of high-speed, long-range bomber that the Ju 288 had been unable to. The pressurized crew compartment, although still extensively glazed, did not feature the dorsal turret or other defensive armament of the Ju 188. It did, however, include the FA15 remote-controlled tail turret that had been trialled on the Ju 188, which was aimed by periscope from the cockpit. The Ju 388L-1 was fitted with additional fuel and reconnaissance cameras in a ventral pannier and powered by the 1,409kW (1,890hp) BMW 801TJ engine. At least some of the 60 or so produced performed operational missions while being evaluated by 3./Versuchsverband – the experimental test unit. Meanwhile, around 15 prototype and pre-production Ju 388K bombers were built, which were distinguished by the expansion of the bomb bay by a ventral pannier. Finally, three prototypes of the Ju 388J night-fighter were produced. These had a more conventional nose and cockpit, allowing either the FuG 220 Lichtenstein or FuG 218 Morgenstern radars to be fitted, while two 20mm (0.79in) and two 30mm (1.18in) cannon were carried in a ventral tray.

The Ju 188, and to some extent the Ju 388, were excellent aircraft with many aircrew rating it more highly

A line-up of Ju 188D-2s at Kirkenes, Norway. The Ju 188D-2 was intended primarily for the maritime strike and reconnaissance roles, and usually carried FuG 200 Hohentwiel radar.

than the Ju 88. Handling at altitude and high all-up mass were better than in the earlier Junkers and it was able to fully utilize the increased power available from the Jumo 213 and BMW 801 engines. They were, however, at least two years later than they needed to be to have had an impact on the war. Only 1,076 Ju 188s were delivered to the Luftwaffe and over half, 570, were of the D and F reconnaissance models, which either failed to spot the largest invasion force ever assembled or were fighting a convoy battle that had already been lost. The bombers, meanwhile, would only equip a handful of Staffeln, production having been started too late to build up sufficient mass before the German Reich was fully on the defensive.

Only 1,076 Ju 188s were completed, compared to around 14,700 Ju 88s. They equipped just two complete Kampfgeschwadern (KG 6 and KG 2), and elements of three more (a few Staffeln from III./KG 26 and I./KG 66, and one Staffel from KG 200). The type also partially equipped elements of 10 Fernaufklärungsgruppen.

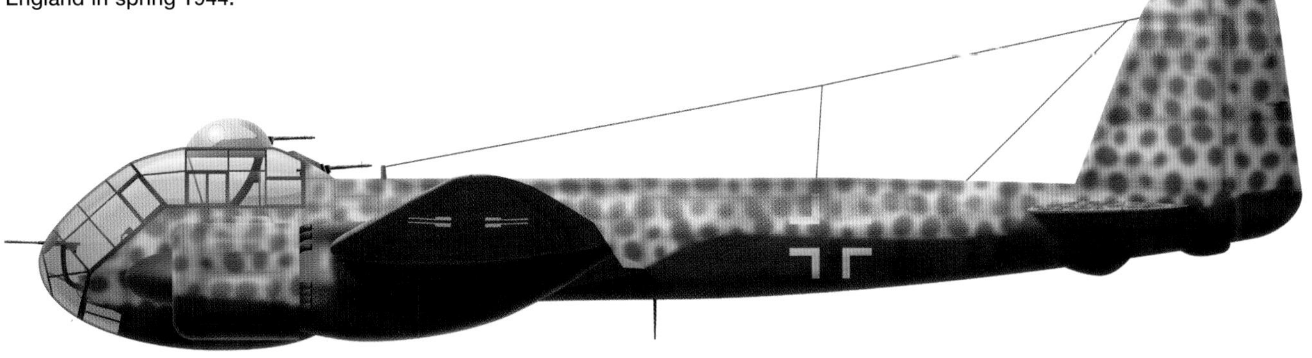

This Ju 188E-2 was flown by Kampfgeschwader 6 during Operation Steinbock, the 'mini-Blitz' that targeted southern England in spring 1944.

Messerschmitt Me 264 'Amerika' bomber (1942)

Designed to cross the Atlantic and attack North America, the Me 264 was an ambitious undertaking. Constantly changing priorities, however, ensured only the one prototype would ever fly.

Primarily a tactical air force, the Luftwaffe had a complicated relationship with strategic bombers: in the 1930s, Oberst Walther Wever, the force's first chief-of-staff, conceived a 'Ural Bomber' able to strike at Soviet industrial targets east of the Ural Mountains. Although this led to the development of the Dornier Do 19, its shortcomings, together with Wever's death and replacement by Oberst Ernst Udet – a proponent of small, fast bombers – led to the requirement being dropped in 1937. At this time, Udet and Göring, head of the Luftwaffe, were opposed to four-engined bombers on the grounds that unless they were capable of dive-bombing, the greater inaccuracy of level bombing would make them less effective than aircraft such as the Ju 88 that could

A Messerschmitt 264 V3. The third prototype, Me 264 V3, wore the registration RE+EP, and represented the standard for the pre-production A-0 series. It was incomplete when the programme was cancelled in 1944.

Engines
Although nominally equipped with four engines, various types being trialled during the development programme including the BMW 801D and Jumo 211J, plans were drawn up for a six engined model to meet the challenging range requirement.

Cockpit
Initially intended to be pressurised this was ultimately deemed too complicated to develop in the time available. Given the cruising height needed to achieve the desired range this would leave the crew on oxygen masks for around 30 hours.

Bomb bay
Bombs were carried in a single bay that took up the lower half of the aircraft's mid-section. It was designed to holdup to 14,000kg of weapons depending on the range to the target including SC 2500 2,500kg bombs.

do this comfortably, despite their smaller bombload. The logical extension of this was that the resources involved in manufacturing one four-engined bomber would be better used making two or three two-engined aircraft. This, though, failed to allow for the mercurial changes in the Führer's military priorities.

In a pattern of vacillation that would be repeated for the duration of the war, by 1939, long-range bombers were back under consideration with the Reichsluftfahrtministerium (RLM) asking for proposals for a four-engined aircraft able to attack the USA from Germany.

Olympic service

Messerschmitt proposed the Me 264 based on earlier work they had done for a long-range bomber, the Bf 165, and the Me 261, an improved design that was originally intended to fly the Olympic torch to Japan for the 1940 Games. The aircraft proposed was a high-wing monoplane powered by four engines, either 990kW (1,328hp) Jumo 211Js or 1,290kW (1,730hp) BMW 801Gs. An internal bomb bay would carry up to 6,000kg (13,228lb) while the defensive armament would be in low-profile remote-controlled barbettes to minimize drag and increase range. By early 1941, Messerschmitt was in a position to submit its proposal together with performance and weight data and assurance that a first flight could take place by May 1942. Sufficiently impressed, the RLM placed an order for 30 of the aircraft with sufficient material to be supplied to begin construction of the first six, subsequently reduced to three, airframes.

Problems with external suppliers led to the already optimistic target date for the first flight being missed; it wasn't until 23 December 1942 that the first prototype, Me 264 V1, would take to the skies for the first time. Fitted with Jumo 221J engines, the V1 conducted a 22-minute test flight from the grass airstrip at Augsburg. The only major setback was the failure of the brakes on landing, which led to the aircraft having to be recovered from a field beyond the end of the

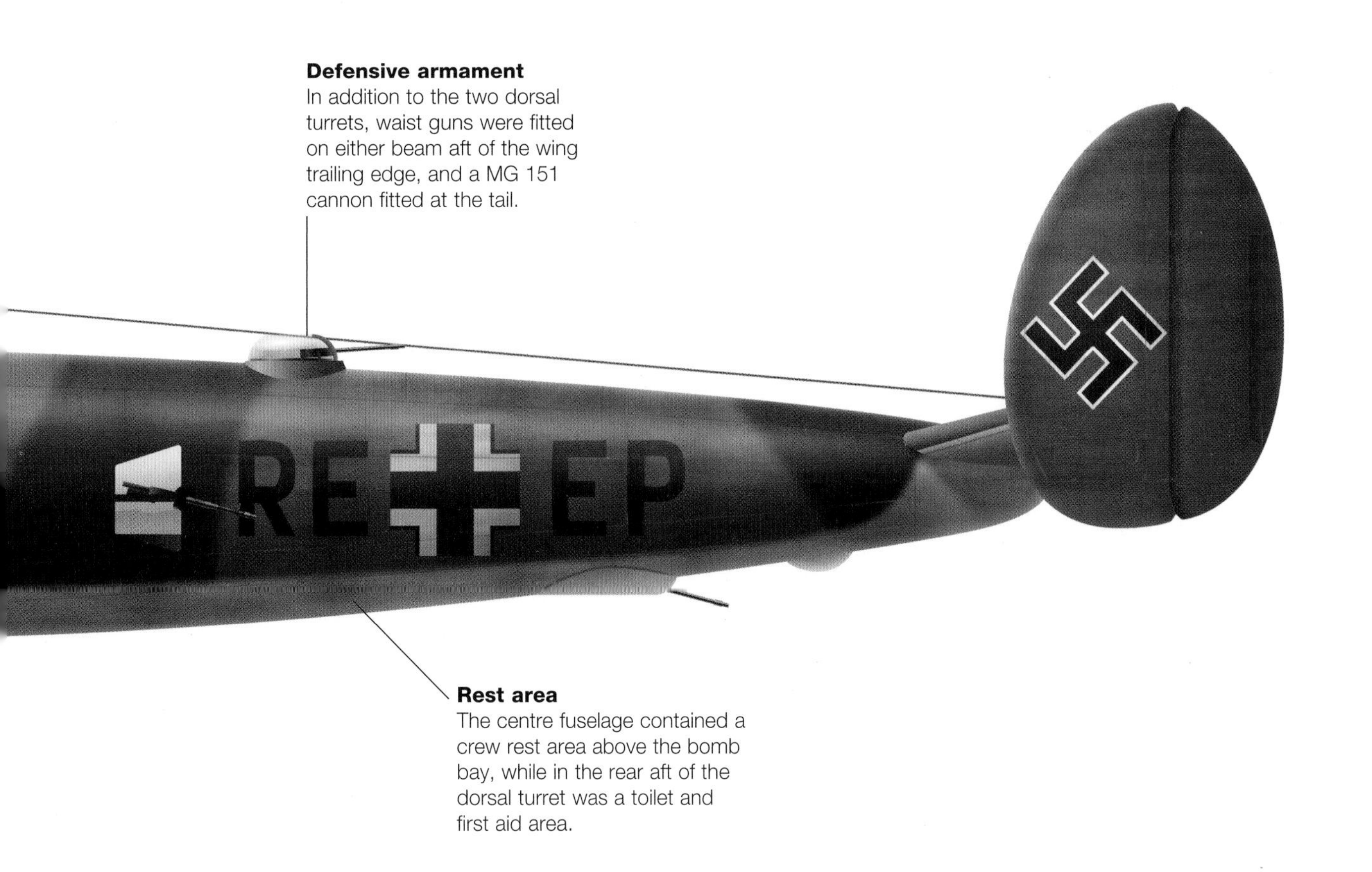

Defensive armament
In addition to the two dorsal turrets, waist guns were fitted on either beam aft of the wing trailing edge, and a MG 151 cannon fitted at the tail.

Rest area
The centre fuselage contained a crew rest area above the bomb bay, while in the rear aft of the dorsal turret was a toilet and first aid area.

Specifications: Me 264H-3 (proposal)

Type:	Long-range heavy bomber
Dimensions:	Length: 20.55m (67ft 5in); Wingspan: 43.1m (141ft 5in); Height: 4.3m (14ft 1in)
Weight:	50,000kg (110,231lb) maximum take-off
Powerplant:	4 x 1,290kW (1,730hp) Daimler-Benz DB 603H liquid-cooled V-12 engines
Maximum speed:	546km/h (339mph)
Range:	14,500km (9,010 miles)
Service ceiling:	8,000m (26,247ft)
Crew:	8
Armament:	2 x 13mm (0.51in) MG 131s in a front remote-control barbette; 2 x MG 151s in a rear remote-control barbette; 1 x MG 151 on either side; a bombload of up to 8,400kg (18,519lb)

landing strip. There were also some concerns about the rudder forces required, which would plague the Me 264 for much of its test programme, and some repositioning of some instrumentation was also recommended by the test pilot. Although the Me 264 was originally intended to be pressurized, the V1 was not and also lacked any armament, mock gun barbettes being installed instead for the trials programme.

The cockpit was at the very front of the fuselage with extensive glazing, not dissimilar to the He 111. Behind the pilots' seats were the positions for the radio operator, navigator and flight engineer. The centre section of the aircraft was divided horizontally, the lower half comprising the bomb bay while the upper gave access to the forward dorsal turret, radio equipment, and at the rear held the crew rest area and a galley. The rear of the airframe contained the waist guns, rear dorsal turret and tail gun, as well as oxygen tanks and the crew toilet.

Massive task

As the test programme continued, uncovering further issues as it went, interest in the Me 264 grew. The entry into the war of the USA after the Japanese attack on Pearl Harbor raised increased interest in carrying out raids on the Eastern Seaboard. Despite this, the Luftwaffe's competing

This photo shows the crew access ladder and gives an indication of how much of the cockpit would have been taken up by the compartment for the nose gear. This led to the crew being sat some way back in the cockpit, one of them just being visible in the image.

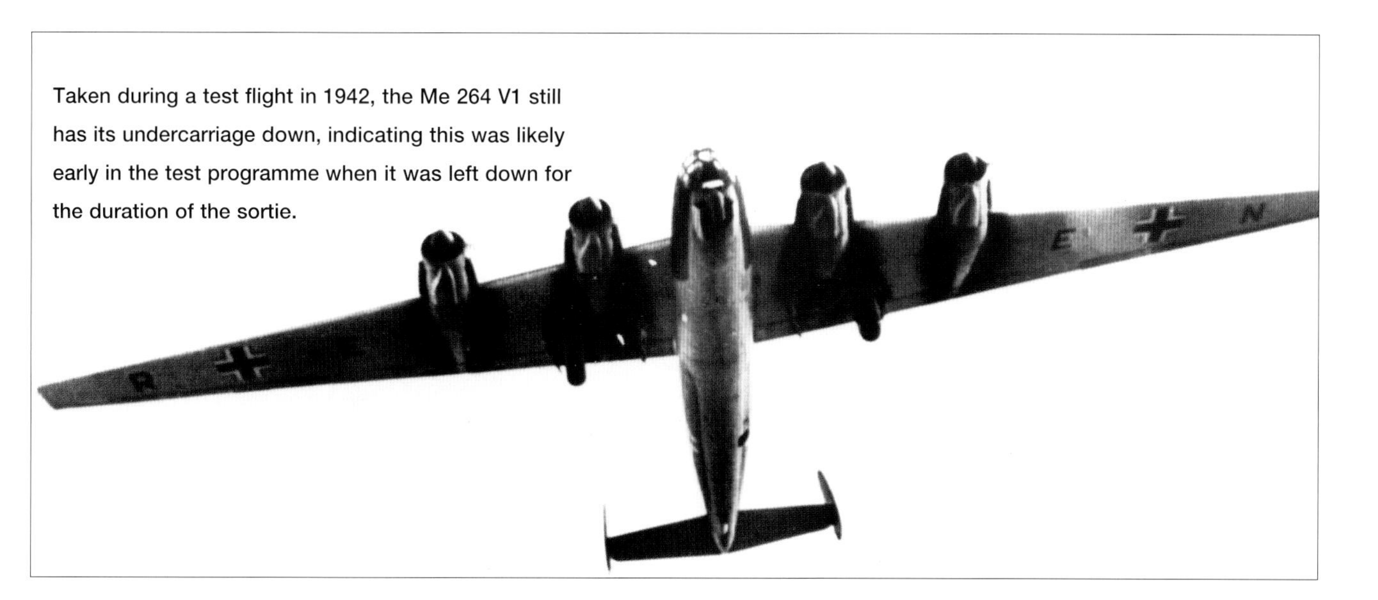

Taken during a test flight in 1942, the Me 264 V1 still has its undercarriage down, indicating this was likely early in the test programme when it was left down for the duration of the sortie.

Mission profiles

Initial planning for the Me 264 raids on North America anticipated using Brest in France as the starting point, which would save as much as 1,000km (621 miles) each way compared to a launch site in Germany. The return journey would cover just over 12,000km (7,456 miles), taking around 30 hours at the Messerschmitt's cruising speed of 350km/h (218mph) at 8,000m (26,247ft). At this range, a mixed load of 3,000–4,000kg (6,614–8,819lb) of general-purpose and fragmentation bombs could be carried depending on the target. The earliest planning had focused on New York as the nearest major city. As the Me 264's range estimates improved, other cities on the Eastern Seaboard came under consideration to force the USA to erect more anti-aircraft defences at the expense of providing arms to the UK. Further refinement of the plans added 21 potential industrial and strategic targets as far inland as Tennessee and stretching into Canada and Greenland. All these missions would have required multiple crews, while any aircraft damaged over North America would have to be nursed back across the unfriendly North Atlantic for over 10 hours. In reality, it is highly doubtful the results would have merited the level of investment required to bring the Me 264 to fruition.

requirements and change of focus to tactical aircraft meant construction of further aircraft was continuously delayed, the V2 and V3 airframes lagging far behind the V1. Meanwhile, between August 1943 and April 1944, the V1 was re-engined with the BMW 801 air-cooled radial engines, ensuring no test flying could be done.

The scale of the task Messerschmitt had undertaken should not be underestimated; an aircraft with a transatlantic radius of action, some 12,000km (7,456 miles), while carrying 2,000kg (4,409lb) of bombs, was at the cutting edge of what was achievable. The nearest comparison was the Boeing B-29. This had started life in March 1938 with a request for proposals from the United States Army Air Corps (USAAC), with the development proper starting in January the following year. In a programme that rivalled the Manhattan Project in scope and cost, the first Superfortresses would enter operational service in 1944. For Messerschmitt to have achieved the same with the limited resources of the Third Reich would have required a miracle and that they managed what they had with the V1 is impressive.

The V1 would be the only Me 264 to fly, achieving 38 hours and 22 minutes over 52 flights before being damaged beyond repair in a bombing raid on 18 July 1944. Despite this, the worsening situation for Germany, and the cancellation of everything but the emergency fighter programme, Hitler himself was still fantasizing over production of the Me 264 as late as August 1944 in order to strike back at the USA. The order went out to industry to hasten its production, only to be rescinded in mid-October when the stark reality of the war's progress led to the final cancellation of the programme.

Arado Ar 234 Blitz (1943)

Originally intended as a reconnaissance aircraft, the Arado twinjet would also be developed as the world's first jet bomber. However, delays in delivering the early jet engines prevented it entering service in sufficient numbers to impact the course of the war.

The Luftwaffe's, and indeed the world's, first jet bomber was originally meant to be a reconnaissance aircraft, design work starting in 1940. Much like the British Mosquito, its primary form of defence was intended to be speed, which led to a very clean design with a cylindrical fuselage mounting a high wing from which the two Jumo 004 turbojets would hang. The pilot sat in a plexiglass glazed nose with good visibility in all directions apart from below and directly behind. To achieve the target 2,150km (1,336-mile) range, the fuselage was almost completely put aside for fuel apart from a camera compartment. This gave the Arado designers some difficulty when deciding where to place the undercarriage; consequently, on the prototypes and Ar 234A models it was decided to use a detachable trolley for take-off with recovery made on to skids mounted on the belly and engine pods.

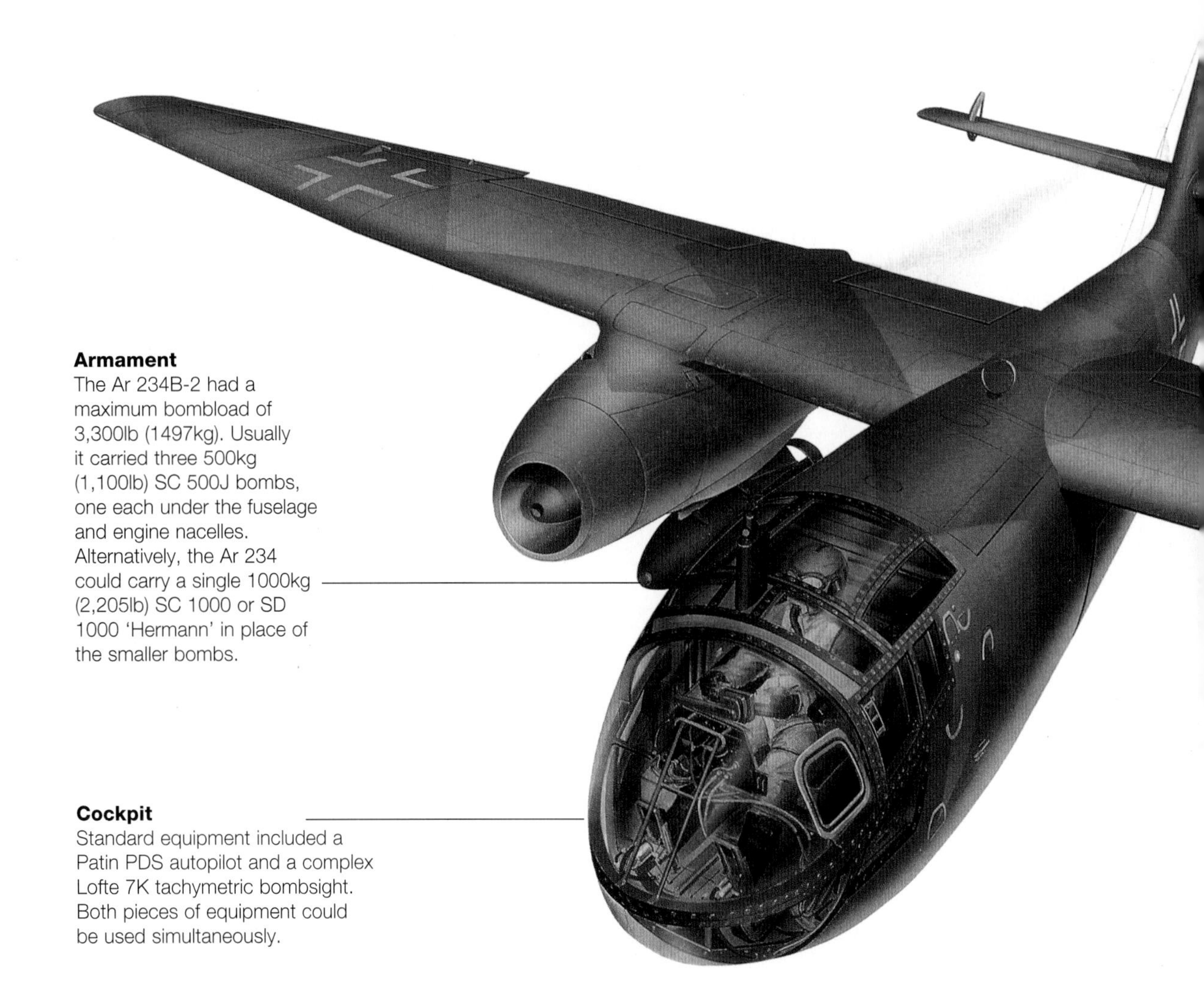

Construction of the first two prototypes was nearing completion by the end of 1941; the only element missing was the Jumo engines. Unfortunately, these would not begin flight testing until the following March, while with priority apparently given to the Me 262 programme, Arado would not receive any for the Ar 234 until February 1943. These engines, however, weren't cleared for flight and it would only be possible to carry out taxi trials later in the year. The intervening period did see consideration given to a bomber variant of the Ar 234 although the design of the undercarriage did not allow for an obvious location to carry bombs.

Engine choices

The prototype finally flew on 30 July 1943. The only issue encountered was with the take-off trolley whose parachute failed to operate when jettisoned at 60m (197ft), destroying it in the resultant landing. After the second test flight suffered the same problem it was decided it would be better to release the trolley while still on the ground. While further prototypes entered the testing programme, an order was placed for the Ar 234B. This would feature a tricycle undercarriage, ultimately allowing for the carriage of bombs under the fuselage and each engine pod. Meanwhile two of the prototypes, the sixth and eighth, would be completed with four engines, before the ninth would emerge as the first with a conventional undercarriage. This was achieved by sacrificing some of the fuel capacity in order to house the main gear in the central fuselage retracting forwards. The nose gear was under the cockpit and retracted to the rear.

By June 1944, the first 20 Ar 243B-0s had been completed. It would be the Ar 234A reconnaissance aircraft that would be urgently needed, however. The

This Arado Ar 234B-2 Blitz flew with the 9th Staffel, III Gruppe, Kampfgeschwader 76, during the defence of the Reich in March 1945.

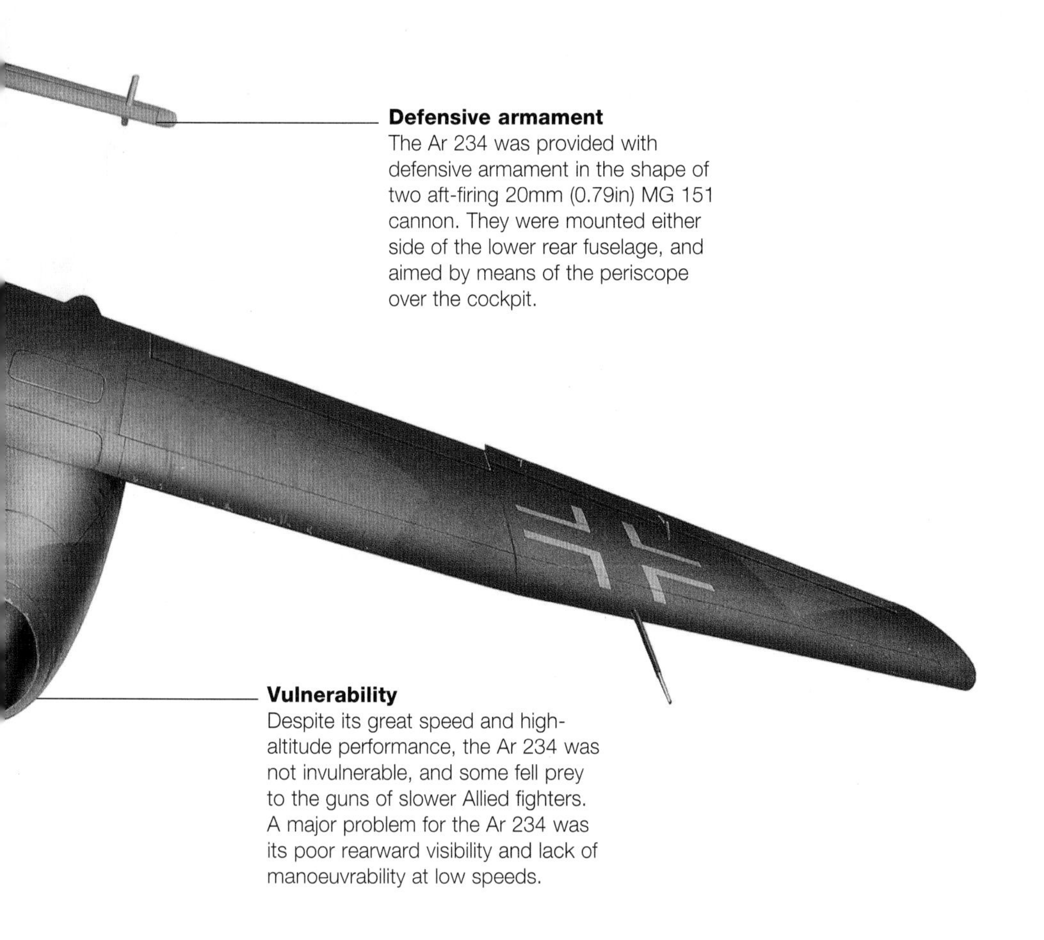

Defensive armament
The Ar 234 was provided with defensive armament in the shape of two aft-firing 20mm (0.79in) MG 151 cannon. They were mounted either side of the lower rear fuselage, and aimed by means of the periscope over the cockpit.

Vulnerability
Despite its great speed and high-altitude performance, the Ar 234 was not invulnerable, and some fell prey to the guns of slower Allied fighters. A major problem for the Ar 234 was its poor rearward visibility and lack of manoeuvrability at low speeds.

Specifications: Ar 234B-2

Type:	Light jet bomber
Dimensions:	Length: 12.64m (41ft 6in); Wingspan: 14.41m (47ft 3in); Height: 4.29m (14ft 1in)
Weight:	9,800kg (21,605lb) maximum take-off
Powerplant:	8.83kN (1,984lb) thrust Junker Jumo 004 turbojets
Maximum speed:	742km/h (461mph)
Range:	1,556km (967 miles)
Service ceiling:	10,000m (32,808ft)
Crew:	1
Armament:	2 x 20mm (0.79in) MG 151/20 cannon fixed firing aft; a bombload of up to 1,500kg (3,307lb)

This captured Ar 234B-2 is seen being examined by Allied personnel near Stavanger, Norway.

Allied invasion force in Normandy were resisting attempts by the Luftwaffe's propeller-powered aircraft to be photographed, greatly complicating attempts to repulse it. The fifth and seventh prototypes were fitted with cameras and deployed to Reims on 25 July, one aircraft having to turn back with engine failure shortly after leaving Oranienburg. Upon reaching Reims, the remaining aircraft demonstrated the shortcomings of the original design by promptly being placed on a low-loader and stored in a hangar until its undercarriage arrived a week later. It wasn't until 2 August that a mission would be flown, the first by a jet-powered reconnaissance aircraft. Using liquid rocket boosters to reduce the take-off run, the Ar 234A was at 10,000m (32,808ft) within 20 minutes and heading towards the Cotentin Peninsula. Making the photographic run from west to east at 740km/h (460mph), the Arado was untroubled by Allied aircraft or ground fire while it made three parallel runs above them. On its return to Reims, the sortie had achieved more than all

View of Arado Ar 234 B Blitz (w. nr. 140312) on display in the Boeing Aviation Hangar, Smithsonian National Air and Space Museum's Udvar-Hazy Center in Virginia, USA.

the other reconnaissance sorties flown by the Germans over the area to that date. If the mission itself had been a success, what it revealed of the scale of the Allied enclave in northern Europe could only depress German field commanders as they realized the extent of the forces arrayed against them.

Frontline role

September 1944 saw the first Ar 234B-1 becoming available to the front line and the last A model was retired, the first having been written off in a landing incident. Reconnaissance missions continued, the aircraft now being operated under the designation Kommando Sperling and operating over liberated Europe and the United Kingdom. The Arado suffered far fewer problems with its engines than did the Me 262 units, the operating profile being far more sympathetic to the first-generation jets with few throttle changes during a sortie.

The speed provided by the Jumo turbojets, meanwhile, made the Ar 234 almost invulnerable to interception, the first not being shot down until 11 February, when a Tempest of 274 Squadron caught up with an aircraft of Kommando Sperling as it was making its approach to land at Rheine air base. Despite this, most aircraft were fitted with a pair of 20mm (0.79in) MG 151 cannon in the rear fuselage, firing directly aft and sighted using the distinctive

Automatic bombing

In addition to the preferred shallow-dive or low-level bombing profiles, the Arado Ar 234B-2 could also use a highly advanced bombing system integrated with the automatic pilot for a medium-level attack.

After routing to a distinctive feature within 30km (18.6 miles) or so of the target, the autopilot was engaged and the control column moved to one side to allow the pilot to operate the Lofte 7K bombsight. The pilot then used controls on the sight to keep the target centred in the crosshairs, which in turn updated the autopilot on the course to steer. At the critical point, the bombs were automatically released, after which the pilot could move the flying controls back into place and disengage the autopilot to make his escape.

The only downside to using this state-of-the-art system was the requirement to fly straight and level over enemy territory for 30km (18.6 miles), while vulnerable to flak or fighters diving on the Arado from behind. Consequently, it was rarely if ever used.

periscope above the cockpit. It's not clear if these were ever used in anger and guidance on being intercepted tended to emphasize using the Arado's superior speed rather than attempting to shoot the other aircraft down.

The Arado's cockpit was well arranged, being larger than the similarly powered Me 262's, and with the glazing providing the pilot with a view past his feet more typical of a helicopter. Fitted with a three-axis autopilot during a bombing run, the pilot was able to move his control yoke out of the way and fly the aircraft using the Lofte 7K bombsight's control knobs. Handling was good at all speeds and the aircraft was fully aerobatic when not carrying bombs. With a bombload, the maximum speed was reduced by up to 95km/h (59mph) and the handling was less crisp but otherwise there were no serious handling issues reported. However, to assist with take-off, the bombers would frequently use Walter 109-500 booster rockets to get up to flying speed.

Christmas sortie

The first bomber unit to operate the Ar 234B-2 was 9./KG 76, which became operational in December 1944 with the first sorties being flown on Christmas Eve. These were against Liège in Belgium and targeted the railway yards and factories. Carrying a single 500kg (1,102lb) bomb under the fuselage, nine aircraft carried out shallow dive attacks before returning home at a height of 2,000m (6,562ft). With little to fear from interception by Allied fighters, the Arados did at least scare a lone Spitfire, which dived away from the approaching jets, unaware they had no offensive armament. Further missions would be flown against Liège in the coming days using the same attack profile. New Year's Eve would, though, see a change of mission.

To gather weather reports for Operation Bodenplatte, a mass attack on Allied airfields, four aircraft of 9./KG 76 would fly a route over Rotterdam, Antwerp, Brussels and Liège, the final two cities becoming victim to the first night-jet bomber raid to conceal the actual purpose of the mission. Gilze-Rijen airfield in Holland would be attacked by the Ar 234Bs during the day as part of the operation itself but the rest of the month only saw them conduct four further missions due to poor weather. III./KG 76 would reach full strength by the third week of the month with 7 and 8 Stafflen joining 9 at Achmer air base on the 23rd, although this was marred by three aircraft being shot down by 401 (Canadian) Squadron, who had the fortune to be over the base when the new aircraft arrived.

To conserve fuel and engine life, the Ar 234B was regularly towed around their airfields in lieu of taxiing. Here a Sd.Kfz.2 Kettenkrad tracked motorcycle tows an Ar 234B-2 of KG 76.

Bombing attack

Although the Ar 234B-2 was a capable bomber, the small numbers available prevented it from having a major impact on the closing phases of the war. 21 February saw III./KG 76 conducting 37 sorties in one day, the most ever flown, against British troops near Aachen for a total of 18 tonnes (20 tons) of bombs dropped against well-dispersed troops and vehicles. In comparison, Allied medium-bomber formations could deliver 54 tonnes (60 US tons) on a target up to 10 times a day.

March, meanwhile, would see KG 76 conducting eight days of raids on the bridge at Remagen, which was being used by the American Army to cross the Rhine. Although the bridge finally fell on the 17th it was at the cost of five Ar 234s. In addition, US Army engineers had used the time to construct pontoon bridges, ensuring the continued flow of men and materiel.

Despite the imminent collapse of the Third Reich, Arado were still constructing the first Ar 234Cs in March 1945. These featured four BMW 003 engines, the increased total thrust allowing the Walter booster rockets to be dispensed with, although it was now possible to exceed the design limits of the control surfaces in level flight at full power. Fourteen of these aircraft were delivered before the factory was destroyed in the face of advancing Soviet troops. Meanwhile, of the 210 Ar 234Bs built, only 38 were operational a month before the war ended. A technical marvel, the Ar 234B-2 entered full-scale production too late to have an impact on the war. The A and B-1 reconnaissance models had a much greater effect, allowing the German command to have accurate intelligence of Allied positions and the initial success experienced during the Battle of the Bulge in 1944–45 was in no small part due to the Arado.

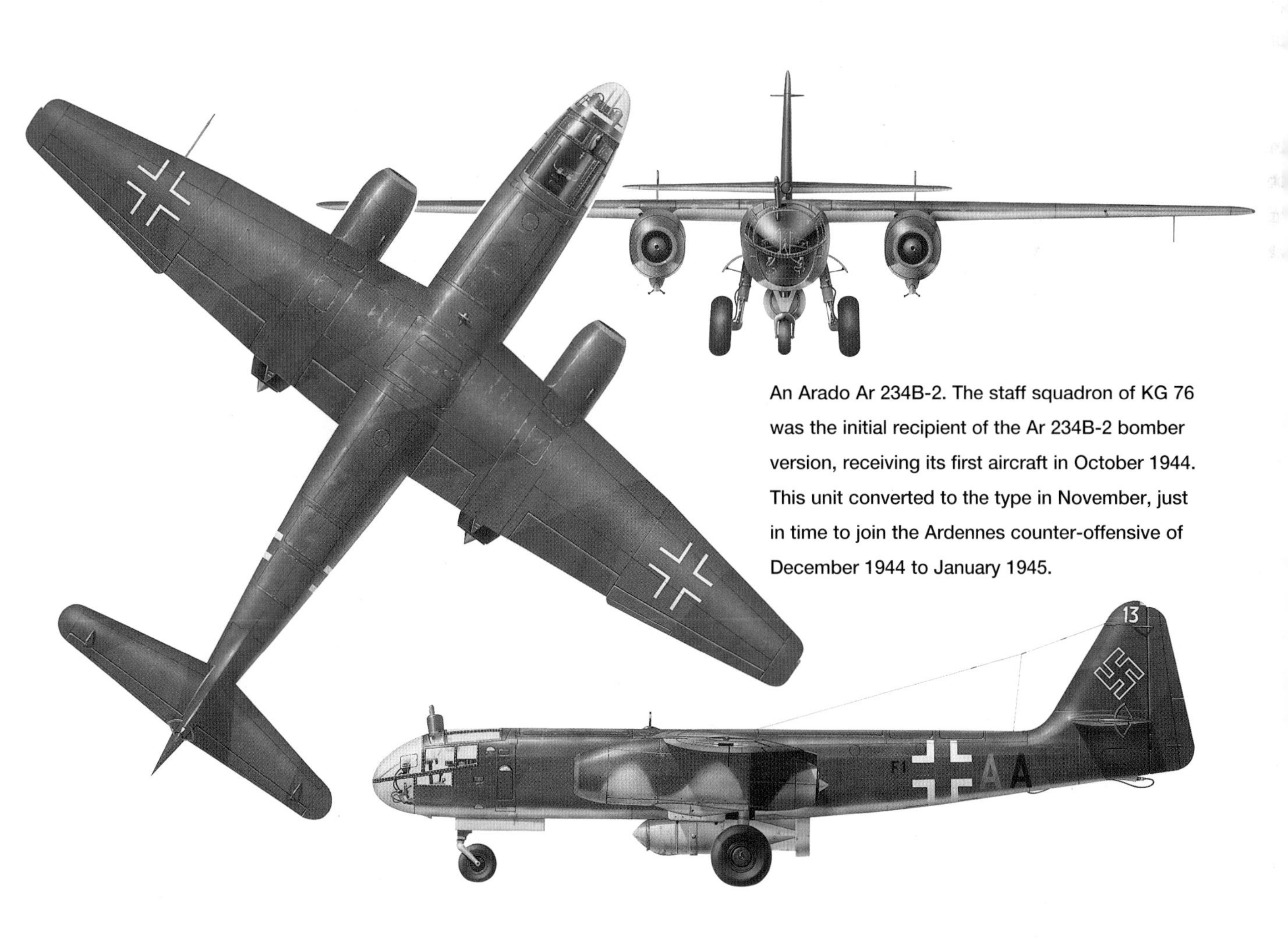

An Arado Ar 234B-2. The staff squadron of KG 76 was the initial recipient of the Ar 234B-2 bomber version, receiving its first aircraft in October 1944. This unit converted to the type in November, just in time to join the Ardennes counter-offensive of December 1944 to January 1945.

E

Transports, Reconnaissance & Seaplanes

The Luftwaffe pioneered air mobility operations during World War II with the Junkers Ju 52/3m being used not only to deploy paratroopers but also to resupply them while they waited for ground forces to join up with them. This would be taken even further on the Eastern Front, where isolated pockets of German forces would be successfully resupplied by air for months on end, ultimately leading to false confidence that the Stalingrad pocket could be maintained. A range of seaplanes were developed for maritime reconnaissance and attack missions, and as spotters for the Kriegsmarine's warships. These would operate in coastal waters from the North Cape of Norway to the Mediterranean and Black seas.

Opposite: A Blohm und Voss BV 138C flying boat docks somewhere in northern Germany, 1937.

Junkers Ju 52 (1930)

Conceived as an airliner, the Ju 52 would see widespread use as a military transport wherever German forces operated during the war. However, its pioneering capabilities would be hampered by its increasing vulnerability, leading to unsustainable loss rates.

The Junkers Ju 52 started life as a monoplane single-engine airliner powered by a range of powerplants producing around 560kW (751hp), including the BMW IV inline 6 and the Armstrong Siddeley Leopard 14-cylinder radial engines. As with Junkers' previous designs, it featured extensive use of a corrugated skin, which increased the stiffness in the axis perpendicular to the corrugations. This was, however, at the cost of increased parasitic drag due to the greater surface area. The type also featured Junkers' patented double wing or Doppelflügel, where the flaps and control surfaces are effectively a second narrow chord wing trailing the main lift surface. The prototype first flew in 1930 but was considered underpowered. The eighth airframe would therefore feature an additional two engines, one on either

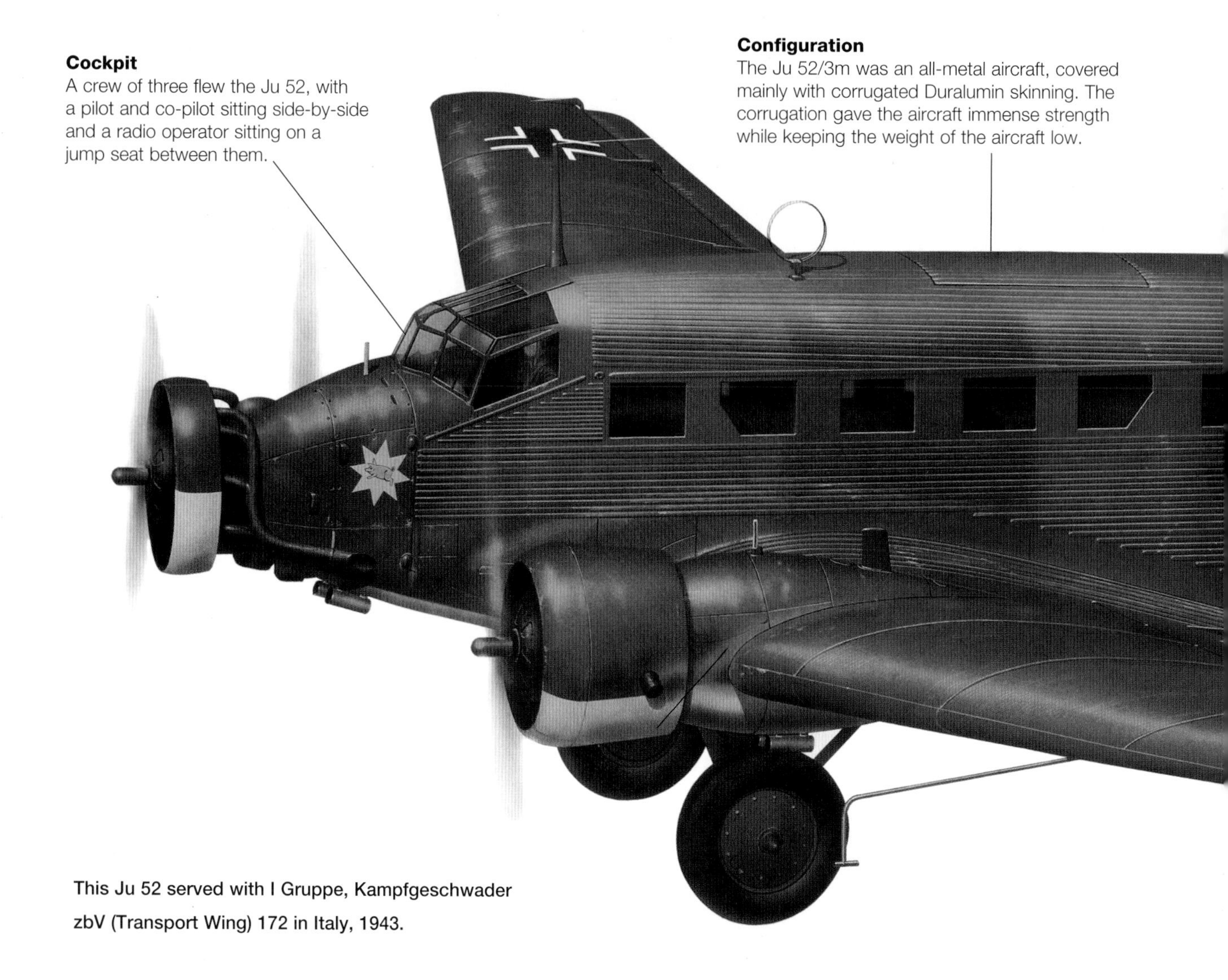

This Ju 52 served with I Gruppe, Kampfgeschwader zbV (Transport Wing) 172 in Italy, 1943.

wing, to become the Ju 52/3m, or drei Motoren for 'three engines'. This allowed a doubling or even tripling of the installed power depending on the engines used, improving the load-carrying ability of the type and providing a level of redundancy. Able to carry 17 passengers, the type was an almost immediate success with airlines from South America, Africa and Europe ordering it. Meanwhile, the Colombian and Bolivian air forces would procure float-equipped examples in 1932–33 for use as transports in the Colombia-Peru and Chaco wars respectively.

Airliner

While Lufthansa was expanding its passenger network through Europe with the Ju 52, the secretly reformed Luftwaffe was looking for a suitable aircraft to help it build its capabilities. This led to the Ju 52/3mge, which first flew in 1934. In addition to its role as a transport, the 3mge could carry a load of six 100kg (220lb) bombs in two internal bomb bays, which displaced the internal seating. Self-defence was provided by a 7.92mm (0.31in) MG 15 machine gun on a mount between the cabin door and the tail plane and a second one in a retractable dustbin that lowered beneath the aircraft. When acting as a transport, the Ju 52 could carry 17 paratroopers, or by removing the seats, 2.7 tonnes (3 tons) of cargo. Air dispatch of both was possible via the main cabin door, or in the case of cargo, through the bomb bay doors, while it was also possible to carry loads externally between the main gear.

First deliveries of the Ju52/3mge began in 1934 and 450 were on strength by the end of the following year.

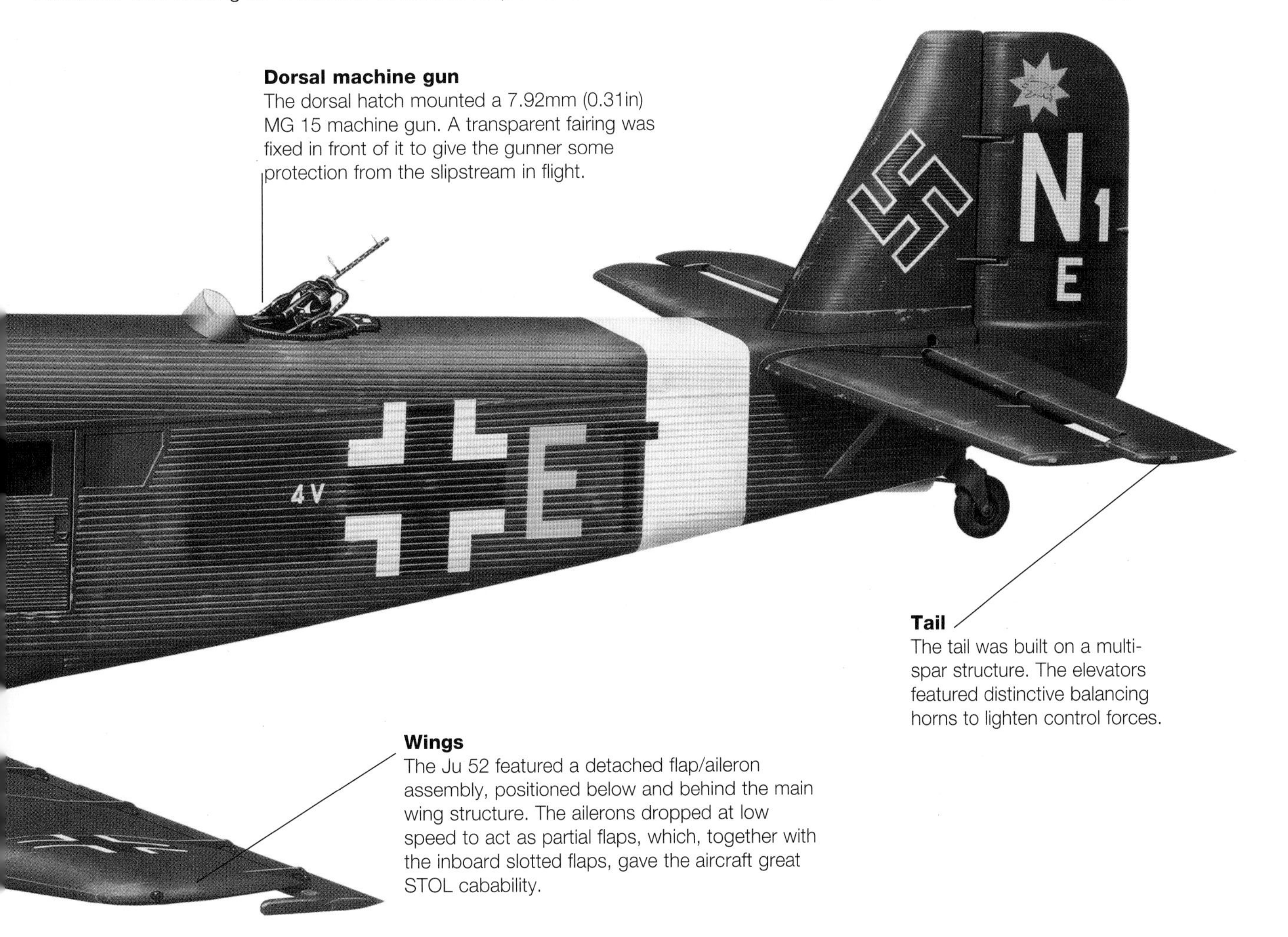

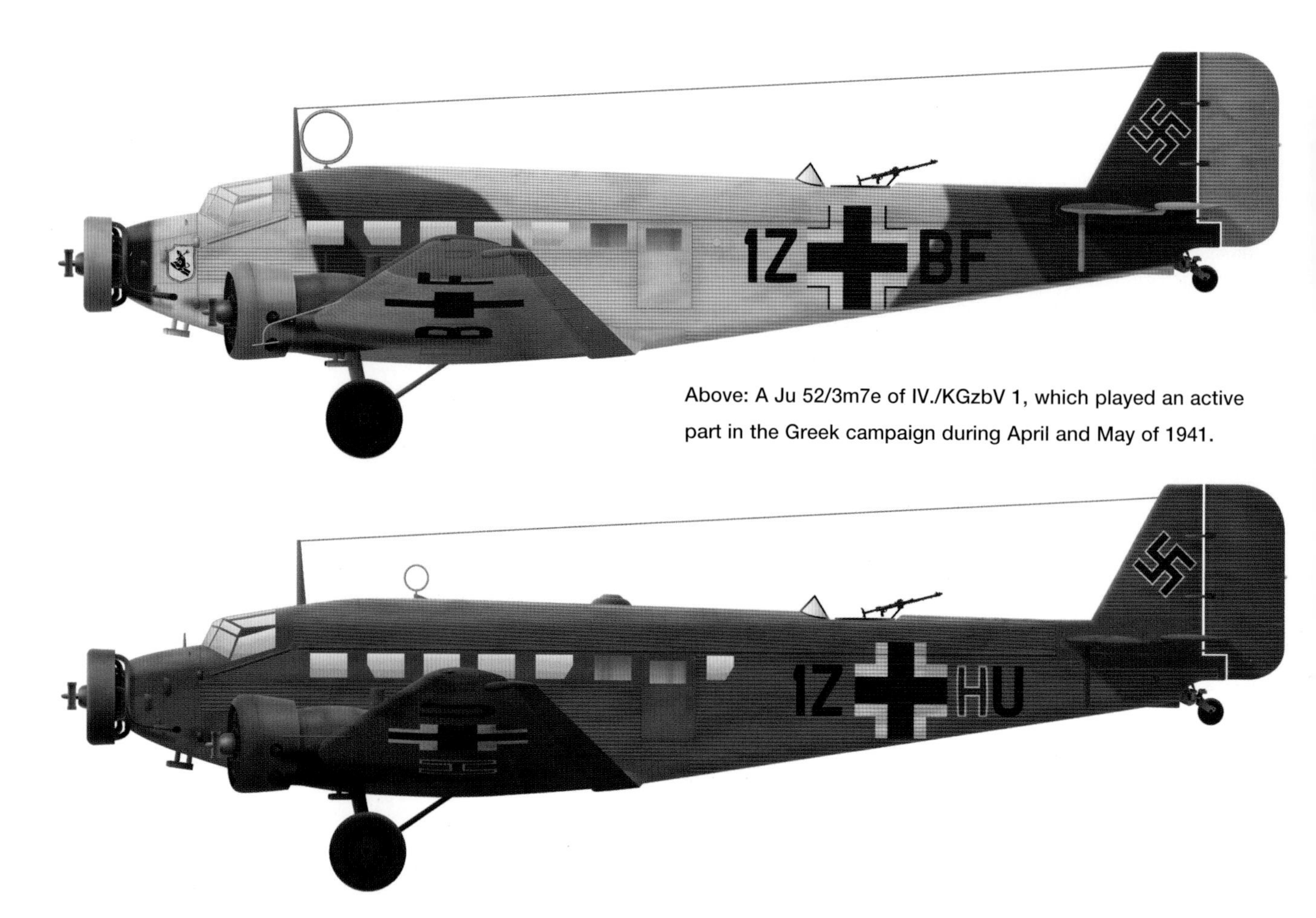

Above: A Ju 52/3m7e of IV./KGzbV 1, which played an active part in the Greek campaign during April and May of 1941.

Above: Another Ju 52/3mg7e of 10./KGzbV, this time while operating in North Africa in 1942. Above the cabin is an additional air intake for the hotter climate.

Specifications: Ju 52/3mg7e

Type:	Tactical transport
Dimensions:	Length: 18.8m (61ft 8in); Wingspan: 29.25m (95ft 12in); Height: 4.5m (14ft 9in)
Weight:	10,515kg (23,182lb) maximum take-off
Powerplant:	3 x 619kW (830hp) BMW 132T-2 nine-cylinder air-cooled engines
Maximum speed:	295km/h (183mph)
Range:	1,290km (802 miles)
Service ceiling:	5,500m (18,045ft)
Crew:	3
Armament:	1 x 7.92mm (0.31in) MG 15 in a dorsal position; 1 x 7.92mm (0.31in) machine gun on each side

Initially, they served with Kampfgruppe 152 although in 1937 the title was changed to Kampfgruppe zur besonderen Verwendung (KGrzb V), or 'special operation bomber group' – a designation intended to indicate their dual bomber and transport capability. In reality, the Ju 52/3mg3e would only see significant use as a bomber during the Spanish Civil War, although even there its first action would be in the transport role. In the summer of 1936, the 20 Ju 52s of the Condor Legion airlifted around 10,000 troops of the Nationalist-siding Army of Africa from Morocco to Spain, bypassing the Spanish Navy's blockade. After this, the aircraft of Kampfgruppe 88 would take part in air raids, principally along the Mediterranean coast but also infamously as part of the raid on Guernica. However, as the He 111 and Do 17 came on to the Luftwaffe's strength, the Ju 52 would increasingly be used purely in the transport role.

The Ju 52/3mg4e would introduce a tailwheel in place of the original skid and would be the standard equipment of the KGrzb V at the outbreak of the war in Europe, the

Transportverband having 547 on strength for the invasion of Poland. During this campaign, it would conduct 2,460 sorties transporting troops and cargo, although 59 aircraft would be lost – more than 10 per cent of the force. The invasion of Norway the following year saw the use of the Ju52/3mg5e, which could be fitted with conventional landing gear, floats or skis. A handful equipped with floats would conduct transport operations to the fjords in addition to the wide-ranging activities by the Ju 52/3mg4e models, which saw 29,000 men, 118,000 litres of fuel and 2,155 tonnes (2,376 tons) of supplies moved.

As part of the simultaneous invasion of Denmark, nine Ju 52s conducted the first operational drop of paratroopers on 9 April 1940, unleashing 96 Fallschirmjäger to capture the Storstrøm Bridge connecting islands in the south-east of the country. Losses were again high with 150 of the 573 aircraft available at the start of the campaign lost – just over a quarter.

High attrition would continue to be a feature of the Ju 52s'

Developments: Ju 252 & Ju 352

Work started on a replacement for the Ju 52 in 1938, Lufthansa wanting greater range, payload and speed. The proposed Ju 252 was powered by three 1,000kW (1,341hp) Jumo 211F engines and featured a smooth stressed-skin fuselage. With the outbreak of war, the Luftwaffe took over the programme, which saw the introduction of the Trapoklappe – a hydraulic loading ramp at the rear of the aircraft. As the Ju 252 remained a tail-dragging design, this had to be powerful enough to lift the rear off the ground. Light vehicles and stores could be loaded via the Trapoklappe and, if fitted with parachutes, dispatched in flight. First flying in 1942 and clearly superior to the Ju 52, the grave need for transport aircraft meant it was felt to be too risky to interrupt production of the older type; consequently, only 15 Ju 252s were produced.

The Ju 352 was a further development that was predominantly constructed from wood to relieve the demand on strategic materials. Although both were improvements over the Ju 52, it was a mark of the Nazi regime's desperate situation that it could not devote the resources to construct either type in significant numbers.

The Ju 52/3mg6e mine-clearance aircraft of the Minensuchgruppe were used extensively in the Mediterranean, equipped with large dural hoops below the fuselage, energised by an auxiliary motor in order to explode Allied mines.

operations as German forces struck to the west. Despite holding aircraft back for the planned invasion of the United Kingdom after the attacks on the Netherlands and Belgium were completed, a further 167 were lost from the 475 that had been on strength after the Norwegian campaign – over a third of the available force in just five days.

Further upgrades were developed in 1941, with the Ju 52/3mg6e improving the radio fit, and the 3mg7e adding an automatic pilot, room for an 18th troop, and an additional 7.92mm (0.31in) machine gun on either side for self-defence. Some Ju52/3mg6es would also be equipped with a large horizontal hoop placed under the wings and fuselage. This created a magnetic field powered by a dedicated motor that could be used to detonate magnetic mines and was used in coastal waters around Europe. The Mediterranean theatre would see the Ju 52 suffering heavy losses, primarily during the invasion of Crete in May 1941. Although this was a victory for the Germans and influenced Allied thinking on the use of airborne forces, delays and confusion during the initial airborne attack resulted in the loss of another 175 Ju 52s out of the 493 that had been assigned to the operation. Bad enough in their own right, these losses would have a knock-on effect on Operation Barbarossa, the invasion of the Soviet Union, with only 238 aircraft available at the start on 22 June 1941.

With the vast scale of the Eastern Front, air transport was vital to the Wehrmacht's campaign and production attempted to keep up with demand, with 1,892 aircraft being built between 1941 and 1943. The majority of these would follow the example of the Ju 52/3mg8e, which did away with the wheel spats to ease operations in the muddy airfields in the east. It also upgraded the dorsal gun to a 13mm (0.51in) MG 131 machine gun. The 9e would increase the maximum take-off weight to 11,500kg (25,353lb) with strengthened undercarriage and glider towing equipment. The 10e was another float-capable variant, while the 12e was powered by 597kW (800hp) BMW 132L radial engines. Late-1943 would see the introduction of an MG 15 machine gun in a turret

Two aircraft of KGzbV 1 in Greece shortly after the German invasion in May 1941. They were often employed to carry fuel to the various garrisons scattered around the Greek islands.

Paratroopers jumping from a Ju 52/3m – the static lines that automatically deployed the parachutes can be seen trailing from the cabin door. The Luftwaffe would be the first to deploy troops in this method during the *blitzkrieg* through Europe.

above the cockpit on the Ju 52/3mg14e. Losses through this period remained high, however. The evacuation of the Afrika Korps from Tunisia in April 1943 saw hundreds of Ju 52s lost, 24 being destroyed and a further 35 crash-landing in Sicily in one action alone on 18 April. The Eastern Front, meanwhile, saw an equivalent scale of loss as desperate battles were fought to supply isolated German troops by air. The Demyansk Pocket would be successfully defended for three months at the start of 1942 with more than 22,045 tonnes (24,300 tons) of supplies and 15,000 men being airlifted in, though 262 aircraft were lost in the process. It would ironically also convince the Nazi High Command that the tactic could be repeated on an increased scale at Stalingrad, despite the reduction in airlift capacity.

By the end of the war, only around 50 of the 4,835 Ju 52s built were still in operational service with the Luftwaffe, though this would not be the end. Surviving examples were taken over by the victorious Allied powers, British European Airways briefly operating 10 examples on domestic routes between 1946 and 1947. Meanwhile, production continued in France as the AAC.1 Toucan by Avions Amiot, whose factory had been co-opted by Junkers during the war. In all, 415 were built and operated by the French Air Force and Navy and would see service in Algeria and Indochina. They were also operated by Air France and a range of other airlines in Europe and North Africa. CASA in Spain would also produce the type as the CASA 352 and 352L, with 170 being built primarily for the Air Force. More surprisingly, in 2022, the Junkers Aircraft Company announced plans to produce a Ju 52NG featuring the same basic design as the original 1930s aircraft but powered by 410kW (550hp) diesel V-12 engines.

The Ju 52 was a dependable aircraft that allowed the Luftwaffe to pioneer the use of tactical airlift, moving men and materiel at speeds previously unimaginable. Its main shortcoming was its vulnerability, which led to loss rates that were unsustainable and prevented it repeating its earlier successes as the war rolled on.

Henschel Hs 126 (1936)

An evolution of the Hs 122, the Hs 126 was designed to fulfil the Army Cooperation role and despite its early success fading as the Luftwaffe faced serious aerial opposition, it would continue to find employment until the end of the war.

The Henschel 126 was developed to replace the Heinkel He 46 in the Army Cooperation and reconnaissance roles. It was based on the earlier Hs 122 that had been originally developed to replace the He 46 but was found to be too slow. All three types were parasol monoplanes with the single wing mounted above the cockpit; this configuration had been adopted by the He 46 when it was found that the lower wing of its original bi-plane configuration obstructed the downwards view to the extent it was hampering it in the reconnaissance role. The Hs 122 first flew in 1935 with the prototype using a Rolls-Royce Kestrel engine, with subsequent aircraft using the 490kW (657hp) Siemens Sh 22B radial engine.

Although the handling and slow-speed characteristics were considered more than adequate, the top speed was barely better than that of the Heinkel it was intended to replace. In the process of redesigning the aircraft to take the 620kW (831hp) BMW-Bramo Fafnir 323 radial engine, there were sufficient changes to warrant a new designation.

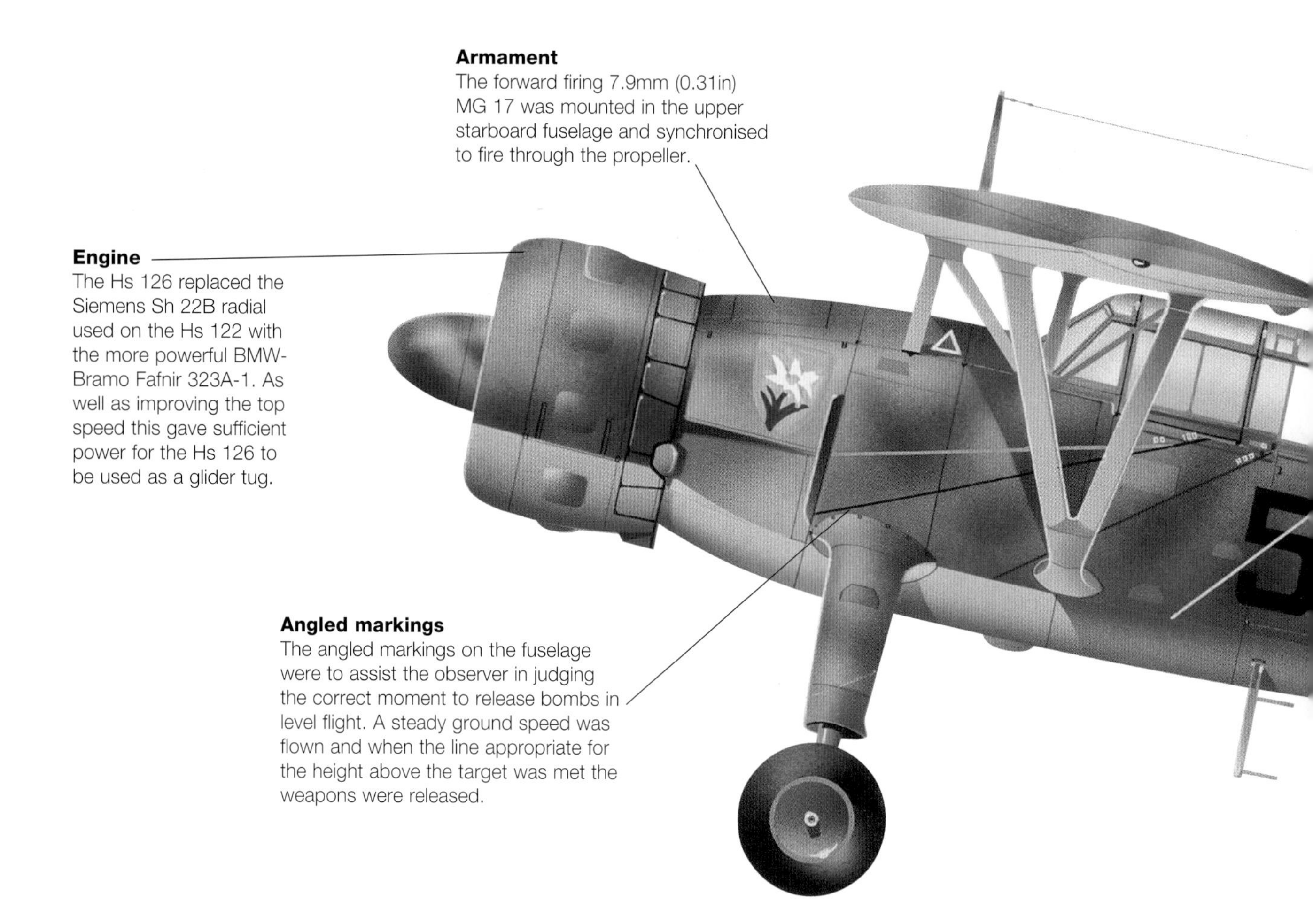

Armament
The forward firing 7.9mm (0.31in) MG 17 was mounted in the upper starboard fuselage and synchronised to fire through the propeller.

Engine
The Hs 126 replaced the Siemens Sh 22B radial used on the Hs 122 with the more powerful BMW-Bramo Fafnir 323A-1. As well as improving the top speed this gave sufficient power for the Hs 126 to be used as a glider tug.

Angled markings
The angled markings on the fuselage were to assist the observer in judging the correct moment to release bombs in level flight. A steady ground speed was flown and when the line appropriate for the height above the target was met the weapons were released.

One of the ten Hs 126A-0 pre-production aircraft D-ODBT is seen here pre-war; the spats on the main gear were frequently removed in service.

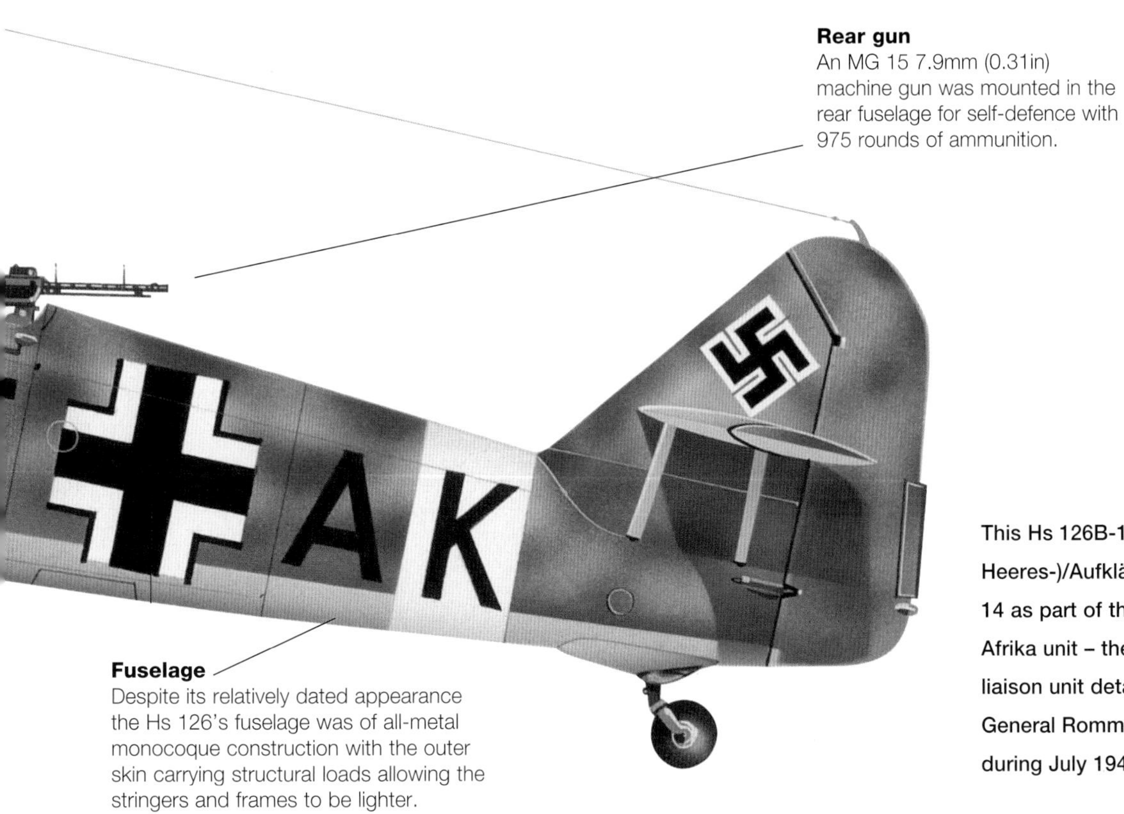

Rear gun
An MG 15 7.9mm (0.31in) machine gun was mounted in the rear fuselage for self-defence with 975 rounds of ammunition.

Fuselage
Despite its relatively dated appearance the Hs 126's fuselage was of all-metal monocoque construction with the outer skin carrying structural loads allowing the stringers and frames to be lighter.

This Hs 126B-1 flew with 2.(Heeres-)/Aufklärungsgruppe 14 as part of the Fliegerführer Afrika unit – the scout and liaison unit detached to General Rommel's HQ – during July 1941.

In comparison to the Hs 122, the new aircraft was longer, with a less curved trailing edge to the wing and a simplified main undercarriage design with a single strut on each side. The pilot sat in a covered cockpit with a sliding canopy. The rear cockpit, meanwhile, was open to the elements, allowing the gunner a clear field of fire. For the reconnaissance role, a fixed camera was carried in the fuselage while the gunner was also equipped with a hand-held model. The Hs 126 V1 was converted from one of the Hs 122 prototypes and featured a Jumo 210 as the BMW engine was not yet available. The second two aircraft gained the BMW radial engines and a slightly taller, more pointed vertical tail.

The Hs 126 V1 first flew in 1936 and retained the good low-speed handling and short take-off characteristics of the Hs 122 while increasing the maximum speed to 310km/h (193mph) – 45km/h (28mph) more than the earlier aircraft. Ten pre-production Hs 126A-0s were completed for trials and evaluation purposes while the first production Hs 126A-1 aircraft were delivered in the beginning of 1938. These would use the 625kW (838hp) BMW 132Dc engine, the 323 still not being available in the required quantity.

Specifications: Hs 126B-1

Type:	Army Cooperation and tactical reconnaissance aircraft
Dimensions:	Length: 10.85m (35ft 7in); Wingspan: 14.5m (47ft 7in); Height: 3.75m (12ft 4in)
Weight:	3,270kg (7,209lb) maximum take-off
Powerplant:	1 x 634kW (850hp) BMW-Bramo Fafnir 323A-1 air-cooled radial engine
Maximum speed:	310km/h (193mph)
Range:	580km (360 miles)
Service ceiling:	8,000m (26,247ft)
Crew:	2
Armament:	1 x 7.92mm (0.31in) MG 17 in the upper forward fuselage; 1 x 7.92mm (0.31in) MG 15 in the rear cockpit; plus a bombload of 10 x 10kg (22lb) bombs in the fuselage bay and 1 x 50kg (110lb) bomb on the fuselage side rack

As with most Luftwaffe types then in service, the Hs 123 was deployed to Spain to serve with the Condor Legion in the Civil War. Six Hs 126A-1s were deployed in 1938 to serve with 5. Staffel of Aufklärungsgruppe 88, the reconnaissance Gruppe. The type also proved successful in the light bomber role, being able to carry up to ten 10kg (22lb) bombs in the camera bay. At the end of the Condor Legion's involvement, the remaining five aircraft were handed over to the Spanish Air Force.

With Hs 126 production ramping up through 1939, 16 aircraft were sold to Greece for the Hellenic Air Force. They saw action during the Greco-Italian War in 1940 but after the decisive German intervention of 1941, they saw no further service. In 1939 production of the Hs 126B-1 included the BMW-Bramo Fafnir 323, which was finally available in sufficient numbers and producing 634kW (850hp). The B-1 aircraft also had an improved radio fit and would be the definitive variant of the Hs 126.

Polish campaign

The invasion of Poland saw 13 Staffeln operating the Hs 126 in the Army Cooperation, tactical reconnaissance and artillery spotting roles. The absence of a significant aerial opposition again saw them strafing ground targets and acting as a light bomber, carrying a 50kg (110lb) bomb on the side of the aircraft in addition to the 10kg (22lb) bombs in the camera bay. During the Phoney War, Hs 126s conducted reconnaissance along the Maginot Line on the Franco-German border. However, when the Battle of France began and the Luftwaffe faced fighters of equal capability to their own, the impunity the Henschel had previously enjoyed was over. Remarkably, one did manage to damage a Spitfire of the RAF's 74 squadron, which was forced to land at Calais before being abandoned for the Germans to capture, although the Hs 126 was also lost in the action.

Despite the losses, it would be some time before the Fw 189 would fully replace the Hs 126 in its primary role and 47 Hs 126 Staffeln were involved in the opening stages of Operation Barbarossa, the invasion of the Soviet Union. Here, the losses would again rapidly mount, and the type was withdrawn from the daylight role by the end of 1942 as its safety could not be guaranteed even with an escort of Bf 109s. The Eastern Front also saw the

An Hs 126 overflies support elements of Rommel's 7th Panzer Division as it advances through France in May 1940. Such close cooperation with ground forces was the raison d'être of the Henschel.

Glider tugs

While some Hs 126s would continue operating in an offensive role until the end of the war, others would see service as glider tugs. Some of these would tow training gliders for future Me 163 Komet pilots while others were involved in Operation Eiche, the mission to free Italian dictator Benito Mussolini, who had been deposed in July 1943. He was held captive in a hotel 2,000m (6,562ft) above sea level in the Gran Sasso mountains east of Rome. On 12 September, 10 Hs 126s launched from Pratica di Mare air base, each towing a DFS 230 carrying nine soldiers in addition to the pilot. During the approach to the target area, the lead section of Henschels had to orbit to gain height, but by 14:00 hrs they had released their gliders to arrive on the plateau. Mussolini was subsequently evacuated by Fieseler Fi 156 and installed as dictator in northern Italy, though that would only delay his inevitable downfall.

Henschel used to help relieve encircled ground forces at Kholm during the first few months of 1942, towing transport gliders and themselves carrying 200kg (441lb) of cargo. That autumn would see them used in the night ground-attack role – an idea copied from the Soviets, both sides using relatively low-performance aircraft to harass enemy troops during the hours of darkness. Hs 126 would continue to be used by the Nachtschlachtgruppen through to the end of the war.

Around 810 Hs 126s were produced between 1938 and 1941. As well as with Spain, Greece and Germany, they also saw service with the Estonian Air Force, five of whose aircraft were captured and used by the Soviet Union before being destroyed in German bombing raids.

Fieseler Fi 156 Storch (1936)

A private-venture light observation aircraft, the Fi 156 excelled at low-speed flying and would serve throughout the war. Its qualities were such that captured examples were often pressed into service by the Allies and it would remain in use long after the war ended.

The Fieseler Storch was the brainchild of World War I fighter pilot Gerhard Fieseler, who after the war had developed an interest in short take-off and landing aircraft. First flying in May 1936 as a private venture, the Fi 156 was an ungainly-looking aircraft with a high wing mounted over a heavily glazed cockpit that extended over the sides of the fuselage to improve the view downwards. The wing was fitted with an array of high-lift devices, including a fixed leading-edge slat. The stalky undercarriage, meanwhile, hung far below the aircraft in flight as a result of its long stroke, designed to absorb the shock of near vertical landings. These features all combined to produce an aircraft that, powered by a 179kW (240hp) Argus engine, could fly as slow as 51km/h (32mph), allowing it to descend almost vertically. Indeed, with a moderate breeze it was possible to land on the lift of an aircraft carrier, as proved by a British test pilot after the war, while at full load the take-off distance was at most 45m (148ft). Realizing the type's potential as an observation

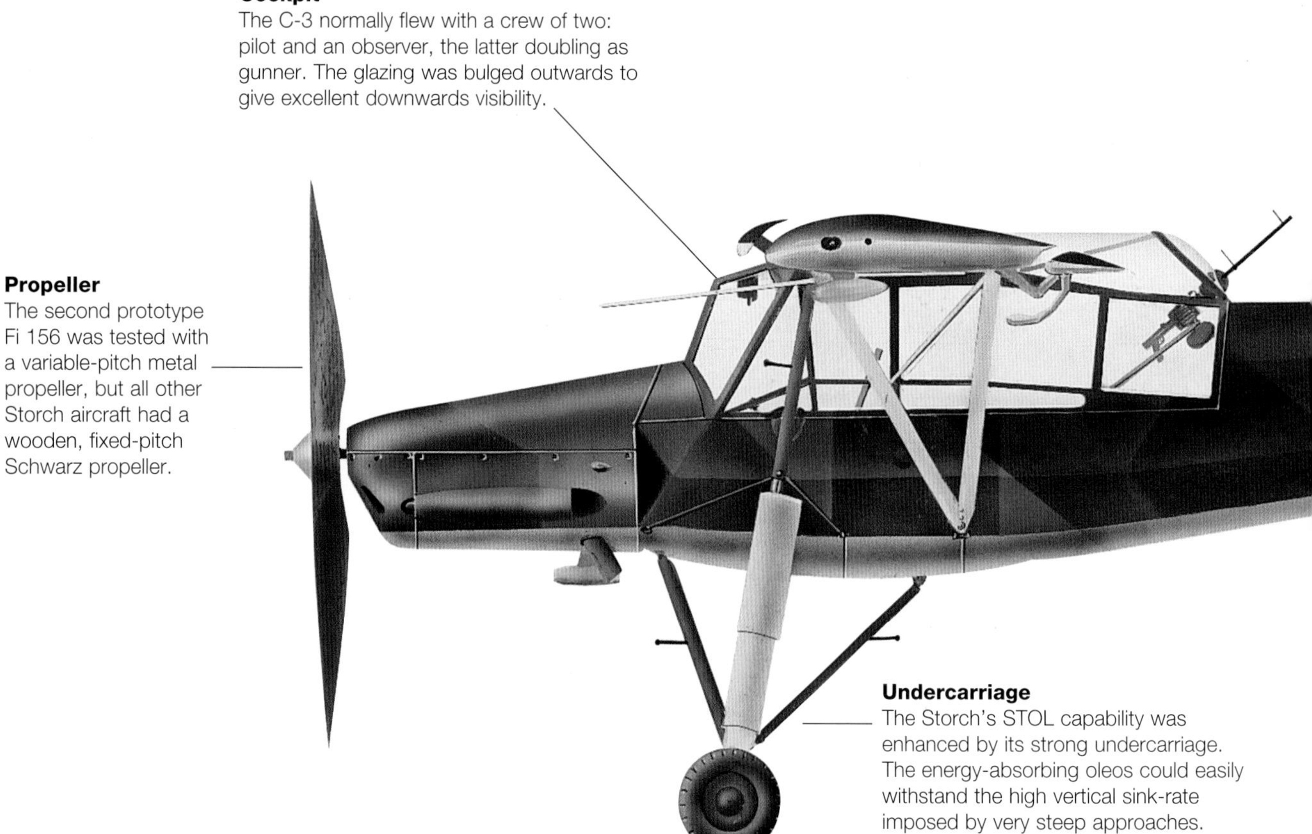

and Army Cooperation aircraft, the Luftwaffe ordered further prototypes, with deliveries beginning in late 1937.

The Fi 156A-1 was relatively unchanged from the prototypes and was constructed from steel tubes covered in fabric while the fixed slat was made from aluminium and the flaps from wood. Fuel was carried in the inner wings while an overload tank could be fitted in place of the passenger seats. Trials indicated that the slow speed made the Storch a difficult target for fighters as it hung in the air at 55km/h (34mph), while it also demonstrated an ability to accurately drop 50kg (110lb) bombs and 135kg (298lb) depth charges.

While Fi 156A-1s were used by the Condor Legion supporting the Nationalists in the Spanish Civil War, the Fi 156C-1 entered production in 1938. This added a mount for a 7.92mm (0.31in) MG 15 machine gun at the rear of the cabin. The C-2 gained a vertical reconnaissance camera, while both could be fitted with skis or attachments for a stretcher. Serving wherever the German Army went, the Storch had a remarkably low loss rate despite regularly operating in full view of the enemy. This included while operating as an air ambulance tasked to recover casualties and aircrew who had been shot down. While the C-5 could carry additional fuel in an under-fuselage drop tank, enabling it to fly recovery missions deep into the deserts of North Africa, the D-1 would gain a loading hatch on the right-hand side. This hinged upwards and extended the opening for the rear cockpit, allowing a stretcher to be easily loaded on board.

Ground support role

The opening stages of the war in the West saw the Fi 156 pressed into an unusual role. As part of the advance through the Ardennes Forest, it was planned to capture two critical crossroads in the towns of Nives and Witry. This would be done using Fallschirmjäger. However, there

Surfaces
The low-speed qualities of the Fi 156 were provided by full-span leading-edge slots and large slotted inboard flaps, allied to a high-lift aerofoil section. The ailerons were slotted and enlarged to ensure roll authority at speeds down to 32 mph (51 km/h).

Fieseler Fi 156C-3, Eastern Front, 1943. This anonymous aircraft retains its four-letter factory codes instead of its unit identification.

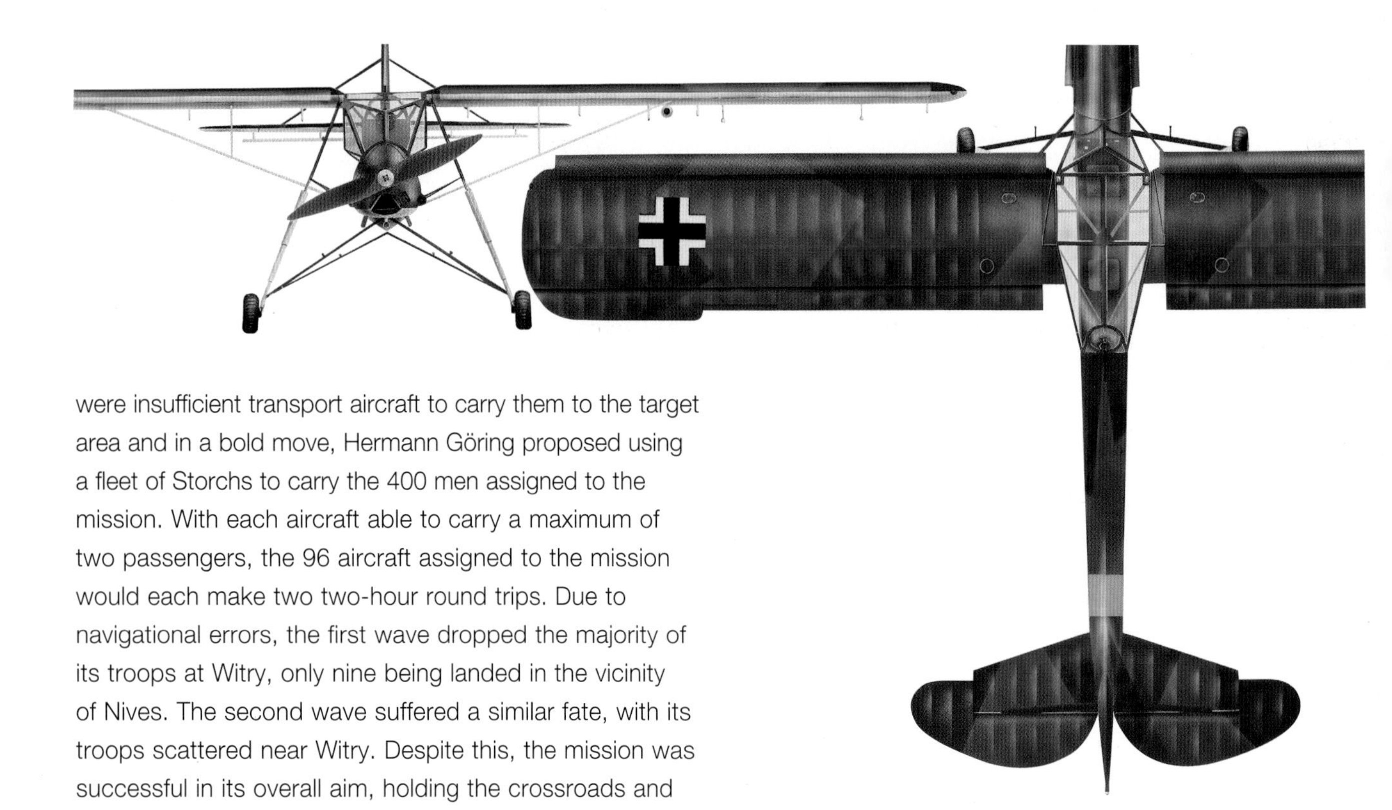

were insufficient transport aircraft to carry them to the target area and in a bold move, Hermann Göring proposed using a fleet of Storchs to carry the 400 men assigned to the mission. With each aircraft able to carry a maximum of two passengers, the 96 aircraft assigned to the mission would each make two two-hour round trips. Due to navigational errors, the first wave dropped the majority of its troops at Witry, only nine being landed in the vicinity of Nives. The second wave suffered a similar fate, with its troops scattered near Witry. Despite this, the mission was successful in its overall aim, holding the crossroads and allowing the 1st Panzer Division to advance towards the River Meuse. However, eight Fi 156s were lost to ground fire and a further 14 in crash landings.

Ungainly, spindly, even gawky – such adjectives come to mind when viewing the Storch, yet the aircraft became the standard by which all other wartime army co-operation/observation types were measured.

Last flight out of Berlin

As the war in Europe neared its conclusion, the Nazis were forced to improvise an airstrip in the heart of Berlin at the Tiergarten – a large former game park in the centre of the city with a sizeable east–west-running boulevard crossing it. Here, transport aircraft could maintain a tenuous air link as Soviet forces advanced on the capital. As the regime disintegrated, Hitler dismissed Göring as head of the Luftwaffe and called for Generaloberst Ritter von Greim to be brought to Berlin to be appointed his replacement. On the night of 26 April, test pilot Hanna Reitsch flew Greim in to the Tiergarten in a Fi 156, overflying the Soviet 3rd Shock and 8th Guards armies that now surrounded the city. A few days later, on the night of 28 April, the Storch would take off from the improvised airfield to head to Admiral Karl Dönitz's headquarters in northern Germany. Reitsch and Greim would be the last people to leave Berlin before it fell.

As the war progressed, a new operator of the Storch would emerge with the RAF capturing 47, predominantly in North Africa, which were subsequently adopted by various squadrons. As well as hacks, these were often used as personal transport for senior officers, including Field Marshal Montgomery and Air Vice Marshal Harry Broadhurst, who used one captured in Libya in northern Europe during and after the invasion of Normandy.

Outsourcing

While Fi 156 production continued through 1942, Fieseler struggled to fill demand as it was also called on to licence-build Bf 109s and Fw 190s. Consequently, a second line was set up at the Morane-Saulnier factory at Puteaux in 1942 and another at Choceň in Czechoslovakia. From 1943, all Storchs would be produced at these two factories. As the tide turned on the Axis powers, the Fi

Specifications: Fi 156C-3

Type:	STOL liaison, observation and rescue aircraft
Dimensions:	Length: 9.9 m (32ft 6in); Wingspan: 14.25m (46ft 9in); Height: 3m (9ft 10in)
Weight:	1,325kg (2,921lb) maximum take-off
Powerplant:	1 x 200kW (268hp) Argus As 10p air-cooled inverted V-8
Maximum speed:	175km/h (109mph)
Range:	470km (292 miles)
Service ceiling:	4,500m (14,764ft)
Crew:	2
Armament:	1 x 7.92mm (0.31in) MG 15 machine gun in the rear cabin

156 would take part in the daring raid to rescue Mussolini from his mountain prison east of Rome. Once the guards had been subdued by glider-landed paratroopers, a Storch landed on the hotel's terrace. Although the plan had been only to take the former dictator, the raid's leader also insisted on riding in the Fieseler, on the grounds that if anything happened to Mussolini, Hitler would have him killed. Overloaded, the plane took off from an 80m (263ft) runway, before making it to Pratica di Mare, where the passengers transferred to a He 111.

Production figures

More than 2,900 Fi 156s were built, while the Soviet Union produced its own version, the OKA-38, copied from one presented to Stalin by Hitler in 1939 after the signing of their non-aggression pact. France would continue to build and operate the type after the war as the M.S. 500, 501 and 502, depending on the engine installation, with some machines eventually ending up with the Vietnamese forces.

Shortly after the fall of Paris to the German *blitzkrieg* in 1940, this Storch landed in the Place de la Concorde, demonstrating its excellent STOL characteristics.

Heinkel He 114 & He 115 (1936 & 1937)

Floatplanes were somewhat anachronistic by the late 1930s and Heinkel's single-engined He 114 would see scant service. The He 115, however, would prove itself useful during the first half of the war, operating in the far north.

The Heinkel He 114 was designed to replace the earlier He 60 that equipped the Kriegsmarine's larger warships. As with the earlier aircraft, the He 114 was a twin-float biplane, although in its case the lower wing was of much shorter span than the upper one. The crew of two sat in tandem cockpits aft of the wing with the gunner's cockpit open to the rear. First flying in 1936, the He 114 was powered by a 716kW (960hp) BMW 132K radial engine, but performance was not considered satisfactory either in the air or while operating on the water. Changes to the wings and floats failed to totally address these issues and the He 114 was only issued to one unit, 1./Küstenfliegergruppe 506, in an attempt to generate export sales. This resulted in sales of 12 to both Sweden and Romania, the latter's aircraft being taken on by the Luftwaffe prior to delivery to assist with the push through the Baltic states at the start of Operation Barbarossa. Romania finally received them at the end of 1941 and operated them on the Black Sea until 1943.

The specification for what would become the He 115 was issued in mid-1935 and called for a floatplane patrol

Crew
The He 115B was flown by a crew of three, comprising pilot, radio operator and navigator. The radio operator sat in the rear part of the dorsal glazed area and operated the aft-facing gun. The navigator had a glazed nose area with an upper step for observation and a lower portion for bomb/torpedo aiming.

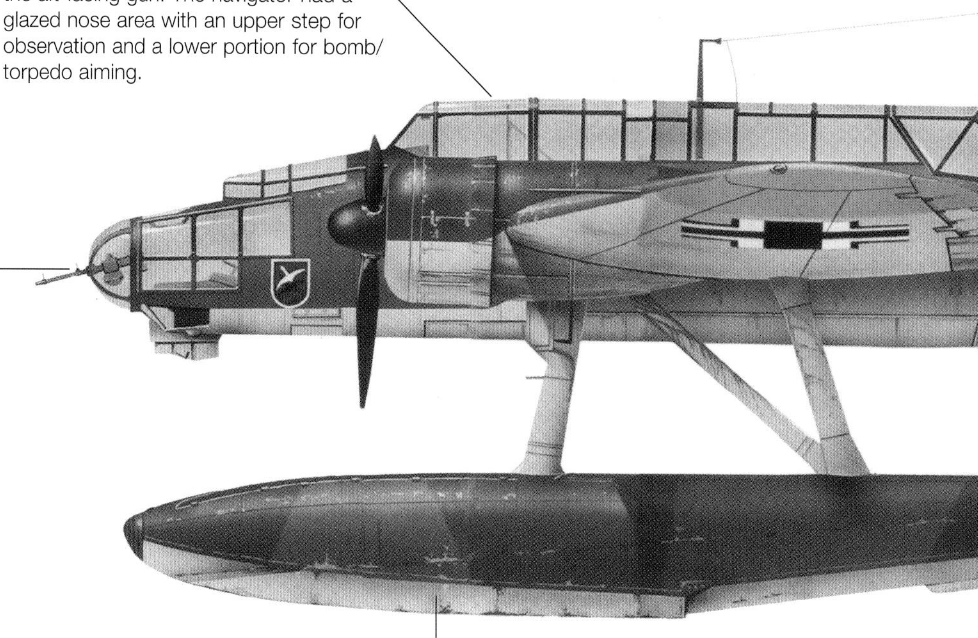

Armament
The navigator had an MG 15 machine gun for defensive fire in addition to the one in the rear cockpit. The He 115C would gain a fixed forward firing 20mm (0.79in) MG 151 cannon for use in the anti-shipping mission.

Floats
A true seaplane with no option to replace the floats with a conventional undercarriage, the He 115 was something of an anachronism even when it entered service.

aircraft capable of conducting torpedo and bombing attacks. First flying in 1937, the He 115 used two of the BMW 132K engines seen on the He 114, fixed to the leading edge of a mid-mounted monoplane with tapering outer sections. Two floats were positioned under the engine nacelles and braced to them and the fuselage. The crew of three comprised a pilot, observer and gunner. The observer's cockpit was in the nose with an upper conventional windscreen for lookout and a bomb aiming section in the underside of the fuselage; at the extreme nose was a cupola for a MG 15 machine gun. The pilot sat in line with the leading edge of the wing and above the observer, allowing him an unrestricted view forwards. The glazing extended aft along the fuselage to the wing's trailing edge with the aft section accommodating the gunner/radio operator, who had a rearward-facing MG 15. A bomb bay was incorporated in the central fuselage and was able to accommodate a torpedo or up to 1,250kg (2,756lb) in

A trio of Heinkel He 114s. Although generally considered unsatisfactory, the outbreak of war led to the Luftwaffe taking over orders destined for other countries. These aircraft were primarily used in the Baltic.

Kill tally
This aircraft carries kill markings for two ships and an aircraft on its tail, despite the He 115's limited performance it provided useful service as a maritime patrol aircraft during the first years of the war.

Markings
This aircraft wears standard tactical codes 'K6' for KüFlGr 406, 'L' for the individual aircraft, in white for the first staffel in the gruppe, and H for the 1. Staffel. Yellow theatre bands are worn as the war in Norway was regarded as being broadly part of the Eastern Front.

He 115B-1 1./Künstenfliegergruppe 406 Sørreisa, Tromso, 1942. By comparison with the He 115A-1 the He 115B-1 had a fuel capacity increased by 65 percent. This resultant increase in range could, in turn, be traded for a larger bomb load and soon after the outbreak of the war He 115s found themselves dropping magnetic mines in British waters.

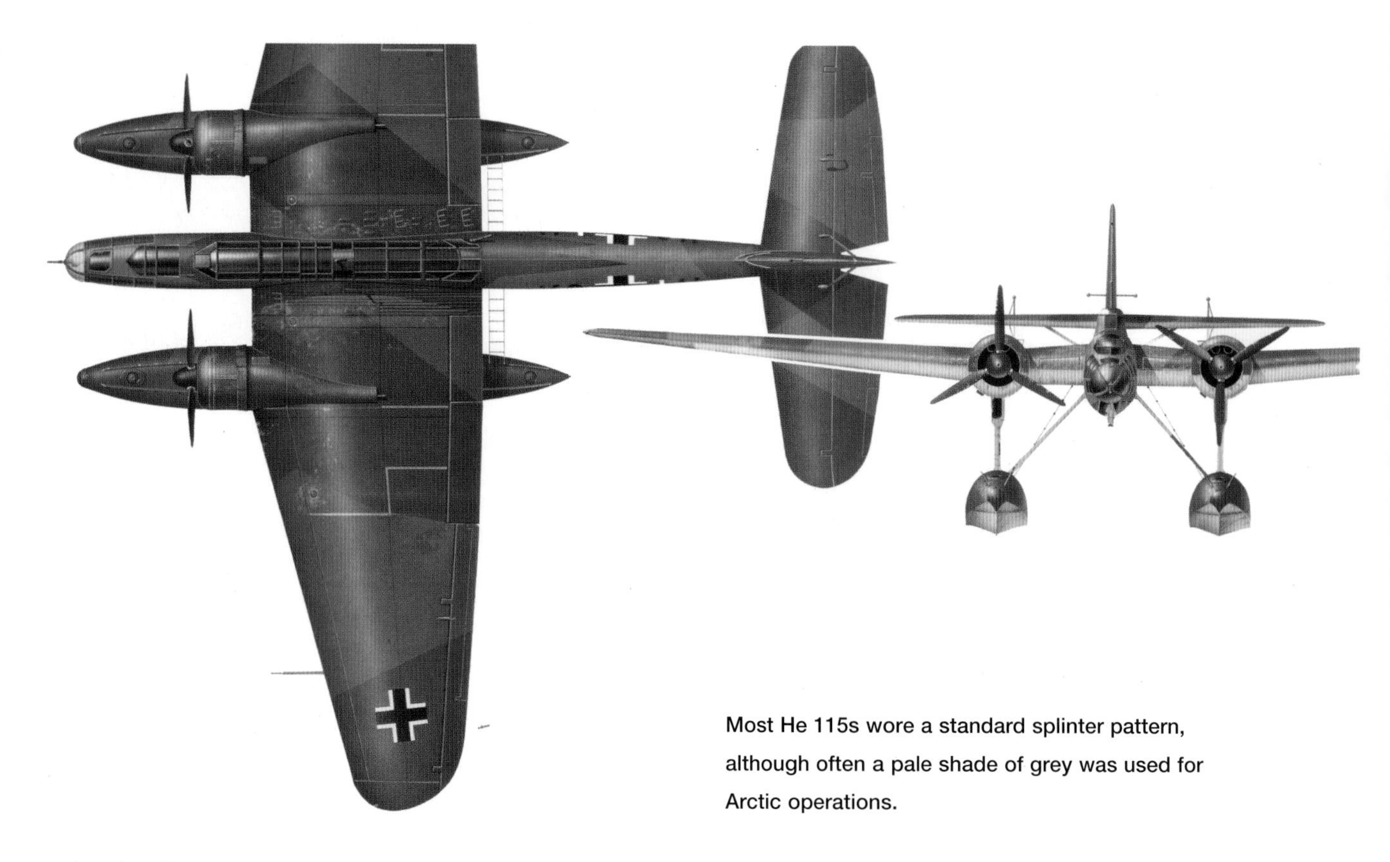

Most He 115s wore a standard splinter pattern, although often a pale shade of grey was used for Arctic operations.

bombs. From the third prototype, the radio operator was also provided with basic flying controls to give him some chance of recovering the aircraft if the pilot was injured.

Into production

Production of the He 115A series comprised 10 A-0s, 34 A-1s and A-3s, and 18 A-2s for export to Norway and Sweden. The A-1 gained underwing racks for two SC 250 bombs while the A-3 had improved avionics. 1./ Küstenfliegergruppe 106 took first delivery of the He 115A-1 as the war started and commenced coastal patrols. During the morning of 5 September 1939, one of these patrols sighted an Avro Anson of Coastal Command and the two aircraft engaged in a 15-minute battle that resulted in the Anson crashing into the sea. The He 115 then landed to rescue the sole survivor of the crew of four, Pilot Officer Hugh Edwards RNZAF, who became the first RAF officer captured. He 115 would also be used for laying mines in the waters around Great Britain, although one aircraft of 3./Küstenfliegergruppe 106 would inadvertently drop their magnetic mine on the mudflats near Shoeburyness, handing the Royal Navy the information they needed to counter its trigger mechanism.

Covert operations

With a small squadron worth's of He 115s gathered in Great Britain after the fall of Norway, the Allies soon looked to use them for clandestine work. Accounts vary but some modifications were carried out, including adding British weapons, and missions were flown across the North Sea, presumably ferrying agents and members of the Norwegian resistance to and fro.

The primary danger was from the RAF and British anti-aircraft fire, since they could not be informed of the missions in advance; one aircraft was attacked by Spitfires as it was returning and had to make a forced landing on the sea. More surprisingly, there are claims that two aircraft were sent via Gibraltar to the Mediterranean, where there were no Luftwaffe He 115s. Although one only managed to conduct a single mission to North Africa before being destroyed in an air raid, the other was operating in the theatre for around six months before being destroyed by Bf 109s.

This He 115B-1 is being loaded with a torpedo, note the use of a small flat-bottomed boat to position the weapon under the bomb bay while the aircraft is moored.

He 115B-1s began entering service in 1940, with an increased fuel capacity and the ability to carry two 500kg (1,102lb) mines. Eighteen He 115B-2s were also produced with skate-runners along the base of the floats for operation from ice. The invasion of Norway that year saw the He 115-equipped Gruppe moving north to operate from the fjords, conducting reconnaissance and anti-submarine missions. At the same time, the Norwegian He 115s would operate against the invading German forces. By the end of the invasion, the Luftwaffe had captured one of the Norwegian aircraft, while two Luftwaffe examples had themselves been captured and flown to Britain along with four Norwegian He 115s. The He 115-equipped Staffeln would spend the majority of their service life in Norway, operations where there was any significant fighter cover proving too dangerous for the relatively slow floatplane. From here, they conducted reconnaissance and attack missions against Russian-bound convoys around the North Cape. They achieved their greatest success against the ill-fated Convoy PQ 17 in July 1942 with units from across northern Norway conducting torpedo attacks and guiding other units to their targets. Ultimately, 23 of the convoy's 36 vessels were lost. After this high point, however, the next Arctic convoy – PQ 18 – was the first to be accompanied by an escort carrier, the embarked Sea Hurricane fighters making life far more dangerous for the Heinkels.

The He 115C-1 was an attempt to improve the type's defensive capabilities, carrying a 20mm (0.79in) MG 151/15 forward-firing cannon under the nose and two MG 17 machine guns firing from fixed positions behind the engine nacelles. However, this proved of limited use, the He 115 being unable to turn with a fighter to bring the forward-firing gun to bear and the fixed rearward-firing guns being virtually useless. Consequently, the aircraft were gradually withdrawn from frontline service with the only two Staffeln still using them in November 1943.

Specifications: He 115B-1

Type:	Coastal reconnaissance and torpedo-bomber floatplane
Dimensions:	Length: 17.3m (56ft 9in); Wingspan: 22.00m (72ft 2in); Height: 6.6m (21ft 8in)
Weight:	10,400kg (22,928lb) maximum take-off
Powerplant:	2 x 716kW (960hp) BMW 132K air-cooled radial engines
Maximum speed:	355km/h (221mph)
Range:	3,350km (2,082 miles)
Service ceiling:	5,500m (18,045ft)
Crew:	3
Armament:	1 x 7.92mm (0.31in) MG 15 machine gun in the nose and dorsal positions; up to 1,250kg (2,756lb) of bombs in the fuselage bomb bay

Arado Ar 196 (1937)

The last combat seaplane to be built in Europe, the Ar 196 was intended to serve as the primary aircraft for the Kriegsmarine's growing fleet of cruisers and battleships. However, it would make as great, or even greater, a contribution while operating from shore bases.

With the rapid pace of aircraft development through the 1930s, by 1936, requests were made for a replacement for the Heinkel He 60 floatplane that had only entered service three years earlier. Although this was initially intended to be the Heinkel He 114, dissatisfaction with its performance led to Arado and Focke-Wulf being asked to produce proposals for the role. The designs had to be capable of operating from both the sea and catapult-equipped warships as maritime patrol aircraft.

Although both companies received orders for prototypes, the Arado monoplane was clearly preferred to the Heinkel biplane from the outset. Powered by a 656kW (880hp) BMW

Powerplant
The Ar 96 prototypes were powered by an 656kW (880hp) BMW 132Dc radial, driving a two-bladed variable-pitch propeller. The pre-production Ar 196A-0 changed to the 716kW (960hp) BMW 132K, which remained the standard powerplant for all subsequent variants. A nine-cylinder air-cooled radial, the BMW 132K drove a three bladed variable-pitch propeller on all production versions of the aircraft.

An Arado Ar 196A-5 from 4./ Seeaufklärungsgruppe 126 that operated from Vukovar, Croatia, in January 1944.

132K radial engine, the Ar 196 featured a low wing with a crew of two in a tandem cockpit, the aft canopy being open to the rear to ease operation of the 7.92mm (0.31in) MG 15 machine gun. First flying in June 1937, two prototypes featured a single float under the fuselage with stabilizing floats under the outer wings, while the other two had two main floats braced to the fuselage and inner wings. Although both layouts were reasonably matched in performance, cost and complexity, the latter scheme was eventually chosen for production as the stabilizing floats could cause significant drag on one side during the take-off run if they became submerged. The Ar 196 was relatively traditional in its construction, with a fuselage made from steel tubes clad in aluminium alloy and fabric. The floats were also made from aluminium alloy with each containing a 300-litre (79-gallon) fuel tank and in production, aircraft survival aids. For shipboard stowage, the wings folded aft on an angled hinge to lie alongside the fuselage with the lower surface outermost, in a similar fashion to Grumman's STO-wing system used on the Wildcat, Hellcat and Avenger.

Testing

Twenty production Ar 196A-1s were delivered to the Luftwaffe on 20 July 1939 for testing and evaluation. As well as the rear-mounted MG 15 machine gun, these aircraft could be armed with two 50kg (110lb) bombs under the wings. From November, the Ar 196A-2 began to be delivered, which also carried a fixed forward-firing 7.92mm (0.31in) MG 17 machine gun firing through the engine cowling and a 20mm (0.79in) MG FF cannon in each wing.

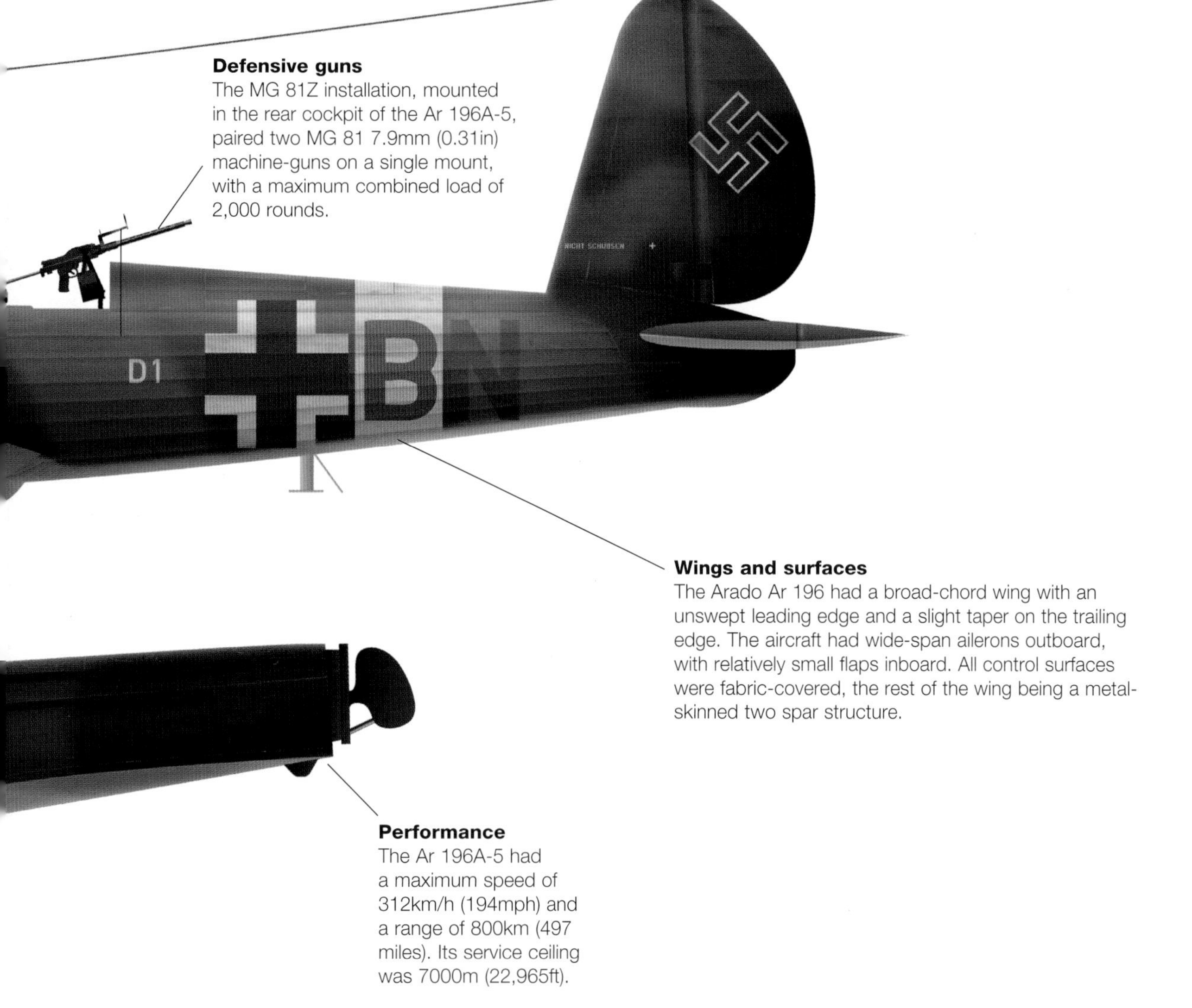

The engines for the A-2 were also uprated to 716kW (960hp). An unusual disruption to the production and delivery programme was the freezing of the River Warnow by the factory in Warnemünde, effectively removing the floatplanes' runway.

Despite having been in existence for two years, knowledge of the Ar 196 outside Germany was almost non-existent at the outbreak of the war, the first report in *Flight* magazine not being published until October 1939. By this point, the first operational aircraft had embarked in the *Admiral Graf Spee*, which was on its way to the South Atlantic and Indian oceans to raid merchant shipping. In this role, the Ar 196 played a vital part in searching for targets and warning of enemy surface combatants, which the *Graf Spee* was at pains to avoid. Serviceability issues would ensure it was not all plain sailing with the engine suffering cracking of its lower cylinder, requiring engine changes and repairs that kept the one embarked aircraft grounded for several days. A rough landing on 11 December while off the coast of South America finally rendered the engine beyond repair. Two days later, the Royal Navy's Force G – comprising the cruisers HMS *Exeter*, *Ajax* and *Achilles* – found the *Graf Spee* and prepared to attack. As a ruse, the three warships manoeuvred as if they were the vanguard of a larger force that would outgun the German pocket battleship. Unable to launch the Arado, the *Graf Spee*'s captain had no way of knowing this wasn't the case and after a running battle sought refuge in neutral Uruguay. Only permitted to remain there for 72 hours, the *Graf Spee* was scuttled in the River Plate on 17 December.

While this action demonstrated the importance of the Ar 196 to naval operations, the difficulties of keeping fragile aircraft serviceable at sea would remain. Indeed, while the *Graf Spee* had been in the South Atlantic, the Ar 196s on the *Gneisenau* and *Scharnhorst* had been damaged from the shock of their own ships' guns being fired without ever getting airborne. There would be better news for the Ar 196 force in the new year, though, when aircraft of 1. and 5. Staffeln of Bordfliegergruppe 196 were involved in the capture of HMS *Seal* – the only Royal Navy submarine to be captured during the war. Damaged by a mine in the narrow seas between Denmark and Scandinavia, the *Seal* was attempting to make for neutral Sweden on the surface when she was attacked by two Ar 196A-2s early in the morning of 5 May. With his situation untenable, the captain

Around the world

As well as operating from warships of the Kriegsmarine, Ar 196s served on board auxiliary cruisers – essentially merchant ships fitted with a range of large-calibre guns to attack enemy merchant shipping. The embarked aircraft scouted for potential targets and warned of approaching enemy warships. They were regularly painted in fake markings to conceal their identity and ranged through the Atlantic, Indian and Pacific oceans. With the auxiliary cruisers occasionally docking in Japan, some spare aircraft appear to have been stored there.

In early 1944, two of them were dispatched to Penang in Malaya to join the U-boat detachment established there. Although operated and maintained by Germans, the aircraft were painted in Japanese markings and conducted patrols along the Malacca Strait, where Royal Navy submarines were increasingly active. They were involved in the rescue of survivors from the former Italian U-boat UIT-23 when it was sunk by HMS *Tally-Ho* in February 1944 and also held off an attack by the same submarine when U-532 was transiting to the docks at Singapore. After the German surrender, the aircraft were handed to the Imperial Japanese Navy Air Service, but their final fate is unknown.

Ar 196A-3

Type:	Maritime reconnaissance
Dimensions:	Length: 10.96m (35ft 12in); Wingspan: 12.44m (40ft 10in); Height: 4.44m (14ft 7in)
Weight:	3,303kg (7,282lb) maximum take-off
Powerplant:	1 x 716kW (960hp) BMW 312K radial engine
Maximum speed:	312km/h (194mph)
Range:	800km (497 miles)
Service ceiling:	7,000m (22,966ft)
Crew:	2
Armament:	2 x 20mm (0.79in) MG FF cannon in wings; 1 x 7.92mm (0.31in) MG 17 machine gun in the nose; 1 x 7.92mm (0.31in) MG 17 in the aft cockpit; plus a bombload of up to 2 x 50kg (110lb) bombs

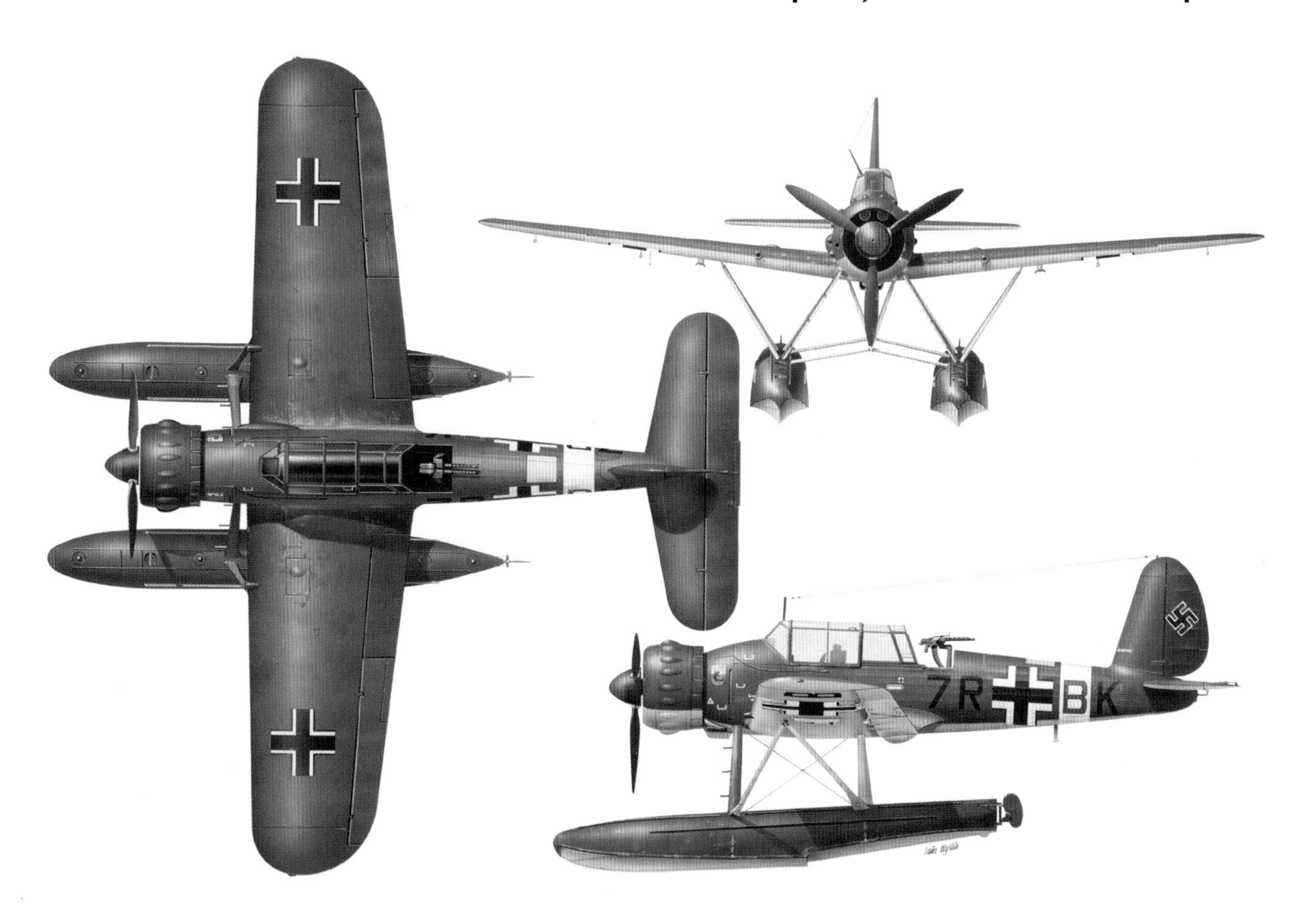

This Arado Ar 196A-5 served with 2./Seeaufklärungsgruppe (SAGr) 125 in the eastern Mediterranean and Aegean Seas during 1943. The unit later became 4./SAGr 126 under the control of Luftwaffenkommando Südost.

of the stricken vessel was forced to surrender. Having been taken away in one of the two aircraft, the *Seal* was eventually briefly operated by the Kriegsmarine as a training boat.

Anti-submarine role

With the Arado's anti-submarine capabilities bolstered by sharing in the sinking of HMS *Shark* off Norway, the type would be based around the coast of Occupied Europe from August 1940. As well as anti-submarine patrols and convoy escort, the Ar 196 was also used for search-and-rescue operations in the English Channel as the Battle of Britan raged overhead. Based in Boulogne-sur-Mer by the Straits of Dover, the mission was not without its dangers, the RAF attacking any German aircraft they found. Indeed, 15 August saw six Spitfires of 266 Squadron attacking a single Ar 196A-2. Although the Ar 196A-2 was ultimately shot down and the rear gunner was killed, the encounter took around 15 minutes and saw one Spitfire damaged and another shot down by the German aircraft's 20mm (0.79in) cannon. Rescue missions, meanwhile, could be equally as challenging with the two-man crew called upon to haul downed airmen from the waters of the Channel and then find space for them in the rear cockpit, which was designed to accommodate only one person. Rescues later in the war would see the aircraft taxied on the surface to the nearest point of land to avoid overloading.

The Ar 196A-3 and A-4 entered service in 1940. The latter was essentially an A-2 with enlarged fuel tanks while the former retained the original 300-litre (79-gallon) tanks but had a strengthened structure. The final A-5 model was broadly similar apart from some equipment changes and the fitting of a twin MG 81Z in the rear cockpit. These types, along with the earlier A-1 and A-2, would see service on the Eastern Front with the commencement of Operation Barbarossa in June 1941. Primarily operating in the Baltic, they carried out maritime patrols from shore while aircraft of 1./BoFlGr 196 embarked on *Tirpitz, Admiral Scheer*, *Nürnberg* and *Leipzig*

and acted as spotters when they shelled coastal targets. Other units operated in Finland and northern Norway and supported patrols by Finnish troops and rescued downed airmen, operating from the numerous lakes in the region. They also carried out medical evacuations. At the southern end of the Eastern Front, Ar 196s were operated around the Black Sea, primarily from Bulgaria and Romania. Anti-submarine patrols were again the primary mission, although attacks on surface shipping also took place.

Arado Ar 196A-2, T3+HK of 2./BFlGr 196 in 1939. Aircraft of 1 and 2./BFlGr 196 were embarked on ships of the Kriegsmarine with T3+HK being assigned to the *Admiral Hipper* at this point.

Convoy escort

In the Mediterranean, Ar 196s were tasked with escorting convoys to North Africa, keeping Royal Navy submarines based in Malta at bay. As the Afrika Korps were pushed out of Egypt in 1942 and ultimately all of Africa in 1943, the Arados would increasingly come into contact with RAF fighters and torpedo-bombers. Contrary to expectations, these encounters would not be one-sided, with the floatplanes putting up a spirited defence of their convoys. January 1943 saw a Marauder of the RAF's 14 Squadron shot down by an Ar 196 while attacking a convoy south of Athens. Later that month, another attack by 14 Squadron was interrupted by an Arado, which prevented one of the Marauders from dropping its torpedo. As the course of the war turned against Germany, the Ar 196s became increasingly vulnerable, though they continued to be fully engaged in the war even as they withdrew to Central Europe. Aircraft remained assigned to the *Tirpitz* even while she was under extended repair in northern Norway in preparation for a breakout. The three Ar 196s of the *Prinz Eugen*, meanwhile, were engaged in spotter activity as late as January 1945, while the ship was attacking Soviet troop positions during the evacuation from the Courland Pocket and East Prussia. Coming under attack from Yak fighters, the Ar 196s formed a wagon wheel formation for mutual defence and survived the encounter.

Production

In all, 541 Ar 196s were produced, including 23 examples produced by SNCASO in France and 69 by Fokker in Holland. Although a relatively small total, the type played an important role both in the waters around Europe and further afield when embarked. In addition to serving with the German allies, Bulgaria and Finland, around 37 were taken by the Soviets after the war and remained in service until 1955 as an inshore maritime patrol aircraft.

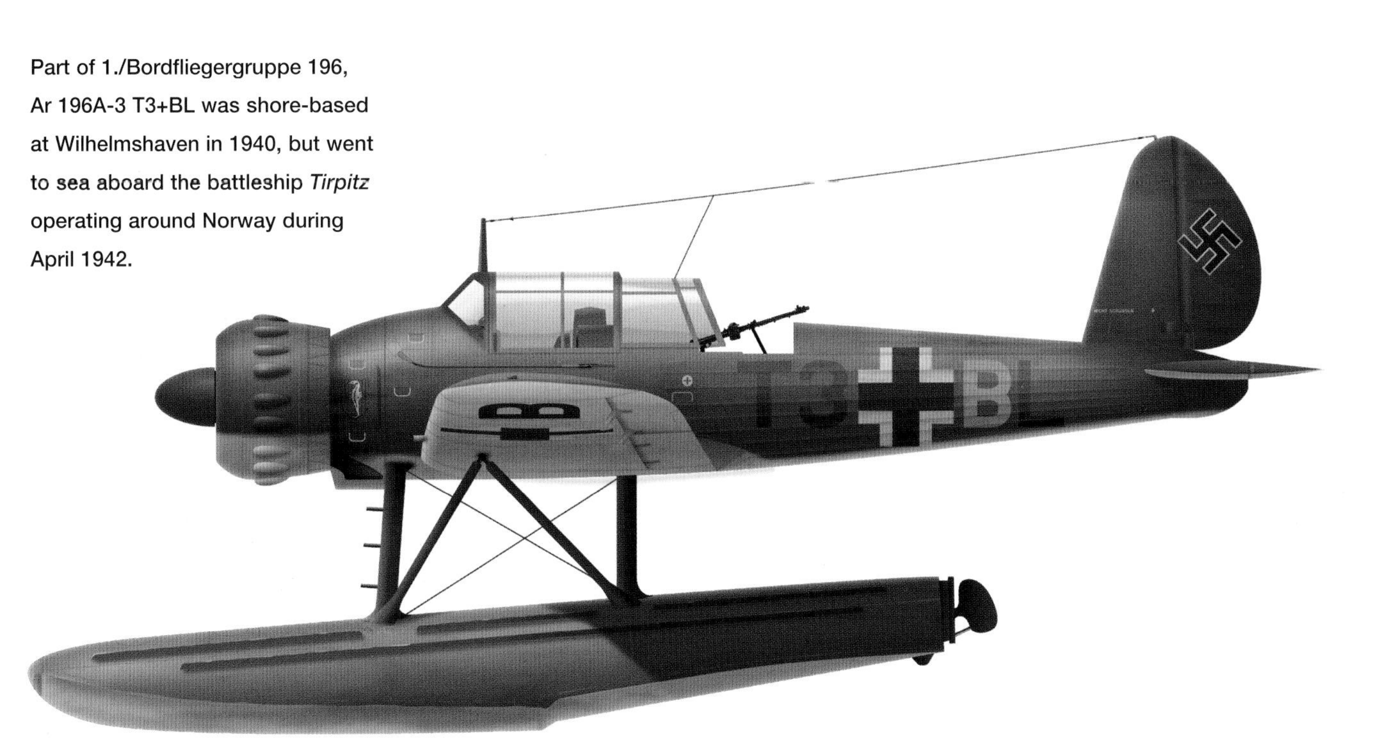

Part of 1./Bordfliegergruppe 196, Ar 196A-3 T3+BL was shore-based at Wilhelmshaven in 1940, but went to sea aboard the battleship *Tirpitz* operating around Norway during April 1942.

An Ar 196, T3+AH, being launched from a catapult onboard the cruiser *Admiral Hipper*. The cradle between the aircraft's floats is propelled along the track by a cordite charge.

Blohm und Voss BV 138 (1937)

A twin-boom, three-engined flying boat, the BV 138 did not look conventional but it would eventually prove to be a rugged patrol aircraft that would play an important role shadowing Allied convoys.

Designed to fulfil a requirement for a medium-range flying boat capable of operating in the open ocean, Blohm und Voss's first flying boat featured distinctive looks that earned it the nickname of the Der Fliegende Holzschuh or the 'Flying Clog'. Work for what would become the BV 138 started in 1934, the initial design being a gull-wing monoplane powered by two 746kW (1,000hp) engines, the final choice being decided by flying trials. When these failed to emerge, the 485kW (650hp) Jumo 205C two-stroke diesel was chosen, necessitating the addition of an extra engine. The first two prototypes flew in 1937 and proved to be both aerodynamically and hydrodynamically unstable. Although attempts to rectify the issues with a revised tailplane were made, a significant redesign was ultimately required.

Coastal flying boat

First flying in February 1939, the BV 138A had a bigger, deeper hull and larger tail booms. Although a significant improvement on the earlier aircraft, the structure was not sturdy enough to withstand the typical in-service loads experienced by a flying boat. Despite this, the

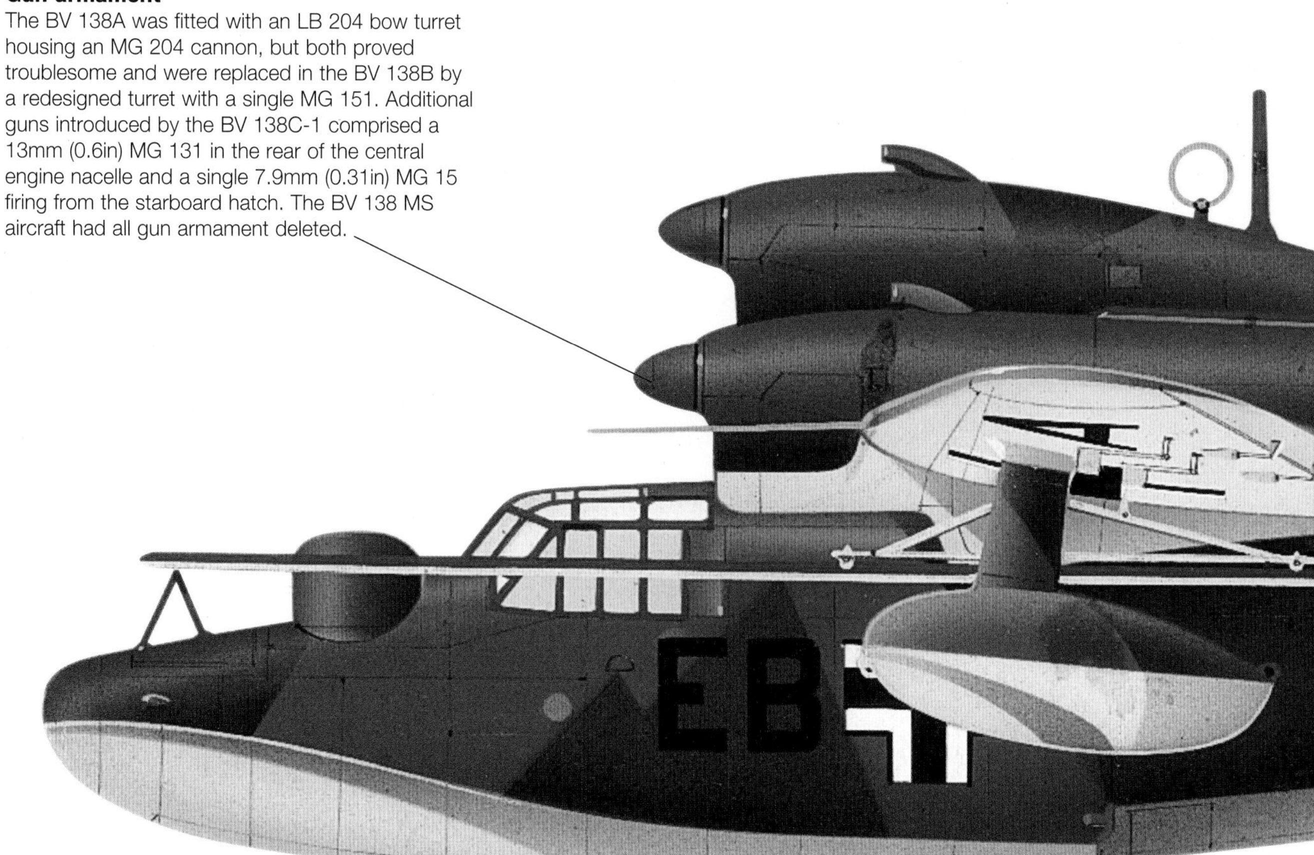

Gun armament
The BV 138A was fitted with an LB 204 bow turret housing an MG 204 cannon, but both proved troublesome and were replaced in the BV 138B by a redesigned turret with a single MG 151. Additional guns introduced by the BV 138C-1 comprised a 13mm (0.6in) MG 131 in the rear of the central engine nacelle and a single 7.9mm (0.31in) MG 15 firing from the starboard hatch. The BV 138 MS aircraft had all gun armament deleted.

pressing need for coastal flying boats saw 25 BV 138A-1s constructed, with two examples seeing service during the invasion of Denmark and Norway, both of which were lost to flak. By October 1940, 1. and 2./Küstenfliegergruppe 506 were equipped with BV 138A-1s and operating in the Bay of Biscay, where the weakness of the hull, which struggled with the rough sea conditions regularly found there, was further demonstrated. At the same time, there were emerging issues with the bow armament, which comprised a LB 204 turret fitted with a 20mm (0.79in) MG 204 cannon, both of which had a tendency to jam.

Strengthened fuselage

The BV 138B-1 addressed these problems with a strengthened fuselage and uprated 656kW (880hp) Jumo 205D engines to compensate for the increase in mass. Armament was also revised, with the nose and stern turrets gaining a 20mm (0.79in) MG 151 cannon while the open position behind the middle engine fielded 7.92mm (0.31in) MG 15. The B-1 was soon followed by the BV 138C-1, on which the fuselage was further strengthened, the centre engine gained a four-bladed propeller, and the rear MG 15 was swapped for a 13mm (0.51in) MG 131. A factory modification was also available that added underwing hardpoints between the fuselage and tail booms that could carry three 50kg (110lb) bombs or two 150kg (331lb) depth charges on each side. For operations from flying boat tenders, the Flying Clog could be fitted with spools that allowed it to be launched from the ship's catapult, saving fuel compared to a conventional take-off. For convoy shadowing work, many aircraft would later gain the FuG 200 Hohentwiel maritime search radar.

The year 1941 saw 1. and 2./KüFlGr 506 re-equipping with the BV 138B-1 and deployed to the Baltic, while after the invasion of the Soviet Union, BV 138 Staffeln would be established for operations in Norway. With an endurance of up to 17 hours when auxiliary tanks were fitted, the BV 138 was ideal for finding and shadowing convoys in

A BV 138 MS. Although the majority of BV 138s served on conventional coastal patrol duties, a small number were converted for mine-hunting tasks. This aircraft served with 6. Staffel/Minensuchgruppe 1 at Grossenbrode in 1945.

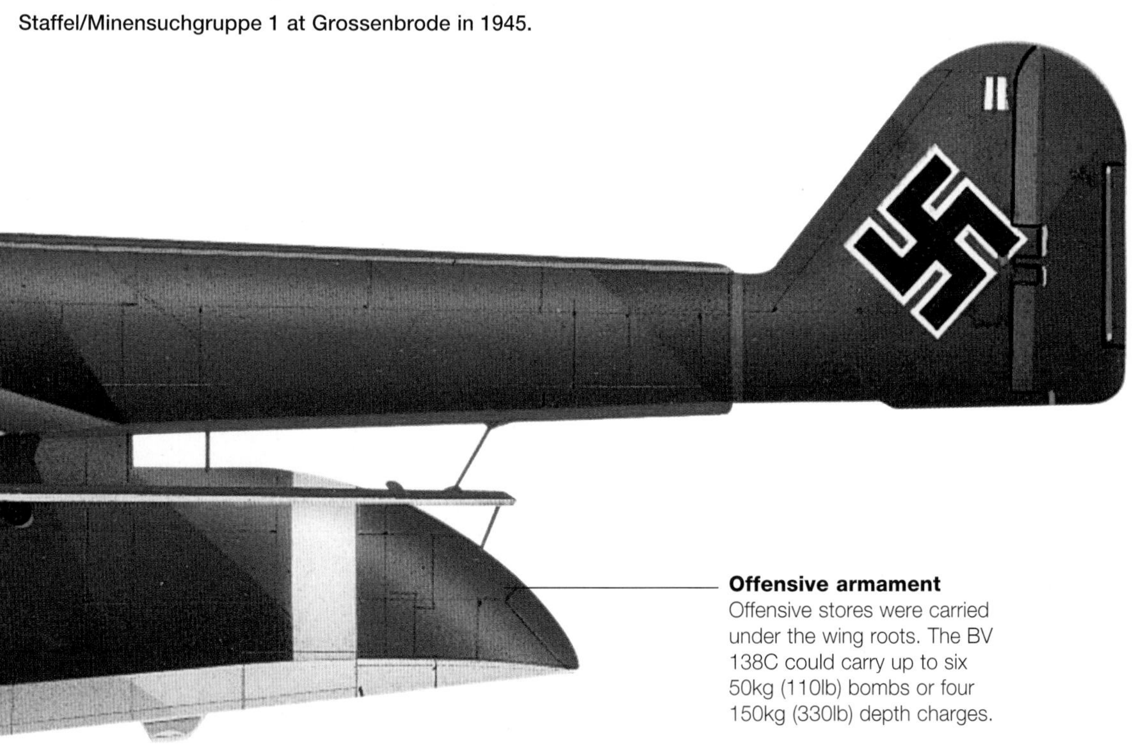

Offensive armament
Offensive stores were carried under the wing roots. The BV 138C could carry up to six 50kg (110lb) bombs or four 150kg (330lb) depth charges.

Specifications: BV 138C-1

Type:	Reconnaissance flying boat
Dimensions:	Length: 19.9m (65ft 4in); Wingspan: 27.0m (88ft 7in); Height: 5.9m (19ft 4in)
Weight:	17,650kg (38,912lb) maximum take-off
Powerplant:	3 x 656kW (880hp) Jumo 205D six-cylinder opposed piston two-stroke diesel
Maximum speed:	285km/h (177mph)
Range:	5,000km (3,107 miles)
Service ceiling:	5,000m (16,404ft)
Crew: 6	
Armament:	2 x 20mm (0.79in) MG 151 cannon in the bow and stern turrets; 1 x 13mm (0.51in) MG 131 behind the centre engine; 1 x 7.92mm (0.31in) MG 15 in the starboard hatch; plus a bombload of 6 x 50kg (110lb) bombs or 4 x 150kg (331lb) depth charges

order to direct strikes by aircraft and U-boats. For even longer-range operations, the aircraft could be refuelled at sea by U-boats, the installation of diesel engines allowing them to use the boats' own fuel. The summer of 1943 even saw a detachment operating from a base at Novaya Zemlya inside the Arctic circle for three weeks, the aircraft supported throughout by U-255, although no suitable targets appear to have been found.

As the war progressed, the Allied Arctic convoys would gain protection from escort carriers of the Royal Navy, the first taking part in Convoy PQ 18 in September 1942. This would make the BV 138s' task more difficult as the defending Sea Hurricanes or Martlet fighters would attempt to drive them off, although the flying boat proved remarkably resilient to attack. The hunted could also become the hunter, BV 138s being able to use their superior speed and armament to fend off attacks on U-boats by the Royal Navy's Swordfish torpedo-bombers. Further south, 3./Aufklärungsgruppe 125 converted from

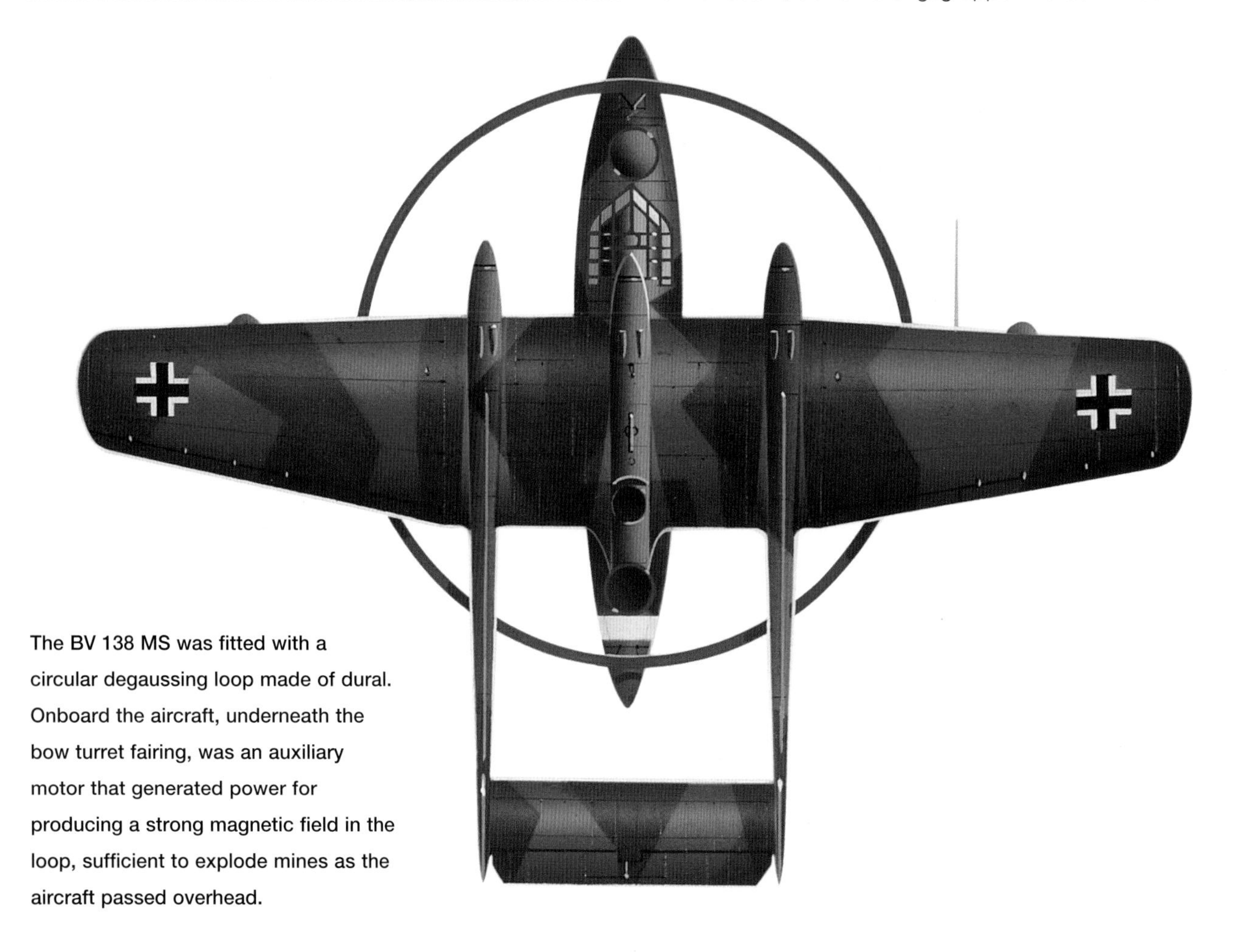

The BV 138 MS was fitted with a circular degaussing loop made of dural. Onboard the aircraft, underneath the bow turret fairing, was an auxiliary motor that generated power for producing a strong magnetic field in the loop, sufficient to explode mines as the aircraft passed overhead.

A BV 138B-1 unusually seen over the countryside. From this angle the open gun position in the rear of the central engine nacelle is clearly visible.

the Arado Ar 95 in 1942 and operated from Constanza on the Black Sea. Here, they were primarily used in the anti-submarine role but also operated as air-sea rescue aircraft when they would carry medical orderlies to tend to rescued airmen. Other units used the BV 138 in the Mediterranean, while 1./SAGr 129 operated from Lake Biscarrosse in southwest France for missions over the Atlantic.

A handful of BV 138s were modified to BV 138MS standard for mine clearance work. These carried a degaussing loop with a diameter equivalent to the fuselage's length attached above the bow and stern and under the wings, the defensive armament being removed. An onboard generator powered the ring and the resultant field would trigger magnetic mines as the aircraft flew overhead.

With production ceasing at the end of 1943 after 276 BV 138s had been built, the last years of the war would see enemy action, a shortage of fuel and the impact of normal operations slowly reduce their numbers. The final units were 1. and 3./SAGr 130 and 3./SAGr 126, who were operating in the Baltic and Norway until the end of the war, one of the latter's aircraft allegedly being the last out of Berlin on 1 May 1945. Although the BV 138 had not had the most promising of starts, the type eventually emerged as a solid maritime patrol aircraft able to shadow Allied convoys with some degree of security.

Jumo 205 engine

The Jumo 205 that powered the BV 138 was an unusual design for an aircraft engine, being a two-stroke diesel. To further complicate it, each of the six vertically mounted cylinders had two pistons that moved in opposite directions. At the top and bottom of each cylinder, exhaust and inlet ports were uncovered by the pistons' movement. With the bottom crankshaft running 11 degrees behind the upper one, the inlet ports were opened slightly after the exhaust ports, helping remove exhaust gases. Meanwhile, an engine-driven compressor would force air into the cylinder as the lower piston opened the inlet port. Consequently, as the pistons moved back together, covering the ports, there were no exhaust gases left in the cylinder, unlike conventional two-strokes. Finally, as the pistons approached their closest point, four injectors around the middle of the cylinder injected diesel for the power stroke. Although the Jumo 205 proved inappropriate for combat aircraft, not being responsive enough, it was ideal for patrol aircraft, having a typical two-stroke's high power-to-weight ratio.

Dornier Do 24 (1937)

Adopted by the Luftwaffe almost as an afterthought, the Dornier Do 24 would see extensive service as an air-sea rescue aircraft – a role to which the type was ideally suited.

The Dornier Do 24 was slightly unusual in originally being designed in response to a requirement from the Royal Netherlands Navy, rather than the Reich Air Ministry, for a maritime reconnaissance aircraft. The Netherlands naval air service, the Marineluchtvaartdienst (MLD), had operated the Dornier Do J Wal in this role since the late 1920s, primarily in the Dutch East Indies, but by 1936 they were due a replacement and a contract was signed with Dornier in August of that year for six Do 24s. This would be a flying boat with the wing mounted on struts above the fuselage, a so-called parasol design, the wing itself carrying the three 661kW (886hp) Wright Cyclone engines specified by the MLD for compatibility with their American-built Martin 139 bombers.

Armament consisted of a 7.92mm (0.31in) machine gun in turrets at the bow and aft of the tailplane, and a 20mm (0.79in) Solothurn cannon in a dorsal turret just aft of the wing. The crew of six were provided with living and sleeping quarters, allowing the Do 24 to make extended patrols away from its main base.

This Do 24T-2 of the 7.Seenotstaffel/SBK XI, flew in the Aegean area in 1942. The legend on the nose reads 'Asbach Uralt', a popular brand of German brandy.

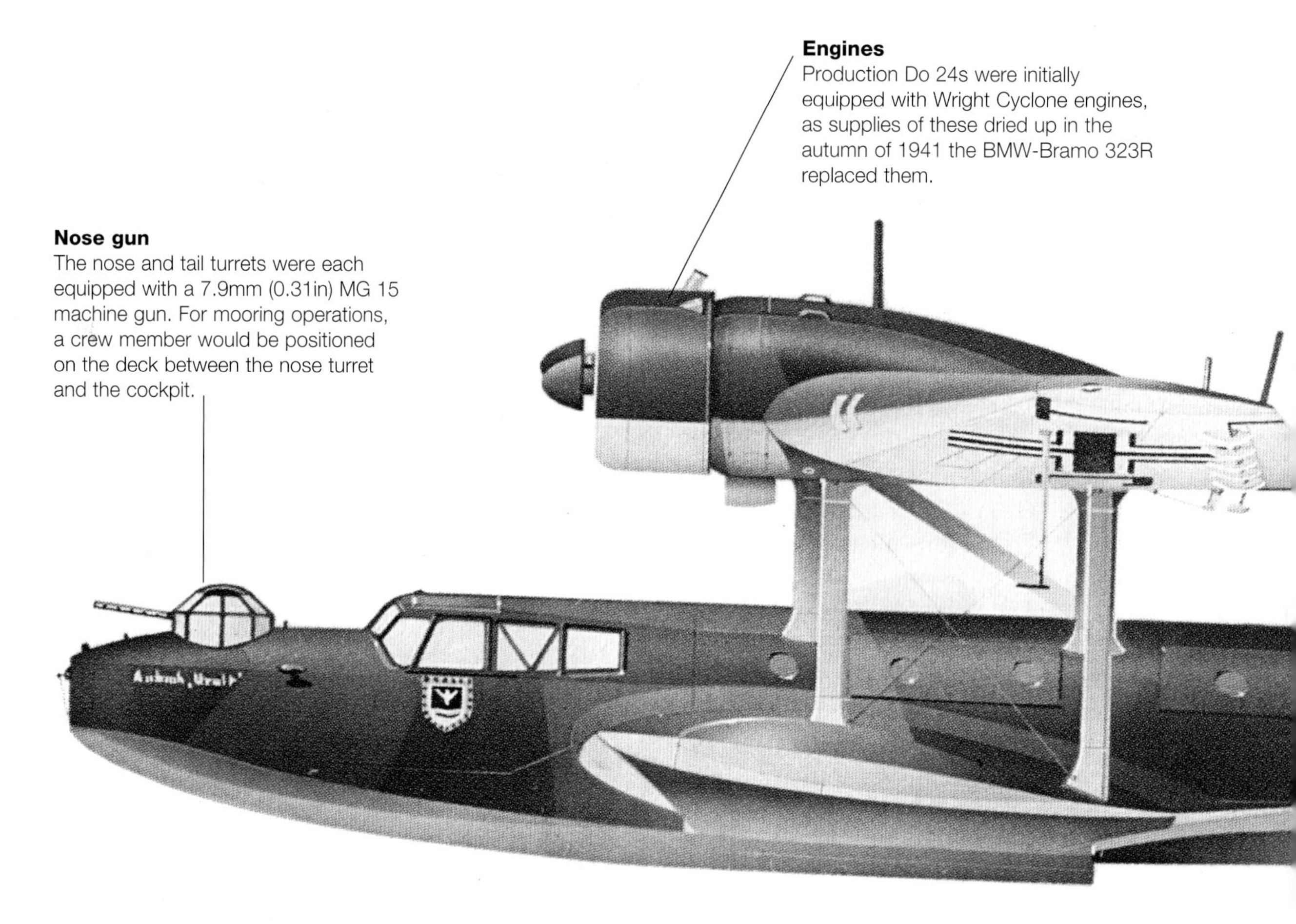

The first of two prototypes flew on 3 July 1937 with a further 12 Do 24K-1s being built by Dornier for the MLD, after which the plan was for 48 Do 24K-2s to be licence-built in the Netherlands by Aviolanda. The K-2 would use a more powerful 820kW (1,100hp) version of the Cyclone and have an increased fuel capacity. At the same time, two examples were built for evaluation by the Luftwaffe, powered by the same 450kW (604hp) Jumo 205C engines as the BV 138. However, with the performance of the latter aircraft already being considered satisfactory, no further interest was shown until after the outbreak of the war.

In advance of the planned invasion of Denmark and Norway, the two Jumo-powered aircraft were taken out of storage and fitted with 7.92mm (0.31in) MG 15 machine guns in the bow and stern, and a 29mm (1.14in) MG 151 cannon in the midship turret. They were then delivered to Kampfgruppe zur besonderen Verwendung 108 (Combat Group for Special Purposes 108) – a transport unit operating an eclectic mix of flying boats and seaplanes to deliver stores and troops into the fjords of Norway. Although the Do 24

Seen during sea trials D-ADLP was the first Do 24K built for the MLD. It would see service in the Dutch East Indies as X-1 before retreating to Australia in the face of the Japanese advance through Southeast Asia.

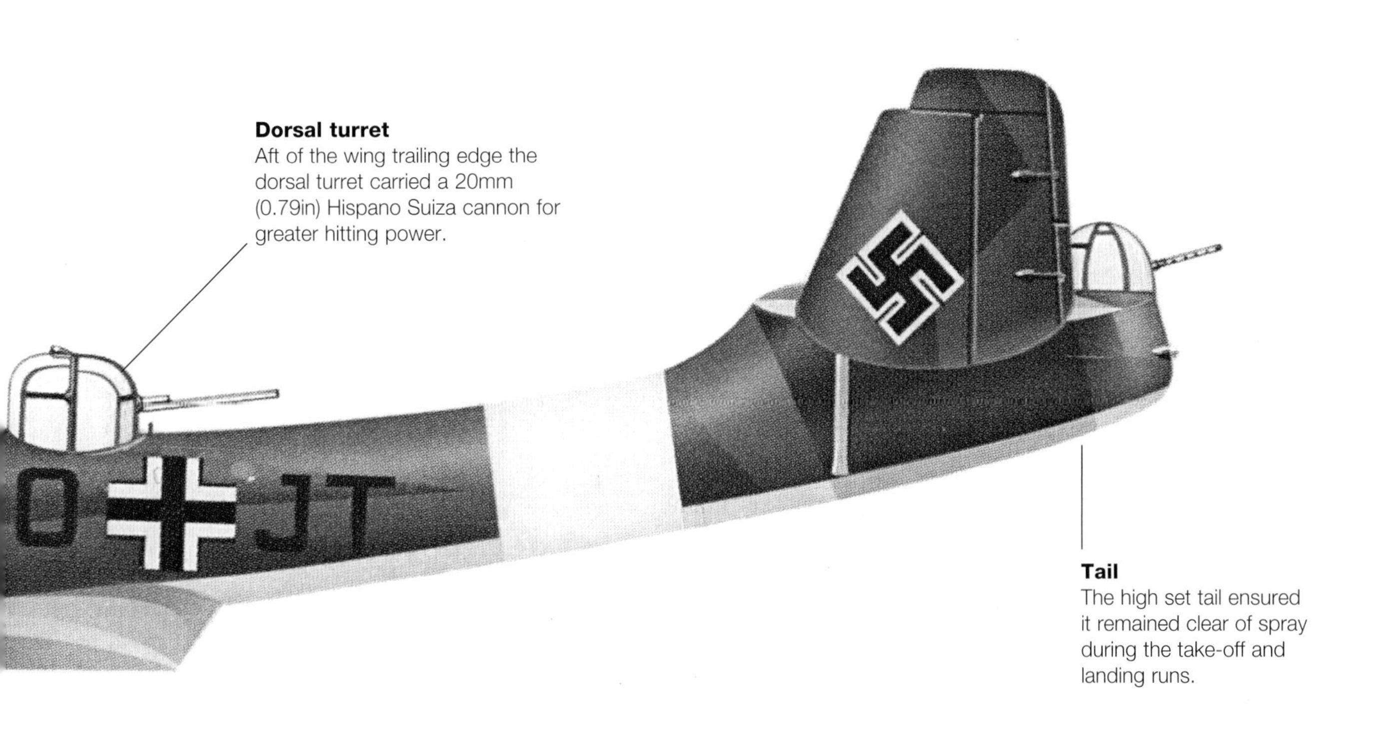

A Do 24 of Seenotstaffel 10 being hoisted ashore in Norway. As with all pure flying boats, bringing the aircraft ashore for hull maintenance was an involved process.

Specifications: Do 24T-1

Type:	Air-sea rescue and transport flying boat
Dimensions:	Length: 22.05m (72ft 4in); Wingspan: 27.00m (88ft 7in); Height: 5.75m (18ft 10in)
Weight:	18,400kg (40,565lb) maximum take-off
Powerplant:	3 x 746kW (1,000hp) BMW-Bramo 323R-2 air-cooled radial engines
Maximum speed:	331km/h (206mph)
Range:	4,700km (2,920 miles)
Service ceiling:	7,500m (24,606ft)
Crew:	6
Armament:	1 x 7.92mm (0.31in) MG 15 machine gun in bow and stern turrets; 1 x 20mm (0.79in) Hispano-Suiza 404 cannon in the dorsal turret

performed well in this role, Dornier were not asked to resume production and it was not until the Aviolanda factory was overrun during the invasion of the Netherlands that additional aircraft would become available. Three completed Do 24K-2s were captured along with a further 20 on the production line. These were fitted with German instruments and radios, MG 15 machines guns in the bow and stern, and captured Hispano-Suiza 20mm (0.79in) cannon in the dorsal turret. The first 11 completed were delivered as Do 24N-1 before production shifted to the Do 24T-1, which used the 746kW (1,000hp) BMW-Bramo 323R radial engine in place of the now unavailable Wright Cyclone. In all, 170 of these would be produced, the majority being used by the Seenotsdienst, the Luftwaffe's air-sea rescue service, which had been relying on the elderly He 59 and Do 18.

Rescue service

With the Seenotsdienst, the Do 24 would carry out rescue missions from the Arctic to the Mediterranean, recovering downed airmen from the waters off the Scilly Isles at the extreme south-west of Britain, and on another occasion,

This Dutch-built Do 24K-2 was completed for the Luftwaffe for air-sea rescue duties.

560km (348 miles) out into the Atlantic Ocean. Able to operate in waters up to a sea state 6, they were ideally suited to the role. Off Norway, a Do 24T landed in seas so severe the tail broke off; by sealing the watertight bulkheads, the crew were able to complete their rescue mission and then taxied back to Kjølle fjord. When not engaged on rescue work, the Do 24s were called on to act as troop transports and convoy escorts, and to conduct maritime reconnaissance. When land-based aircraft were no longer able to resupply the Kuban bridgehead in the spring of 1943, due to the thaw, Do 24Ts shipped in more than a thousand tons of supplies while also evacuating injured troops. A handful of aircraft were also adapted for the mine-warfare role, carrying a degaussing loop powered by an onboard generator. Flying at low level, the influence of the loop would trigger magnetic mines.

French production

In addition to the Aviolanda factory in the Netherlands, a further production line was established at Sartrouville in France in the former CAMS facility. This produced 48 aircraft before being abandoned in 1944 in the face of the Allied invasion; it would produce a further 22 after the war for the Aéronavale. As the war turned against the Axis, Do 24s would be used to evacuate troops and the surviving aircraft gathered in northern Germany. These were eventually captured by the Allies and the majority destroyed, although those used as gunnery targets proved remarkably resilient and, in some cases, had to be scuttled with explosives.

As well as with the Aéronavale, the Do 24 would see post-war service with the Spanish Air Force, which acquired 12 Do 24T-3s in June 1944 for air-sea rescue work. In a testament to the type's suitability for the role, they would remain in service until 1970. Even this was not the end for Dornier's flying boat, with a single PT6 turboprop-powered example equipped with a new wing and retractable undercarriage flying as a technology demonstrator in 1983. Although initially shunned by the Luftwaffe, the Do 24 would prove to be ideal in the air-sea rescue role, while also being able to perform as a transport aircraft when required.

Sinking of the *Shinonome*

The Do 24 would ironically have its biggest impact as a maritime patrol aircraft while operated by the Allies. On 17 December 1941, in response to Japanese landings on Borneo, three Dornier Do 24K-2s of the Dutch MLD mounted an attack on the invasion fleet. Launching from Tarakan in the early hours, the aircraft utilized the overcast conditions to make their approach. While one of the aircraft made a forced landing before reaching the target area, two – X-32 and X-33 – arrived over Miri to find a Japanese warship. Singling her out, X-32 made a bombing run, dropping five 200kg (441lb) bombs. The destroyer *Shinonome* was hit by two while a third exploded in the sea abeam the rear 12.5cm (5in) turret. This triggered ammunition in the turret's magazine, causing a large explosion that sent smoke 1,500m (4,921ft) into the air. Within minutes, the *Shinonome* had disappeared beneath the waves. The Dorniers would subsequently retreat to Australia, where they would be operated by the RAAF.

Focke-Wulf Fw 200 Kondor (1937)

Conceived as a state-of-the-art long-distance airliner, the Focke-Wulf Fw 200 could cross the Atlantic without refuelling. The coming of war saw it re-employed as a maritime patrol aircraft, where its civilian roots were both a blessing and a curse.

The Fw 200 was designed as a long-range airliner able to carry 26 passengers non-stop across the Atlantic, something that had previously been the preserve of flying boats such as the Boeing 314 Clipper. First flying in July 1937, the Fw 200 V1 prototype would subsequently prove its capabilities just over a year later on 10 August 1938 when it flew from Berlin-Tempelhof to New York's Floyd Bennett airport in 24 hours 55 minutes. The return flight was made in 19 hours 47 minutes, giving an average speed of 330km/h

This Focke-Wulf Fw 200C-1 Kondor flew anti-convoy missions with 1./KG 40 IV Fliegerkorps, Luftflotte 3, from Bordeaux-Mérignac, France, in 1940.

(205mph) and proving the endurance of the type. By late November, it had also made the trip to Tokyo, stopping in Basrah, Karachi and Hanoi in only 46 hours 18 minutes – about the same time it took Imperial Airways' flying boat service to reach Karachi.

The Condor's design was aerodynamically clean with a high-aspect-ratio wing holding four engines, initially Pratt & Whitney Hornets providing 653kW (876hp). Subsequent aircraft differed from the V1 in having slight sweepback to the outer wings, modified tail surfaces and BMW 132 engines – essentially licence-built Hornets. However, only a handful of Fw 200A series aircraft were built, two going to the Danish airline DDL and two to a Brazilian airline. A further two examples were built as government transports, with one being set aside for the personal use of Adolf Hitler. This had a special armoured seat equipped with a parachute allowing the Führer to escape via a jettisonable trap door in the event of an emergency, although this was never used in anger. The subsequent Fw 200B with increased all-up mass and more powerful versions of the BMW 132 also failed to gain much interest, Deutsche Lufthansa – the German state airline – having no real requirement for a long-range airliner on its mainly European routes. Only a handful of orders were received, five for Dai Nippon KK of Japan and two for Aero O/Y of Finland, and these would end up being operated by the Luftwaffe after delivery was made impossible by the start of the war in

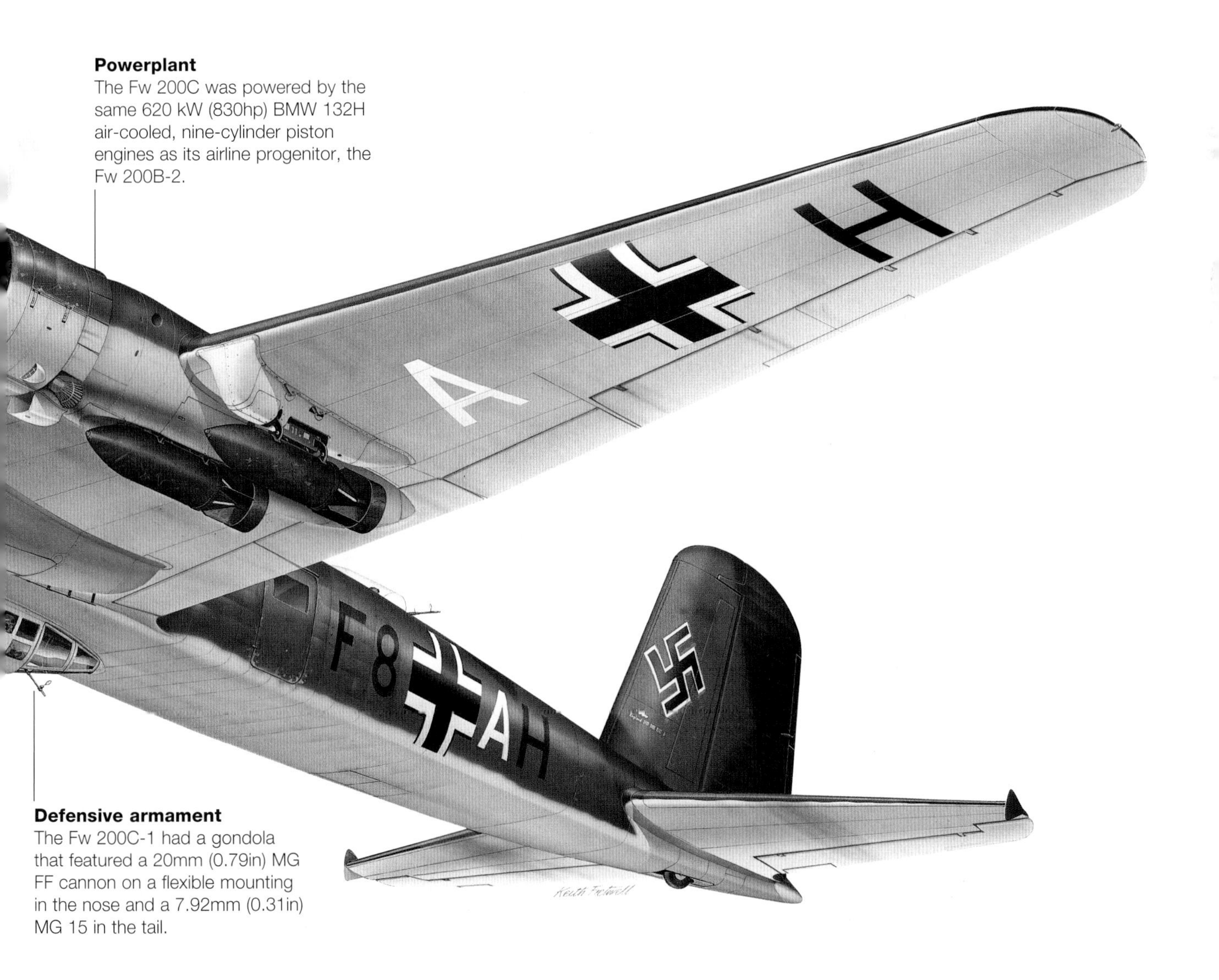

Powerplant
The Fw 200C was powered by the same 620 kW (830hp) BMW 132H air-cooled, nine-cylinder piston engines as its airline progenitor, the Fw 200B-2.

Defensive armament
The Fw 200C-1 had a gondola that featured a 20mm (0.79in) MG FF cannon on a flexible mounting in the nose and a 7.92mm (0.31in) MG 15 in the tail.

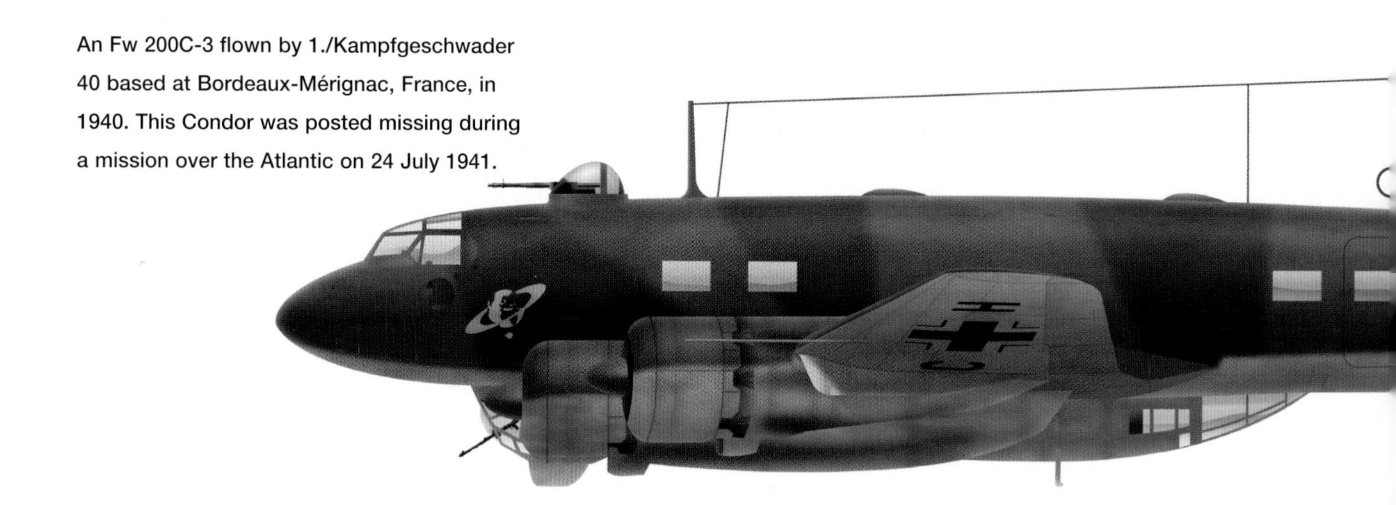

An Fw 200C-3 flown by 1./Kampfgeschwader 40 based at Bordeaux-Mérignac, France, in 1940. This Condor was posted missing during a mission over the Atlantic on 24 July 1941.

Europe. However, with that war appearing ever more imminent, at the start of 1939, belated planning was begun for attacking shipping out in the North Atlantic to break the immense logistics chain that France and especially Britain relied on. Without a maritime patrol aircraft in development, and with limited time, the Fw 200 was chosen as the best option for conversion to the role. Serendipitously, Focke-Wulf had already covertly started work on modifying the Fw 200B prototype into such an aircraft for the Imperial Japanese Navy. This carried 60 per cent more fuel in tanks in the cabin, had provision for 2,000kg (4,409lb) of equipment and three 7.92mm (0.31in) MG 15 machine guns for self-defence, although at this stage there was no bomb bay. The Condor would, though, carry a handicap throughout its life due to its heritage as an airliner. Designed for cruising at altitude, the structure was not suited to the rigours of low-level flight in turbulent air or manoeuvring in tight turns, both of which would be required in the battles ahead. Nor would it stand up well to the rigours of field operations, at least eight breaking their backs on landing in the first year of operations. Focke-Wulf's

Specifications: Fw 200C-3

Type:	Long-range reconnaissance bomber
Dimensions:	Length: 23.45m (76ft 11in); Wingspan: 32.85m (107ft 9in); Height: 6.3m (20ft 8in)
Weight:	24,520kg (50,057lb) maximum take-off
Powerplant:	4 x 895kW (1,200hp) BMW-Bramo 323R-2 air-cooled radial engines
Maximum speed:	360km/h (224mph)
Range:	3,560km (2,212 miles)
Service ceiling:	6,000m (19,685ft)
Crew:	7
Armament:	4 x 13mm (0.51in) MG 131 machine guns; 1 x 20mm (0.79in) MG 151 cannon; a bombload of up to 2,100kg (4,630lb)

engineers attempted to address these issues with local strengthening but structural failures would haunt the type for its entire career.

The first four Fw 200C-0s were essentially transport aircraft converted on the line from Fw 200Bs. However, the later six gained three MG 15s as defensive armament and the ability to carry four 250kg (551lb) bombs, one each under the outer engine nacelles and wings. These aircraft were delivered from April 1940 while production of the Fw 200C-1 commenced. This gained a small amount of armour around the pilot's position and a distinctive ventral gondola that housed a

Long-range rescue

Accurate weather forecasts are vital in wartime, so much so that the Germans manned remote meteorological stations in the Arctic to enable them to be made. In early summer 1944, nine of the 10-man team of one such station in the Franz Josef Land archipelago fell ill. With no alternative, a Fw 200 was dispatched with a medical team to assist them. On 7 July, arriving over the island after an eight-hour flight and with no airfield available, the pilot chose a suitable-looking area and made his approach. Unfortunately, during the landing one mainwheel and the tailwheel were damaged. The aircraft briefly tipped on to its nose and then sank up to its axles in the melting ice. While the medical team set about treating the men from the weather station the Condor's crew assessed their own situation. Fortunately, a Bv 222 flying boat was able to drop the needed spare parts the next day, and over the next 48 hours repairs were made to the aircraft and a take-off strip prepared. Finally, late on 10 July, a successful, if stressful, take-off was made and after 10 hours the aircraft was back in Norway.

Another Kampfgeschwader 40 aircraft, this Focke-Wulf Fw 200C-4 was on strength with 7./KG 40, operating out of Norway in May 1945.

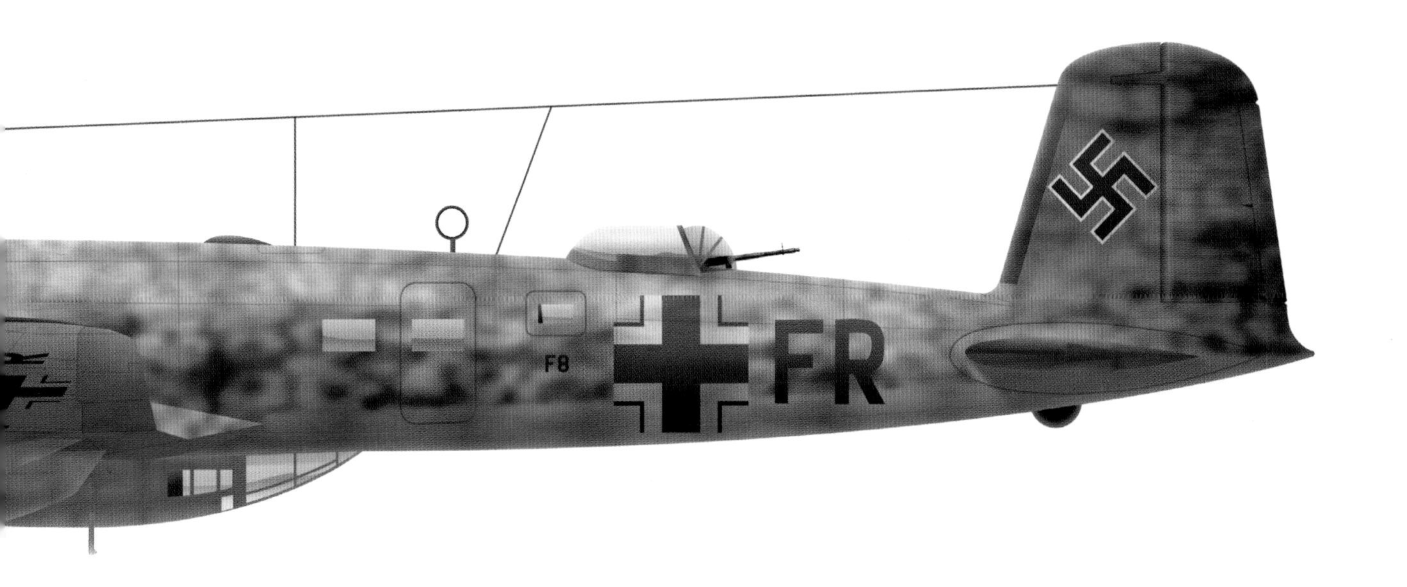

20mm (0.79in) MG FF cannon at the front, a MG 15 at the rear, a bombsight and a bomb bay with capacity for another 250kg (551lb) bomb.

I./KG 40 would be the first Fw 200 unit to see action, and indeed the primary operator of the type throughout the war. For the invasion of Norway, they were initially used to transport troops, but on 10 April they were tasked with their first bombing mission, flying from Germany to Narvik in Norway, where they attacked British warships, before returning to base – a 17-hour mission in total. Although this did not result in any reported casualties, this would soon change with one of I./KG 40's aircraft hitting a 907 tonne (1,000 US ton) freighter with a 250kg (551lb) bomb on 3 May. This success would be repeated throughout the Norwegian campaign, but it was after the fall of France that the Condor would really earn its reputation as the 'Scourge of the Atlantic'.

French air base

Moving to the airfield at Bordeaux-Mérignac on the edge of the Bay of Biscay, the aircraft of I./KG 40 were now able to roam far into the Atlantic to report on and attack convoys headed to and from the United Kingdom. To extend operations further, they would regularly end their sorties at Trondheim in Norway before returning the next day. Despite rarely having more than a dozen aircraft available, the small force of Condors was responsible for up to a third to a half of all shipping sunk in the autumn and winter of 1940. The totals were 81,647 tonnes (90,000 US tons) in the final five months of 1940 and 57,311 (63,175 US tons) in January 1941 versus 115,015 tons (126,782 US tons) for the U-boats, rising to 76,671 tonnes (84,515 US tons) versus 178,519 tonnes (196,783 US tons) in February. With a shortage of aircraft carriers, the Royal Navy would resort to placing single-use fighters on merchant ships to be catapulted off if a Fw 200 was sighted. Remarkably, these did manage a number of successful interceptions, the Hurricane or Fulmar fighters then being ditched alongside a friendly vessel. Fortunately for the British, from mid-1941, to preserve the limited number of Condors, they were ordered not to attack shipping but only to report its location for subsequent U-boat attack.

The year 1941 saw a number of improvements made to the Fw 200C with the C-2 featuring a recess in the outer engine nacelle, allowing a bomb to be carried inside it. The C-3, meanwhile, gained a powered dorsal turret behind the cockpit, more powerful BMW 323R engines producing 895kW (1,200hp) each, and strengthening that at least partially addressed the issue of structural failures. From early 1942, the Fw 200C-4 began deliveries with the FuG 200 Hohentwiel maritime search radar. By this point, though, the vulnerability of the Condor was becoming more apparent with six being shot down by the Grumman Martlets embarked on the escort carrier HMS *Audacity* between September and December 1941. The first of those shot down suffered structural failure of the rear fuselage as a combined result of the fighter's gunfire and the loads imparted by the pilot's evasive manoeuvring. Another was lost due to a mid-air collision when the attacking fighter came too close during a frontal pass, although in a testament to the strength of Grumman's aircraft, the Martlet survived. The Condor was not helped in these encounters by its relative lack of manoeuvrability. Stable in all three axes, the Focke-Wulf was never designed to be flung around the sky and had little choice but to maintain straight and level flight and rely on its defensive armament. The only other option was to evade attackers in cloud cover if it was available, being well equipped for flight in poor weather. Indeed, as well as being at threat from fighters, the Fw 200 was also involved in running battles with Allied patrol aircraft, such as Coastal Command Liberators, Hudsons and even Sunderland flying boats.

Atlantic patrol

Despite this, the Condor would soldier on in the maritime patrol role into 1944 as its planned replacement, the He 177 Greif, suffered continual delays. This also led to the large Focke-Wulf being modified to carry the Henschel Hs 293A missile, essentially a 250kg (551lb) bomb attached to a rocket-powered, radio-controlled plane. Modified C-3s became Fw 200C-6s, able to carry two missiles and the control equipment, while eight Fw 200C-8s were built with modified outer engine nacelles for more streamlined

The second production Fw 200C-1, BS+AG, shortly after completion at Bremen in early 1940.

missile carriage. Although a moderately successful weapon when employed by the Do 217 and the He 177, very few, if any, Fw 200-launched Hs 293s appear to have hit their targets.

The last successful attack by a Condor took place on 10 February 1944, three aircraft of III./KG 40 bombing shipping in eastern Iceland and sinking the tanker El Grillo. However, KG 40 would suffer further casualties before withdrawing from the role and its long-term home at Bordeaux-Mérignac as the Allied invasion of Normandy made their position there untenable. The remaining aircraft would now be used as transports and KG 40 was disbanded in November 1944. The aircraft then operated under the auspices of Transport-Fliegerstaffel Condor right up until VE Day and the end of hostilities in Europe.

Only 276 Fw 200s were built, 252 being C variants. As well as with the Luftwaffe, a handful of captured or interned aircraft were operated post-war by the Soviet and Spanish air forces. Surprisingly, in addition to the Danish and Brazilian airlines already mentioned, the British Overseas Airways Corporation (BOAC) was another civilian operator, taking over a Danish example that had escaped to Britain after the German invasion in 1940.

Blohm & Voss BV 141 (1938)

One of the most unusual-looking aircraft of World War II, the BV 141's performance belied its appearance. However, its widespread adoption was prevented by the choice of engine, not its looks.

In 1937, the Reichsluftfahrtministerium issued a requirement for a tactical reconnaissance aircraft. This was ultimately fulfilled by the Fw 189 Uhu. Where that aircraft was merely unconventional in appearance, the alternative proposal from Blohm und Voss gave the impression that the designer had only heard of aircraft as an abstract concept. Built as a private proposal, the BV 141 featured a straight monoplane, the outer panels of which had a small degree of dihedral, the tips pointing upwards. Attached to the centre section was a cockpit nacelle to the right of the centreline for the crew of three, while to the left was the nacelle for the single engine and tail surfaces. The decision to adopt such an unusual configuration had a number of factors but the prime one

This is the fourth Blohm und Voss BV 141B-0 pre-production aircraft (BV 141 V12), as seen when it was delivered to Tarnewitz for armament trials.

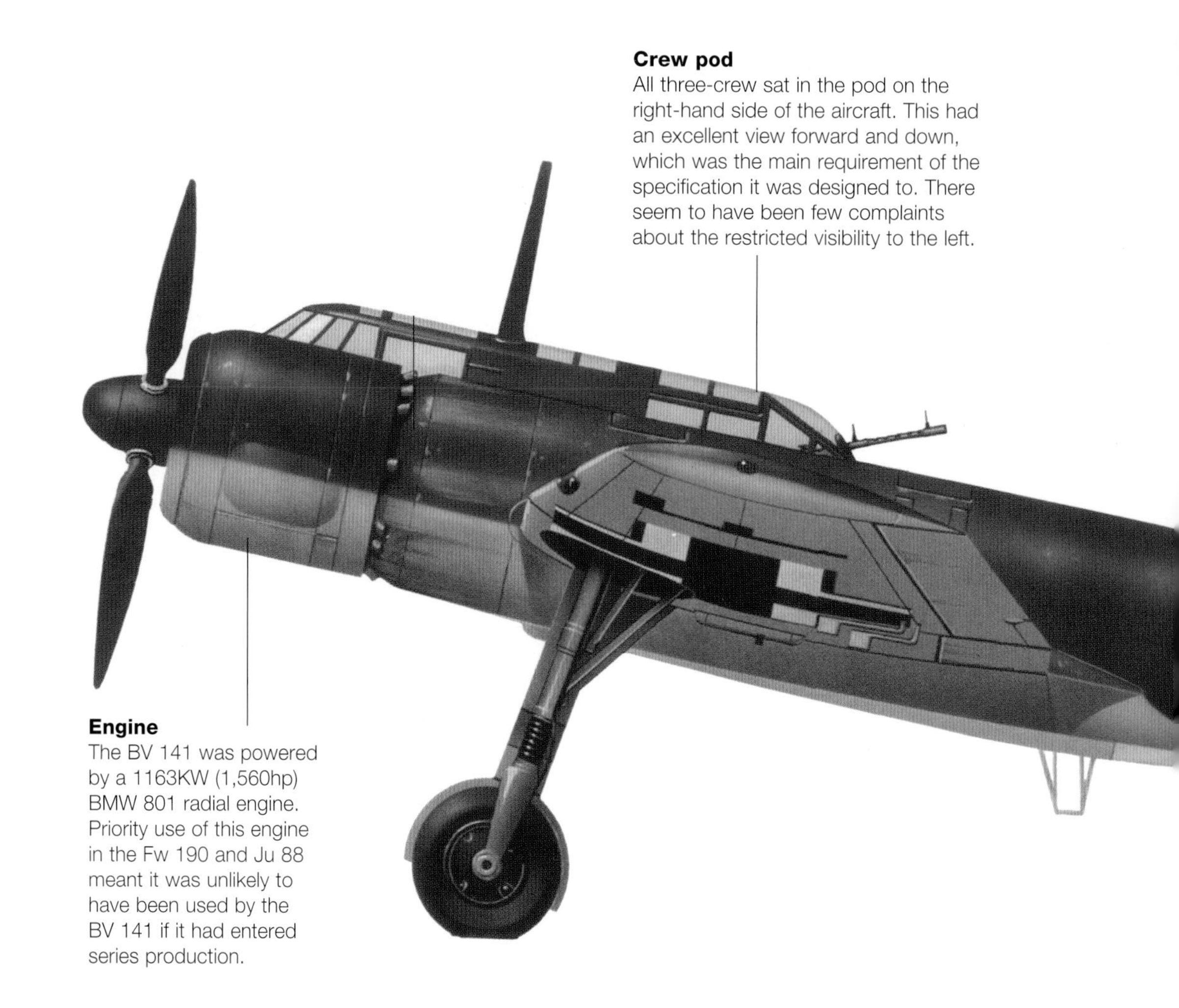

was the requirement for the crew to have a good view down and forwards for the reconnaissance role. Blohm und Voss's designer, Dr Ing Richard Vogt, found that with a conventional layout, the required 25 per cent field of view resulted in an excessively large fuselage. While Focke-Wulf adopted a twin boom design with the cockpit mounted between two low-powered Argus engines to achieve this, Vogt essentially dispensed with one of the booms and used one higher-power BMW engine. In fact, the two aircrafts' cockpit nacelles would end up being remarkably similar to each other.

Unconventional layout

Despite the unusual appearance, when the prototype first took to the air in February 1938 there were no problems with the handling. Shortly after, World War I ace and Luftwaffe technical department head, Ernst Udet, flew the aircraft and was sufficiently impressed for the type to gain official recognition and the designation Ha 141-0, the Ha prefix coming from Hamburger Flugzeugbau, Blohm und Voss's aviation subsidiary. This was soon changed to BV 141-0 as the parent company ensured its mark was present across its product range. Two further prototypes followed, which had increased wingspan and overall length. All three were powered by the 645kW (865hp) BMW 132N radial engine – essentially an improved Pratt & Whitney R-1690 Hornet built under licence. At least one of the prototypes was armed with two 7.92mm (0.31in) MG 17 machine guns and took part in weapons trials.

The prototype trials went well enough to secure an order for five BV 141A-0 pre-production aircraft. These had slightly greater wing area than the prototypes and used the 746kW (1,000hp) BMW Bramo 323 radial engine. The crew compartment also evolved with a fully

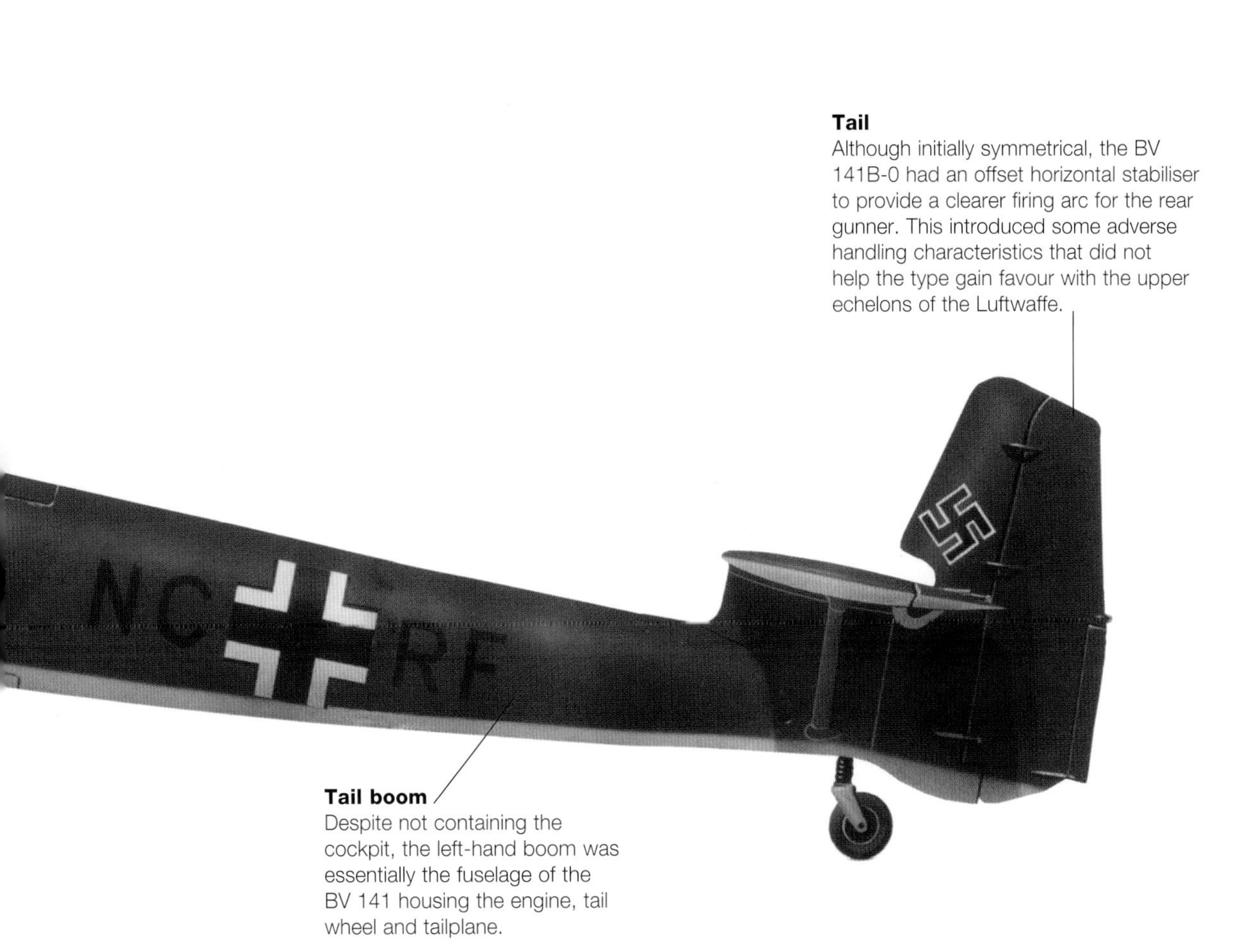

Tail
Although initially symmetrical, the BV 141B-0 had an offset horizontal stabiliser to provide a clearer firing arc for the rear gunner. This introduced some adverse handling characteristics that did not help the type gain favour with the upper echelons of the Luftwaffe.

Tail boom
Despite not containing the cockpit, the left-hand boom was essentially the fuselage of the BV 141 housing the engine, tail wheel and tailplane.

A Blohm und Voss BV 141B-1. This aircraft was found abandoned at the Blohm und Voss works at Wenzendorf in 1945. The finish appears to be overall RLM02 *grau-grün* (grey-green) with standard mid-war markings.

glazed nose section with seating for the pilot on the left-hand side and the bombardier's slightly behind and to the right. The internal layout was almost identical to that of the competing Fw 189, with the pilot's controls on a console to his left, rudder pedals extending over the glazed floor, and the flight instruments on the ceiling just above his line of sight. The similarity to the Fw 189 even extended to the rear gunner's position, which used the same conical turret.

Private venture

Despite a positive evaluation by the Luftwaffe, the BV 141 lost out to the Fw 189 and was officially cancelled in April 1940. Not to be deterred, Blohm und Voss proceeded with the private venture BV 141B, which was intended to address some of the shortcomings noted during its official evaluation.

The horizontal stabilizer, previously symmetrical, was redesigned to lie to the left of the tail fin, which opened up the rear gunner's field of fire. Power was also increased by installing a 1,163kW (1,560hp) BMW 801A radial engine and the undercarriage revised to bear the higher all-up mass of around 6,000kg (13,228lb). Ten BV 141B-0s were built but test flights revealed the new tail configuration had reduced the directional stability while the increased power from the BMW 801A required strengthening of the aircraft structure.

Specifications: BV 141B-0

Type:	Tactical reconnaissance aircraft
Dimensions:	Length: 13.95m (45ft 9in); Wingspan: 17.46m (57ft 3in); Height: 3.6m (11ft 10in)
Weight:	6,000kg (13,228lb) maximum take-off
Powerplant:	1 x 1,163kW (1,560hp) BMW 801A radial piston engine
Maximum speed:	370km/h (230mph)
Range:	1,200km (746 miles)
Service ceiling:	10,000m (32,808ft)
Crew:	3
Armament:	2 x 7.92mm (0.31in) MG 17s firing forwards; 2 x 7.92mm (0.31in) MG 15s in the rear turret; a bombload of up to 200kg (441lb)

Operational trials

Blohm und Voss did what they could to rectify these problems but there was one insurmountable challenge. With the BMW 801 also being used by the Fw 190 and Ju 88, the BV 141 would never be a priority for its supply. At the same time, the Fw 189 was already proving more than adequate for the tactical reconnaissance role. Some BV 141B-1 airframes may have been produced but it is thought these were remanufactured prototypes or B-0s.

The BV 141B-0 appears to have undergone some operational trials with a reconnaissance unit, Aufklärungsschule 1, in the east of Germany but this is as close as the type would come to frontline service.

Asymmetric handling

Although it seemed an unusual design choice, there was method to the lopsided arrangement of the BV 141. A single propeller creates a number of forces as a by-product of producing thrust, all of which cause a rotation around the aircraft's vertical axis, or a yawing moment. For a propeller that rotates clockwise, when viewed from behind, the slipstream's flow over the vertical stabilizer will yaw the aircraft to the left. At the same time, in a climb, or when sitting nose up on the ground, the downwards-going propeller will produce more thrust than the upwards-going one due to the difference in angle of attack. This effect is known as the P-factor. On a clockwise-rotating propeller this will again produce a yawing motion to the left. However, when an engine is offset from the aircraft's centre of gravity, the thrust itself will create a yawing moment towards the opposite side. By placing the BV 141's engine on the left of the aircraft the thrust yawed the aircraft to the right, while the slipstream and P-factor yawed it to the left. The two yawing moments therefore effectively cancelled each other out.

Although the Reich Minister for Propaganda, Joseph Goebbels, took the opportunity to claim the type was making a great contribution to the war against the Soviet Union in the summer of 1942, the planned deployment of two aircraft had in fact already been cancelled.

By this point, around 20 BV 141s of all variants had been built, a number endorsed by the type's designer. No examples survived the war, some being used as decoys at the Blohm und Voss factory in Hamburg, which had switched to licence-building Fw 200s. Although a fundamentally sound design with a performance that matched that of the Fw 189, the BV 141 was hampered by its choice of engine – one that was also needed in large numbers for the Luftwaffe's premier fighter and medium bomber, and which meant it would always face an uphill battle to win an order.

The BV 141's asymmetry is clearly visible in this overhead shot with the crew pod to the right of the centreline while the fuselage is to the left. Used for propaganda purposes by the Nazi regime, the original caption claimed the type was already in large-scale production.

Focke-Wulf Fw 189 Uhu (1938)

Unconventional in appearance for the time, the Focke-Wulf Fw 189 would prove to be a tough, manoeuvrable, multipurpose aircraft that would see the majority of its action on the Eastern Front.

As a primarily tactical air arm intended to support the German Army, the Luftwaffe had a requirement for an army cooperation and reconnaissance aircraft. Originally fulfilled by the Henschel Hs 126 in 1937, the Reichsluftfahrtministerium issued a requirement for its replacement. Arado, Focke-Wulf and Blohm und Voss responded, with only the first submitting what could be described as a conventional design. Blohm und Voss proposing a highly unorthodox design with a fuselage pod to one side of the centreline and the engine and tail to the other. Focke-Wulf's design, meanwhile, placed their fuselage pod on the centreline with twin booms supporting the tail and engines. Both were intended to provide the crew with the best all-round visibility. Focke-Wulf also offered the possibility of using a range of fuselage pods to fulfil a variety of missions with the same basic airframe. Despite institutional reluctance to adopt an unconventional design, the poor performance of the Arado prototype in

Armament
Although intended for the reconnaissance role, the Fw 189 was surprisingly well armed. In addition to the two pairs of MG 81Z 7.9mm (0.31in) machine guns in the crew nacelle, the aircraft had a pair of 7.9mm MG 17 machine guns in the wing roots and bomb racks for up to eight SC 50 50kg (110lb) bombs.

comparison to that of the Focke-Wulf led to its selection, the Blohm und Voss design still being too radical for consideration. Although Focke-Wulf called its design the *Eule* or 'owl', the Luftwaffe preferred *Uhu* or 'eagle owl' (confusingly, this name was also used for the Heinkel 219 night-fighter), while it would be called *das fliegende Auge*, 'the flying eye', in official media.

Novel turret

Unlike its design, construction of the Fw 189 was conventional, with a stressed-skin, flush-riveted structure and fabric-covered control surfaces. Each boom housed one of the main landing gears, a fuel tank and an Argus As 410A-1 engine. These were air-cooled inverted V-12s driving two-bladed, constant-speed propellers, which were controlled automatically by the vaned cap in front of the spinner itself. The horizontal stabilizer was low-mounted between the rear end of the two booms and held the tailwheel, which retracted into it to the left. The central fuselage pod had a typically German glazed nose for the cockpit, a central metal-skinned section and a glazed rear for the tail gunner. The rear gun itself was mounted in a novel 'turret', which consisted of a glazed cone with the apex to the rear. The turret rotated around the longitudinal axis to allow a cut-out section for an MG 15 7.92mm (0.31in) machine gun to move in the required direction. Additional aft-firing protection was mounted at the rear of the cockpit and operated by the bombardier while forward-firing 7.92mm (0.31in) MG 17 machine guns were fitted in the wing roots.

The pilot sat to the left of the cockpit with the rudder pedals extending forwards into the glazed nose. The majority of the aircraft controls were on a console to the pilot's left while the flight instruments were mounted on the

This Focke-Wulf Fw 189A-2 of AufklGr (H)/14 is seen with Eastern Front tactical markings in yellow, as it was when captured by US Forces in Salzburg in 1945.

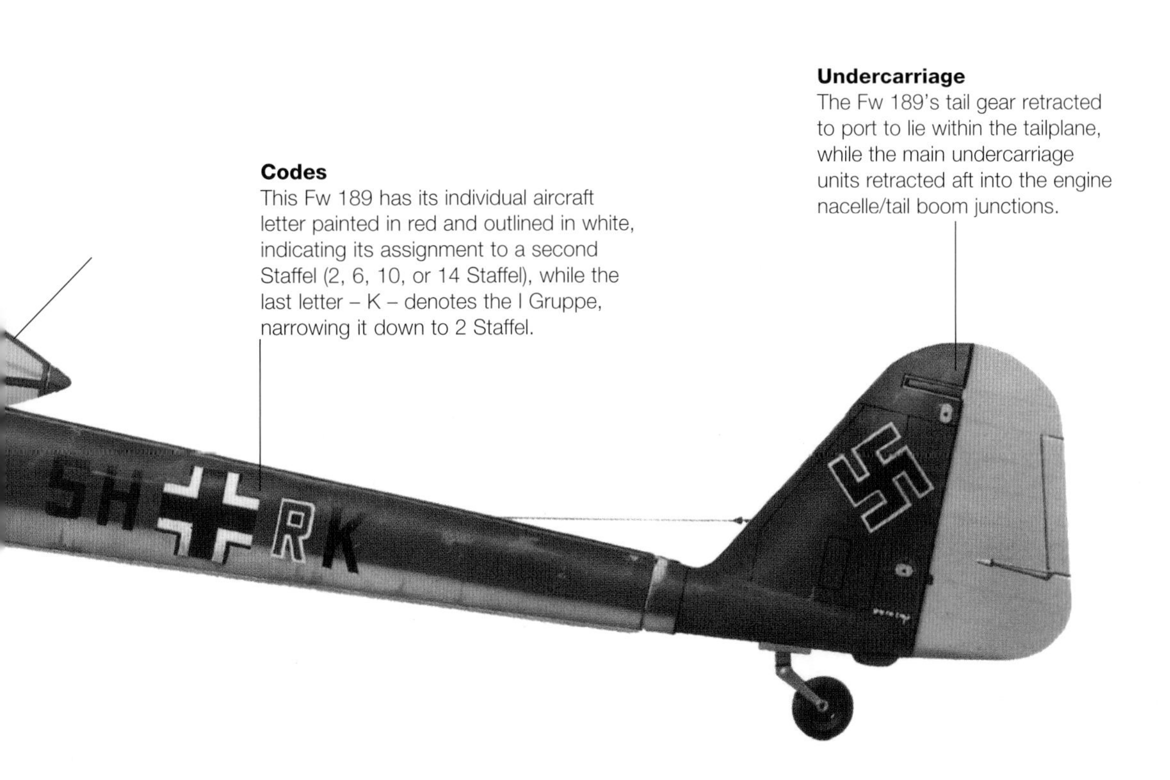

Codes
This Fw 189 has its individual aircraft letter painted in red and outlined in white, indicating its assignment to a second Staffel (2, 6, 10, or 14 Staffel), while the last letter – K – denotes the I Gruppe, narrowing it down to 2 Staffel.

Undercarriage
The Fw 189's tail gear retracted to port to lie within the tailplane, while the main undercarriage units retracted aft into the engine nacelle/tail boom junctions.

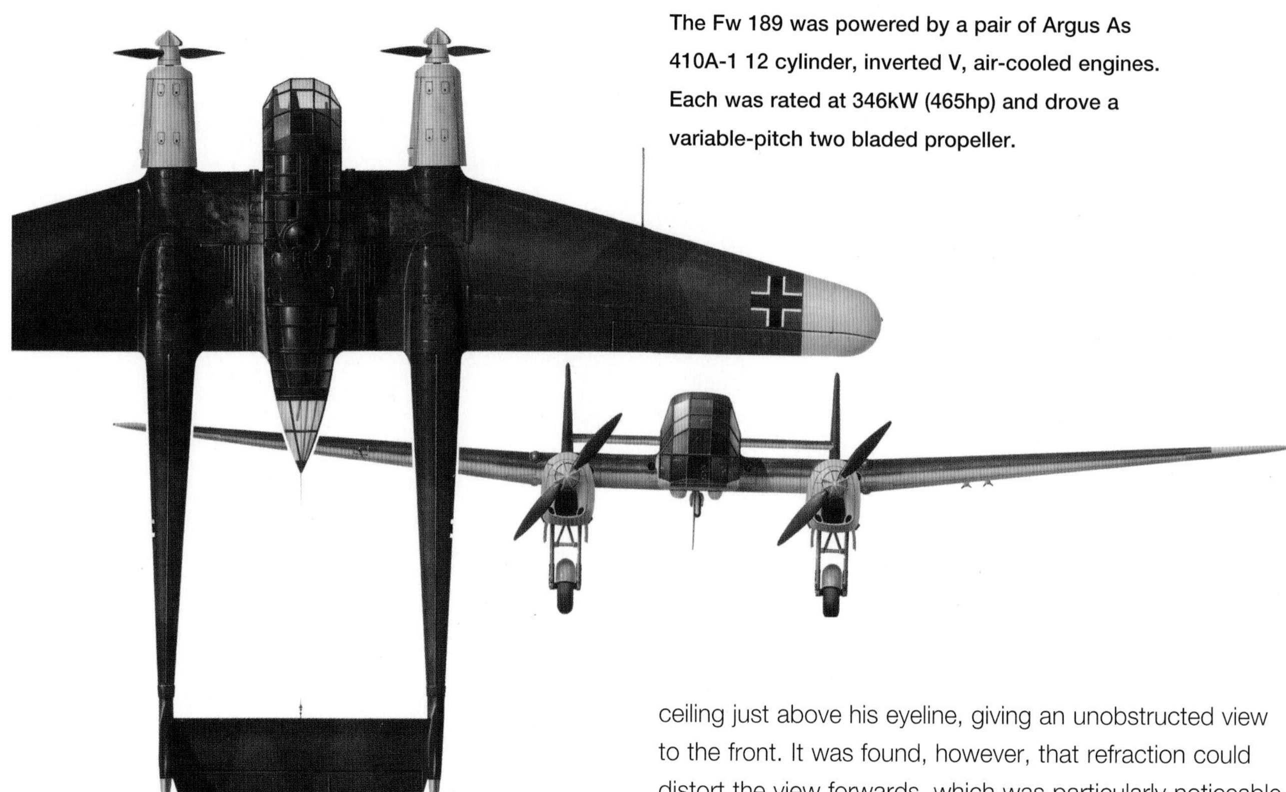

The Fw 189 was powered by a pair of Argus As 410A-1 12 cylinder, inverted V, air-cooled engines. Each was rated at 346kW (465hp) and drove a variable-pitch two bladed propeller.

ceiling just above his eyeline, giving an unobstructed view to the front. It was found, however, that refraction could distort the view forwards, which was particularly noticeable in poor weather. The bombardier sat to the right of and slightly behind the pilot with his bombsight in front of him while the seat swivelled, allowing him to operate the rear-facing machine guns. The bombardier was also responsible for operating the reconnaissance cameras and the radios, making him possibly the busiest of the three-man crew.

Despite successful trials with the pre-production Fw 189A-0s by an Aufklärungsstaffel (reconnaissance squadron), the Luftwaffe was slow to order the Uhu into production on the basis that the Hs 126 was perfectly adequate for its needs. It wasn't until the Henschel's shortcomings became apparent during the invasion of the Low Countries and France that the Fw 189 became a high-priority aircraft. With Focke-Wulf also under pressure to produce the Fw 190, construction was outsourced to Aero in Czechoslovakia and from 1941, SNCASO (Société nationale des constructions aéronautiques du Sud-Ouest) in France. Both occupied countries would frequently outproduce Focke-Wulf's Bremen factory, concentrating as they were on a single type.

The Fw 189A-1 was used extensively on the Eastern Front from 1942, where it demonstrated the ability to outmanoeuvre Soviet fighters at low level and fight them

Specifications: Fw 189A-2

Type:	Army Cooperation and tactical reconnaissance aircraft
Dimensions:	Length: 12.03m (39ft 6in); Wingspan: 18.4m (60ft 4in); Height: 3.1m (10ft 2in)
Weight:	4,170kg (9,193lb) maximum take-off
Powerplant:	2 x 345kW (463hp) Argus As 410A-1 V-12 air-cooled engines
Maximum speed:	350km/h (218mph)
Range:	670km (416 miles)
Service ceiling:	7,300m (23,950ft)
Crew:	3
Armament:	2 x 7.92mm (0.31in) MG 17s in the wing roots; 2 x 7.92mm (0.31in) MG 81s in the dorsal position; 2 x 7.92mm (0.31in) MG 81s in the revolving cone turret; a bombload of up to 200kg (441lb)

off with its defensive armament. More surprisingly, at least two examples survived ramming attacks by Soviet fighters, one boom apparently being sufficient to return to base. The Fw 189A-2 replaced the MG 15s with the MG 81Z twin machine-gun installation and added electrical rotation of the rear cone. From late 1942, a limited number of Fw 189A-4s were introduced with extra armour and replaced the forward-firing guns with 20mm (0.79in) MG FF cannons to give the aircraft a close air support capability.

Typically, the Uhus would operate at around 1,000m (3,281ft), conducting reconnaissance in search of enemy troops, logistics and fortifications. Once found, these targets would be attacked by the Luftwaffe's bomber force. Although successful in the role, and able to outmanoeuvre enemy fighters, as the sheer weight of numbers of the Soviet opposition forced Germany on to the defensive, even the Fw 189 was finding itself increasingly vulnerable. Consequently, from mid-1944, the Fw 189 force began to conduct its operations by night.

In total, 864 Fw 189s were produced. In common with most German aircraft types, production ended in 1944 as the focus was moved to the emergency fighter programme. As well as with the Luftwaffe, the Uhu would also see service with the Bulgarian, Hungarian and Romanian air forces as part of the Axis, while Norway would operate four captured examples post-war.

Fw 189 V1 D-OPVN was the first example to fly, taking to the air in July 1938. Kurt Tank himself took the controls for the first flight. Due to the soundness of the design, the production example differed little from the V-1.

Variants

Focke-Wulf's concept of different fuselage pods for different roles was not widely used, although the Fw 189B featured a more conventional-looking cockpit with a solid nose. Intended as a crew trainer, the B had dual controls and was initially of more interest to the Luftwaffe than the A model as it promised a more cost-effective training aircraft than anything else they had in their inventory. Ultimately, though, only three B-0s and 10 B-1s would be produced. The Fw 189C was proposed as a close air support aircraft for the same requirement as the Henschel Hs 129. The entire fuselage pod was replaced with a small two-person armoured crew section, barely longer than the wing chord line, for the pilot and rear gunner. Much like with the Hs 129, the view was poor, and the cockpit's cramped dimensions caused issues. Although marginally better-performing than the Henschel, the Fw 189C was not pursued primarily on the grounds of cost, the Hs 129 being cheaper.

Arado Ar 232 (1941)

Despite the Ar 232 being in many ways a pioneering design that featured elements that would appear in all post-war military transports, the Luftwaffe failed to order it in sufficient numbers to make use of its capabilities.

The Arado Ar 232 was designed in response to a 1939 request from the Reichsluftfahrtministerium for a replacement for the Junkers Ju 52/3m. Arado's response was, for the time, an unconventional design featuring a high-mounted mainplane on a boxy fuselage equipped with a rear loading ramp. This maximized the internal volume and avoided the wing spars intruding on the cargo deck. The tailplane was mounted on a narrow boom that extended from the top of the fuselage and featured twin vertical stabilizers. The main undercarriage retraced into the wings while the nose wheel partially retracted into the fuselage. In an unusual move, there were also 11 pairs of small wheels mounted under the fuselage that did not retract and from which the type gained the nickname

This aircraft, the ninth B-series Ar 232B, J4+UH, was active at Mühldorf, Germany, in summer 1944. The operating unit was Lufttransportstaffel 5.

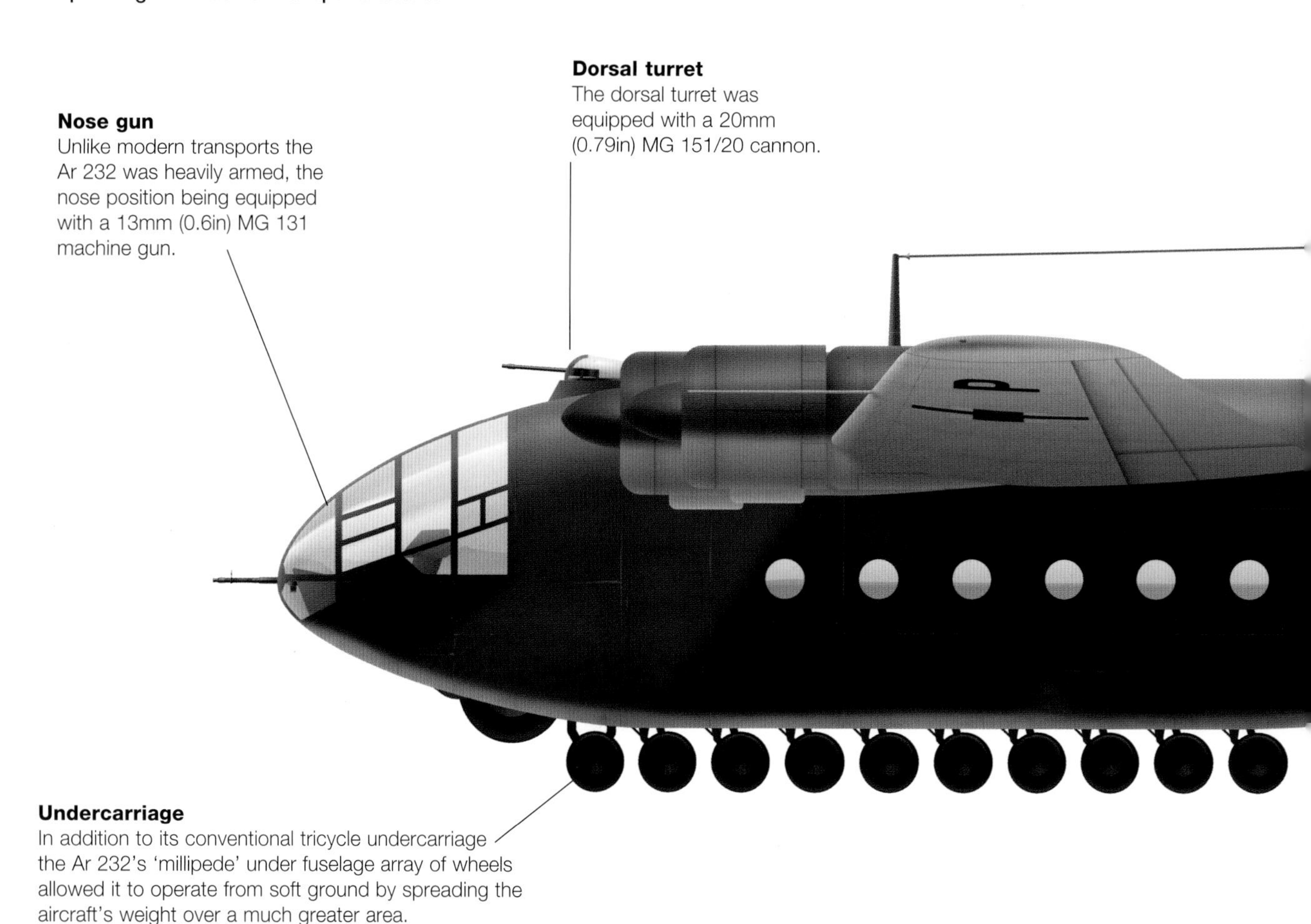

The twin-engined Ar 232 V2 is seen in flight. The type was nicknamed the *Tausendfüssler* ('centipede') on account of its unusual undercarriage.

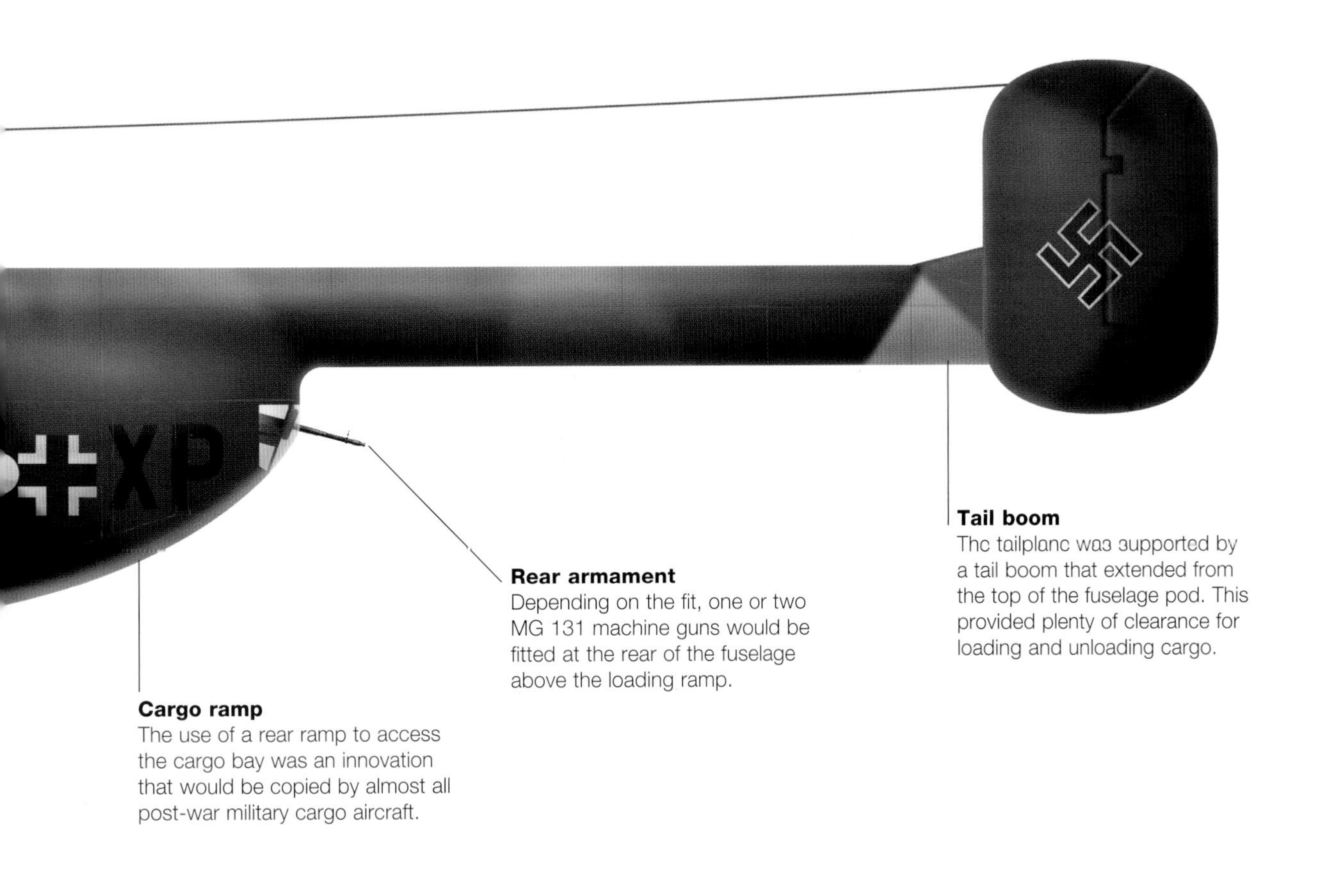

'Tausendfüssler' or centipede. These were intended to allow the Ar 232 to operate on rough ground, the main undercarriage legs having a kneeling function that lowered the fuselage on to the smaller wheels. This eased access to the cargo bay and enabled taxiing over unprepared areas, including the ability to cross ditches up to 1.5m (4.9ft) wide. It would even prove possible to land on the under-fuselage wheels in extreme situations.

The first prototype flew in 1941 powered by two 1,193kW (1,600hp) BMW 801 radial engines and it was on this flight that the ability to land on the so-called 'centipede' wheels was proven when a design flaw led to the main gear stubbornly refusing to extend. While the second aircraft would also use the BMW 801, by this point they were being used in the Fw 190, which had a higher priority. Consequently, Arado carried out redesign work to utilize the BMW-Bramo 323 engine while constructing 10 Ar 232-0s to the original configuration.

With the BMW-Bramo 323 only producing 895kW (1,200hp), it was necessary to use four to provide enough power, which in turn required an increase in the wingspan to allow the necessary spacing. At the same time, the change in centre of gravity led to the fuselage being stretched to compensate. These four-engined models were designated Ar 232B-0 and were arguably more capable aircraft, having greater installed power allowing full use to be made of the enlarged cargo bay, which could now carry a 4,500kg (9,921lb) load. Additionally, the loss of one engine during a critical phase of flight was now less likely to result in an accident, the remaining three being sufficient to maintain controlled flight. For self-defence, both versions were equipped with a 20mm (0.79in) MG 151/20 cannon in a power-operated turret just behind the cockpit, and 13mm (0.51in) MG 131 machine guns in the nose and at the aft of the fuselage pod above the loading ramp.

Blowing in the wind

Another novel feature of the Ar 232 was the inclusion of a system to blow extra air over the flaps. This prevents the airflow separating from the surface and allows controlled flight to be maintained at lower speed, reducing the take-off and landing roll. On the Ar 232, air was sucked from the upper surface of the inner wings and then redirected internally, compressed and exhausted over the rear of the upper wing. The system could reportedly reduce the take-off run to around 200m (656ft) and although it is not clear how heavy the aircraft was when doing this, it was still a significant achievement. The compressor that drove the airflow was powered by high-test peroxide (HTP), a volatile self-oxidizing fuel. However, the bulk of this was earmarked for the Luftwaffe's rocket programmes and use of the blown flaps does not appear to have been a regular occurrence. It was, though, another forerunner of things to come with many post-war jets, such as the F-104 and Buccaneer, which used the technique, albeit with bleed-air from their engines rather than a dedicated compressor.

Specifications: Ar 232B-0

Type:	Heavy transport
Dimensions:	Length: 23.52m (77ft 2in); Wingspan: 33.5m (109ft 11in); Height: 5.69m (18ft 8in)
Weight:	21,135kg (46,595lb) maximum take-off
Powerplant:	4 x 895kW (1,200hp) BMW-Bramo 323R-2 radial air-cooled engines
Maximum speed:	340km/h (211mph)
Range:	1,060km (659 miles)
Service ceiling:	8,000m (26,247ft)
Crew:	4
Armament:	2 x 20mm (0.79in) MG 151 cannon in the dorsal turret; 1 x 7.92mm (0.31in) MG 131 in the nose; and 1 x MG 131 above the loading ramp

Stalingrad aid

With the German Sixth Army encircled at Stalingrad in late 1942, the two prototype Ar 232-0s were sent to the Eastern Front to assist in the doomed efforts to support it by air. Although the first aircraft was lost almost immediately, the second would provide sterling service operating in and out of the pocket and ultimately being among the last to leave before the Soviet victory. Its success was in no small part due to its ability to operate from rougher surfaces than other transports in the Luftwaffe's inventory while also being around 80km/h (50mph) faster than the Ju 52/3m.

Despite this initial success, only 10 Ar 232B-0s would be ordered, the Luftwaffe by this stage of the war prioritizing fighter and bomber production. In a sign of the confusion and lack of planning that crippled the Third Reich as the war progressed, plans would be put in place in mid-1944 to produce a revised design, the partially wooden-winged Ar 432, with production scheduled to start in October 1945. This was at the same time as bomber production was being cancelled to concentrate on the emergency fighter programme.

The Tausendfüssler that were built fulfilled a variety of roles with Arado themselves frequently making use of them to move aircraft parts between factories. The Luftwaffe appear to have used it for general cargo movements. They were also adopted by KG 200 for what would now be known as special operations.

These primarily consisted of attempts to insert agents into the Soviet Union during 1944 and 1945, including two attempts to assassinate Stalin himself. The first of these, in July 1944, was aborted after the Ar 234A-0 was unable to lower its ramp and unload the motorcycle the agents were to use as transport. The Arado's undercarriage subsequently collapsed on landing back in German-occupied Poland. The repeat attempt was made on 4 September with an Ar 232B-0. However, this was damaged by anti-aircraft fire and appears to have made a forced landing around 200km (124 miles) west of Moscow with all involved ultimately being executed.

Although seeing limited production and use, the Ar 232 foreshadowed many of the features that would appear in post-war dedicated transports, including truck-level loading ramps, an internal crane and rough field performance. Of the 22 that were built, two survived the war, being captured and evaluated by the British, who would then use them as transport hacks. One was even flown in front of the public at the post-war German Aircraft Exhibition in autumn 1945, a crewman sitting on the lowered cargo ramp, again foreshadowing modern transport aircraft's capabilities.

Another photo of one of the two twin-engine Ar 232 prototypes in flight. Both these aircraft took part in the attempts to relieve the German Army at Stalingrad, one of them being the last to leave the garrison.

Messerschmitt Me 323 Gigant (1942)

The appropriately named Gigant was the largest land-based transport to fly during WWII and was unusual, although not unique, in being produced as both a glider and a powered aircraft. Able to move anything from troops to tanks, it would serve in Africa and on the Eastern Front.

With Operation Sealion – the planned invasion of the United Kingdom – officially postponed in the autumn of 1940, it was belatedly realized that additional airlift would be required to provide the initial airborne assault with armour and artillery. This would allow the Germans to hold their bridgeheads while waiting for the bulk of the invasion force to make their amphibious landings. Junkers and Messerschmitt were given two weeks to produce designs for heavy assault gliders, which were considered the most efficient way of moving the required loads. Messerschmitt submitted what would be the winning design on time on 1 November 1940. This was initially given the designation Me 261w before briefly becoming Me 263, before finally settling on Me 321. This was a high-wing braced monoplane with a conventional tail, all loading and unloading being done via a clamshell door that formed the nose. The single pilot cockpit was above this, level with the leading edge of the wing. Constructed of steel tubing covered by doped fabric

Defensive armament
The Me 323D introduced two gun positions in the clamshell nose doors, each with a 7.9mm (0.31in) MG 15 machine gun, with two similarly equipped positions on each side of the wing trailing edge/fuselage junction. In the Me 323E the door guns were replaced by less rudimentary gun positions lower down on each side of the centreline accommodating a 13mm

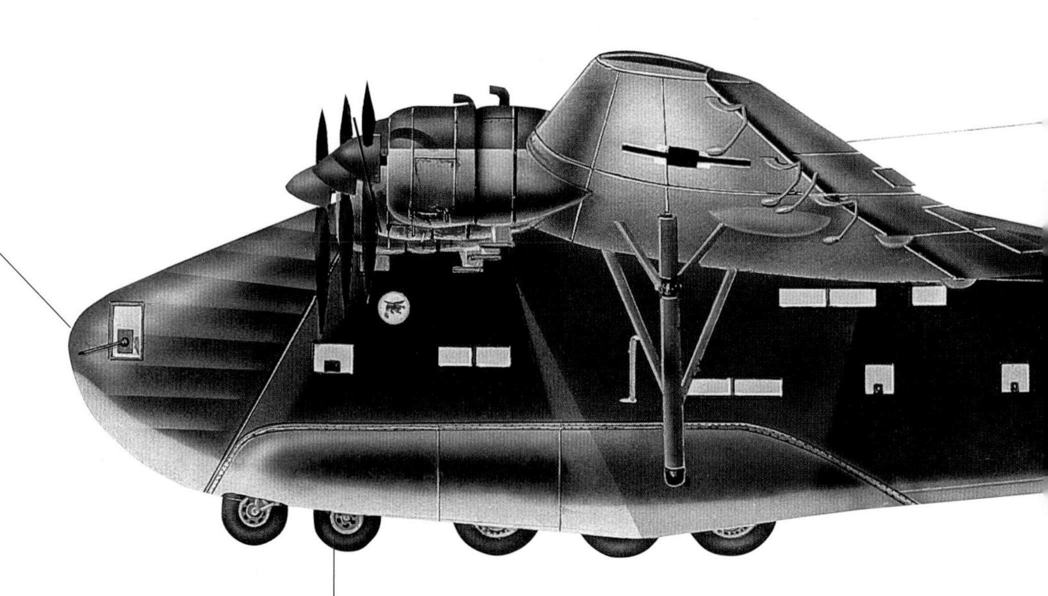

Undercarriage
The Me 321 had a simple skid undercarriage with a reusable, jettisonable take-off dolly weighing 1700kg (3740lb). The Me 323 introduced a new undercarriage with sponson fairings accommodating five tandem mainwheels on each side of the fuselage.

to minimize the empty weight, the Me 321 could carry an impressive 23,000kg (50,706lb), allowing it to lift a Panzer IV medium tank or 200 troops. Another weight-reducing measure was the four-wheel undercarriage that was mounted on a dolly that detached from the aircraft after take-off, landing being made on four skids.

Initial test-flying from February 1941, using a Junkers Ju 90 as the tug, revealed heavy controls requiring some modifications, including servomotors and a second pilot to make it a practical proposition. Consequently, the first 100 aircraft were produced as the Me 323A-1 while the next 100 would be completed as B-1s and feature the required modifications and a side-by-side dual-control cockpit. At the same time, only a month after the first test flight, work began on a powered version to overcome the issues of operating a 55m (180ft)-span transport that was unable to even move itself after landing and would require specialist towing aircraft. The Me 321 would see vital

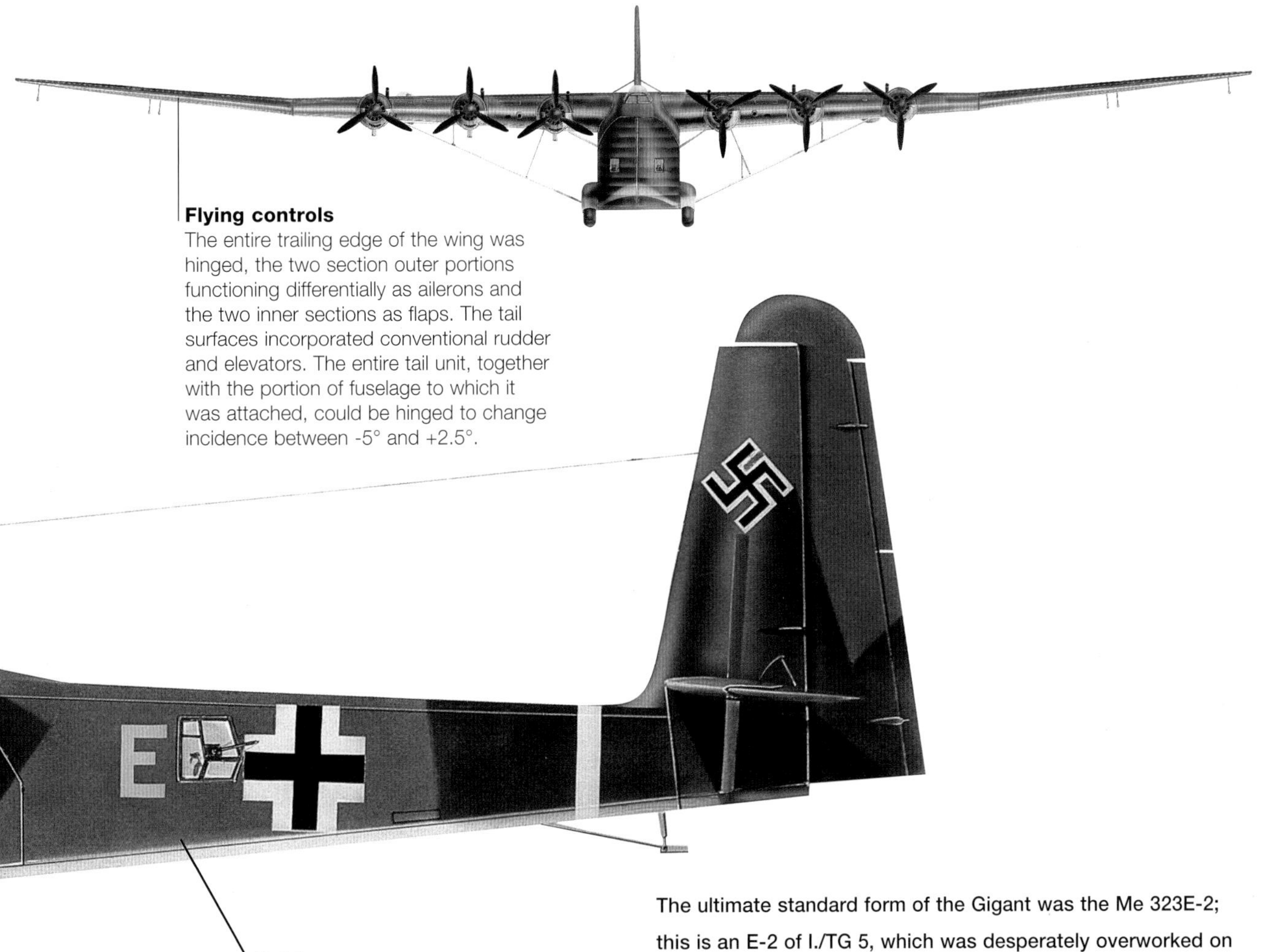

Flying controls
The entire trailing edge of the wing was hinged, the two section outer portions functioning differentially as ailerons and the two inner sections as flaps. The tail surfaces incorporated conventional rudder and elevators. The entire tail unit, together with the portion of fuselage to which it was attached, could be hinged to change incidence between -5° and +2.5°.

Hold
The Me 323D could carry 120 fully equipped troops in its cavernous hold, or 60 stretcher patients with medical attendants. Two 890 litre (195 gallon) auxiliary fuel tanks could be carried in the rear of the cargo hold, increasing the range with a 11,566kg (25,470lbs) load from 750km (466 miles) to 1000km (720 miles).

The ultimate standard form of the Gigant was the Me 323E-2; this is an E-2 of I./TG 5, which was desperately overworked on the Eastern Front from late 1943. This aircraft has a white stripe ahead of the tail instead of the usual yellow theatre band. The E-2 differed from earlier versions chiefly in defensive armament, the normal fit comprising two hand-aimed MG 131s low down in the front doors, another MG 131 firing aft from the radio compartment behind the cockpit, two 20mm (0.79in) MG 151s in low-drag EDL 151 turrets behind the outboard engines, and four single MG 131s firing from front and rear beam positions.

service supporting the German advance into the Soviet Union moving equipment forward as it was needed. However, the glider would never operate in the invasion-supporting role originally envisaged, putative landings on Malta and Sicily being cancelled due to a lack of tugs or suitable landing areas.

For the first flight in January 1942, the Me 323 V1 featured four 850kW (1,140hp) Gnome-Rhône 14N engines that were under production in France. These were attached to a reinforced wing with a flight engineer's cabin between each pair in the wing leading edge. A fixed undercarriage was composed of five wheels in sponsons on either side of the fuselage, each of which was independently sprung and allowed operations from rough forward airfields while keeping the fuselage level. The strengthening and extra weight required of a powered aircraft broadly halved the Me 323's payload from that of the 321, to 12,000kg (26,3446lb), but with only four engines, a tug would still be required for a fully loaded aircraft to get airborne. All subsequent Me 323s would therefore feature six engines with the first Me 323 D-1 entering production almost immediately, and entering service by late 1942. Defensive armament consisted of 7.92mm (0.31in) MG 15 machine guns in each nose door and in sponsons aft of the wing trailing edge. Guns could also be fired from three cabin windows on each side. Experience in the Tunisian Campaign would, though, lead to an uplift in the armament. A 13mm (0.51in) MG 131 replaced the MG 15 in the nose with an additional position fitted to each door. Meanwhile, a dorsal firing position was added for two MG 15s while later variants would gain turrets in the wings aft of the outer engines.

Seen undergoing repairs on the Eastern Front with the port undercarriage sponson removed, this Me 323 has the name *Himmelshaus* (Heavenly House) painted on the nose door.

Specifications: Me 323E-2

Type:	Heavy transport
Dimensions:	Length: 28.5m (93ft 6in); Wingspan: 55.0m (180ft 5in); Height: 9.6m (31ft 6in)
Weight:	45,000kg (99,208lb) maximum take-off
Powerplant:	6 x 850kW (1,140hp) Gnome-Rhône 14N left- and right-handed air-cooled radial
Maximum speed:	253km/h (157mph)
Range:	1,100km (684 miles)
Service ceiling:	4,000m (13,123ft)
Crew:	5
Armament:	2 x 20mm (0.79in) MG 151/20 cannon above the wings; 7 x 13mm (0.51in) MG 131 in the nose, dorsal and beam

Afrika Korps service

With Allied efforts to intercept Axis shipping to North Africa succeeding, the Me 323s were called on to resupply the Afrika Korps, regularly crossing the Mediterranean in formation escorted by fighters. Surprisingly resilient, it was not uncommon for Allied fighters to exhaust their ammunition without destroying the lumbering transport. However, losses were still heavy, with 43 being lost in April 1943 before the German Army was expelled from Africa.

Subsequent Me 323 operations would be concentrated on the Eastern Front, where the ability to carry outsize loads over 1,000km (621 miles) was vital in such a vast area. The year 1943 also saw the introduction of the Me 323 E-1, which had increased fuel capacity, revised nose armament and beam positions for MG 131s in the tail. The E-2 would introduce the wing turrets, while the Me 323 E-2/WT was a one-off attempt at an escort aircraft equipped with eleven 20mm (0.79in) MG 151/20 cannon and four MG 131s. Intended to ward off attacks by Allied fighters on formations of transports, it was soon concluded that the traditional fighter escort was the better approach.

Attempts to improve the performance of the Me 323 with Jumo 211 or BMW 801 engines did not progress due to their need by higher-priority projects. Production continued intermittently into 1944 but only 198 were built in total, some of which were converted from Me 321s. The Me 323's heavy lift capability was impressive for the time; however, its low speed and the limited numbers available meant it was not able to make a decisive contribution in any of the theatres in which it was involved.

Getting airborne

While the Me 323's engines allowed it to get airborne under its own power, at full load, assistance was still sometimes required in the form of high-test peroxide rocket motors. These would provide around 40kN (9,000lb of thrust) of additional thrust for about 30 seconds, at which point the motors would be parachuted to the ground. Compared to the Me 321, however, this was child's play. While initial flights used the Junkers Ju 90 as a tug, at higher weights, more power was required. Ultimately, this would be provided by the five-engined He 111Z but the alternative, and only, option until that was available was using three Bf 110s.

With the central tug 20m (66ft) ahead of the other two, formation take-offs were required and the failure of any element could lead to a catastrophic accident. At least one trial flight in 1941 ended with the Me 321 making a desperate low-level turn away from the snapped tow line with its rocket motors still firing.

An Opel 'Blitz' truck being unloaded from an Me 323D-6. The sheer size of the Me 323 gave the Luftwaffe a capability the Allies would not obtain until after the war.

Junkers Ju 290 (1942)

Evolving from an unwanted long-range bomber programme, the Ju 290 would develop into an excellent transport and maritime patrol aircraft. Lack of numbers would, however, prevent it from reaching its full potential.

The Junkers Ju 290 grew out of the Luftwaffe's Ural Bomber programme of the mid-1930s. When that was cancelled, the Junkers submission – the Ju 89 – was repurposed as an airliner and transport aircraft: the Ju 90. However, this would have to do without the Jumo 211 or DB 600 engines that had been specified for the bomber as they were needed elsewhere. While the Ju 90 initially flew with BMW 132 air-cooled radial engines, these were underpowered and a protracted development would eventually lead to a substantially redesigned aircraft powered by BMW 801s. With provision for defensive armament while operating in the maritime patrol role, the new designation Ju 290 was applied to the subsequent design, which first flew in August 1942.

The prototype was soon followed by two pre-production Ju 290A-0s and five Ju 290A-1 production models, all of which were unarmed. The design was typical for transport aircraft of the time, being a low-wing monoplane taildragger

Cabin
The Ju 290 was flown by a crew of two on the flight deck. Seven additional crew members acted as gunners, observers, navigator, radio operator and radar operator.

Radar
The FuG 200 Hohentwiel radar could detect convoys at ranges approaching 100km (62 miles), a dedicated operator was included in the nine-man crew.

Gondola
The ventral gondola mounted a 20mm (0.79in) MG 151 in the front and a 13mm (0.6in) MG 131 to the rear.

with four 1,300kW (1,743hp) BMW 801 radial engines in nacelles on the wings. With the German Sixth Army at Stalingrad surrounded by the Soviets in late November 1942, the prototype Ju 290 and one of the A-0s were dispatched to help in the airlift of stores and manpower to relieve them. In a reflection of the fate of the Sixth Army, the prototype was lost while the other aircraft was damaged participating in the operation.

Maritime patrol

With the Fw 200's shortcomings as a maritime patrol aircraft all too apparent, the Ju 290A-2 was developed to take over from it. First flying in the summer of 1943, this gained a dorsal turret equipped with a 20mm (0.79in) MG 151 cannon, additional navigation equipment and the FuG 200 Hohentwiel search radar for locating Allied convoys. Three of these were produced before the A-3 variant added

Seen parked in front of a Junkers Ju 88G on a snow-covered airfield, Ju 290A-5 KR+LK was the eleventh and last A-5 to be built.

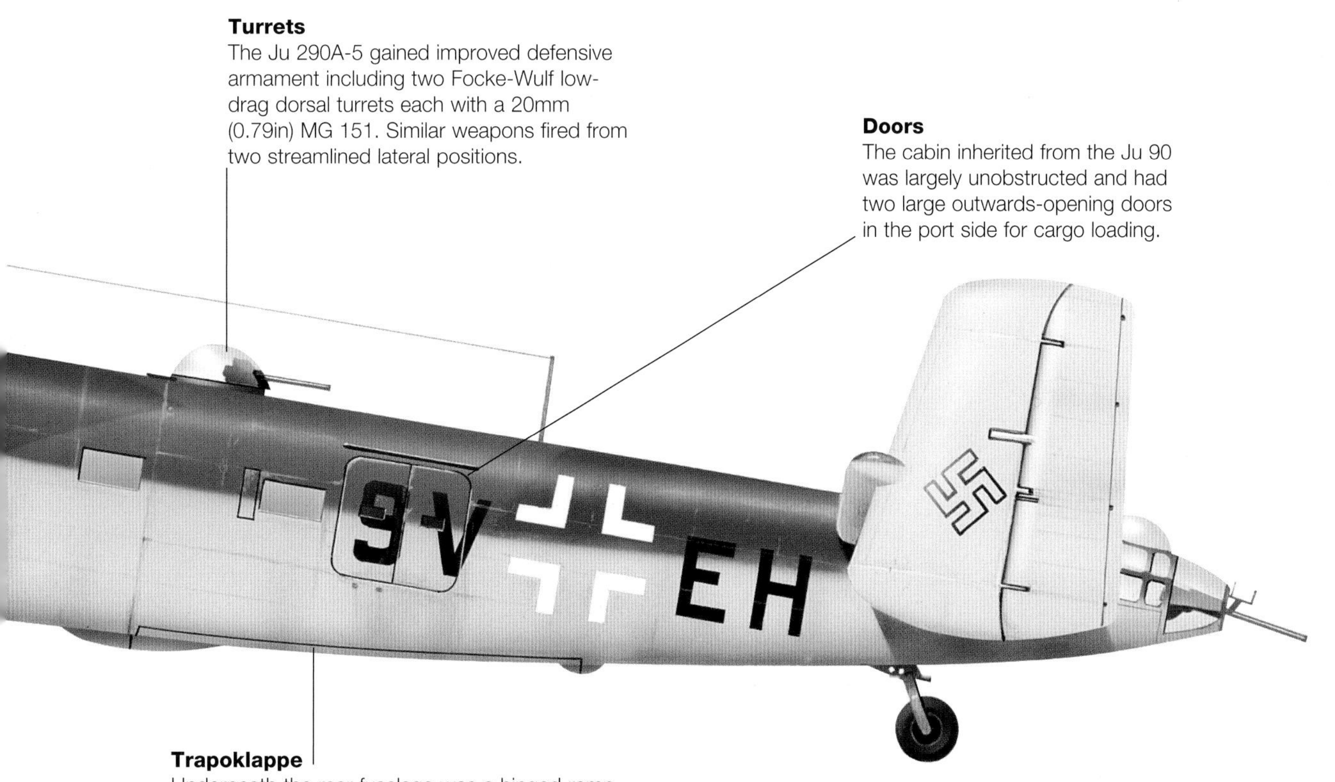

This aircraft flew with Fernaufklärungsgruppe 5 from Mont-de-Marsan in France. With never more than 20 operational aircraft, the unit struggled to fulfil its commitments in the anti-shipping role.

a gondola under the nose equipped with a forward-firing MG 151 and a rearward-firing 13mm (0.51in) MG 131 machine gun. The new variant also had a further MG 151 in the tail and replaced the dorsal turret with a low-drag model designed by Focke-Wulf. Five Ju 290A-3s were produced and together with the A-2s were operated by Fernaufklärungsgruppe 5 from Mont-de-Marsan on the French Atlantic Coast from October 1943. These would soon be joined by the A-4 model that added a second dorsal turret just aft of the cockpit.

By the spring of 1944, the first of 11 Ju 290 A-5s were entering service. These gained armour protection for the cockpit and self-sealing fuel tanks. Although a step up in capability from the Fw 200 and popular with its crews, there were seldom more than 20 aircraft available to FAGr 5. With the whole of the North Atlantic as their search area, they were never able to adequately cover it. At the same time, fighters from the escort carriers now accompanying the convoys were a regular threat. A Ju 290 was even the last aircraft shot down by a Sea Hurricane when an aircraft of 835 Naval Air Squadron from HMS *Nairana* shot one down on 26 September 1944, that ship's fighters having already claimed at least two others that year.

The next model intended for series production was the Ju 290A-7, which was designed as a carrier for the Henschel Hs 293, Hs 294 anti-ship missiles and the Fritz X glide bomb. However, with the Allied invasion of Normandy and the subsequent loss of access to bases in Western France, only around 12 appear to have been built, one of which was taken to the USA for evaluation after the war. Meanwhile, the remaining aircraft of FAGr 5, unable to fulfil their intended role because accessing the Atlantic became too dangerous, were put to task as transports. Helpfully, despite being constructed as patrol aircraft, they retained the hydraulic loading ramp, or Trapoklappe, of the original design. When lowered, this lifted the tail off the ground and allowed vehicles and cargo to be loaded with ease.

Flights to Japan

A direct air route between Germany and its Axis ally, Japan, was a long-term aspiration to overcome the delays and dangers of the sea route, which was the only practical option for the duration of the war. A number of proposed routes were studied with the best appearing to be either Odessa in occupied Ukraine or Nautsi in northern Finland to Manchuria in Japanese-occupied China. Both were a one-way distance of around 6,000km (3,728 miles) – just on the edge of the Ju 290's range. Three Ju 290A-5s were converted for the task in early 1943, with their armour removed and additional fuel tanks fitted, but the plan was then put on hold as the Japanese, not yet at war with the Soviet Union, did not want to provoke them by overflying their territory. A Ju 290A-3 was also modified for the necessary long-range flight in early 1945 to send a replacement air attaché to Tokyo. However, they were instead ultimately sent onboard U-234, which surrendered to the US Navy on 14 May after Germany's unconditional surrender.

Specifications: Ju 290A-5

Type:	Long-range maritime patrol aircraft
Dimensions:	Length: 28.6m (93ft 10in); Wingspan: 42m (137ft 10in); Height: 6.8m (22ft 4in)
Weight:	44,969kg (99,140lb) maximum take-off
Powerplant:	4 x 1,300kW (1,743hp) BMW 801G air-cooled radial engines
Maximum speed:	440km/h (273mph)
Range:	6,150km (3,821 miles)
Service ceiling:	6,000m (19,685ft)
Crew:	9
Armament:	1 x 20mm (0.79in) MG 151 cannon in each dorsal turret; 1 x MG 151 in the tail and each waist position; 1 x MG 151 in the front of the gondola; 1 x 13mm (0.51in) MG 131 in the rear of the gondola

Bomber prototype

While a number of Ju 290 bomber models were proposed, including the pressurized B-1, unpressurized B-2 and Ju 290E with an internal bomb bay, none were completed. One development that did at least reach flying prototype stage was the Ju 390, which was essentially a Ju 290 with a fuselage plug and an extra central wing section allowing an additional engine to be fitted on each side. The first example flew in October 1943, configured as a transport

aircraft. A second slightly longer airframe was built for the maritime patrol role, though there are conflicting accounts of whether or not this flew, and it now appears unlikely that it did. Similarly, post-war claims that one or both Ju 390s had reached the US East Coast from Mont-de-Marsan in 1944 do not stand up to scrutiny.

Limited impact

Only around 45 Ju 290s were built with another 35 cancelled after their production run had started as the war turned against the Axis powers. A definite improvement over its predecessors in the transport and patrol roles, the type arrived too late and in insufficient numbers to materially affect the course of the war. One aircraft believed to be a pre-production Ju 290A-8 would survive the war at the Czech factory where it had been constructed but would fail to find a buyer when offered for sale. Meanwhile, a Ju 290A-5 of Lufthansa was interned in Spain on 5 April 1945 and subsequently put into service by the Spanish Air Force until it was finally scrapped in 1957 after an accident.

This frontal view of a Ju 290A-5 shows the offset nature of the under-fuselage gondola. The A-5 entered service in the spring of 1944 and was a significant improvement over the earlier models.

This Ju 290A-7 was captured intact by Allied forces and flown to the US for evaluation. The A-7 introduced a nose turret: in operational configuration it would have also featured search radar.

Helicopters (1942)

The very first helicopters proved to be difficult to master, and although an initial example was airborne in September 1907, it wasn't until Germany introduced the Flettner Fl 282 *Kolibri* that a truly useful rotorcraft reached the front line. Delivered to the German Navy from 1942, it was followed by the more capable Focke-Achgelis Fa 223 *Drache*, a six-seater with options for various defensive weapons, though the helicopter as a weapon was still immature by the time World War II ended.

Focke-Achgelis Fa 223 *Drache*

The Focke-Achgelis company gained experience of a helicopter with an outrigger-mounted, twin-rotor arrangement with the Fa 61, which was then scaled up to create the six-passenger Fa 226 *Hornisse* ('hornet'). The latter was developed to meet a Deutsche Lufthansa requirement. The prototype completed ground-running and tethered-hovering trials in summer 1940 and a first free

First flown in February 1945, the 51st prototype Fa 223 V23, received the military serial GW+PA. It was used for operational trials before it was captured by US troops.

Accommodation
The crew of pilot, on the left, and observer sat in the cockpit with the four-person passenger compartment immediately behind them and in front of the gearbox.

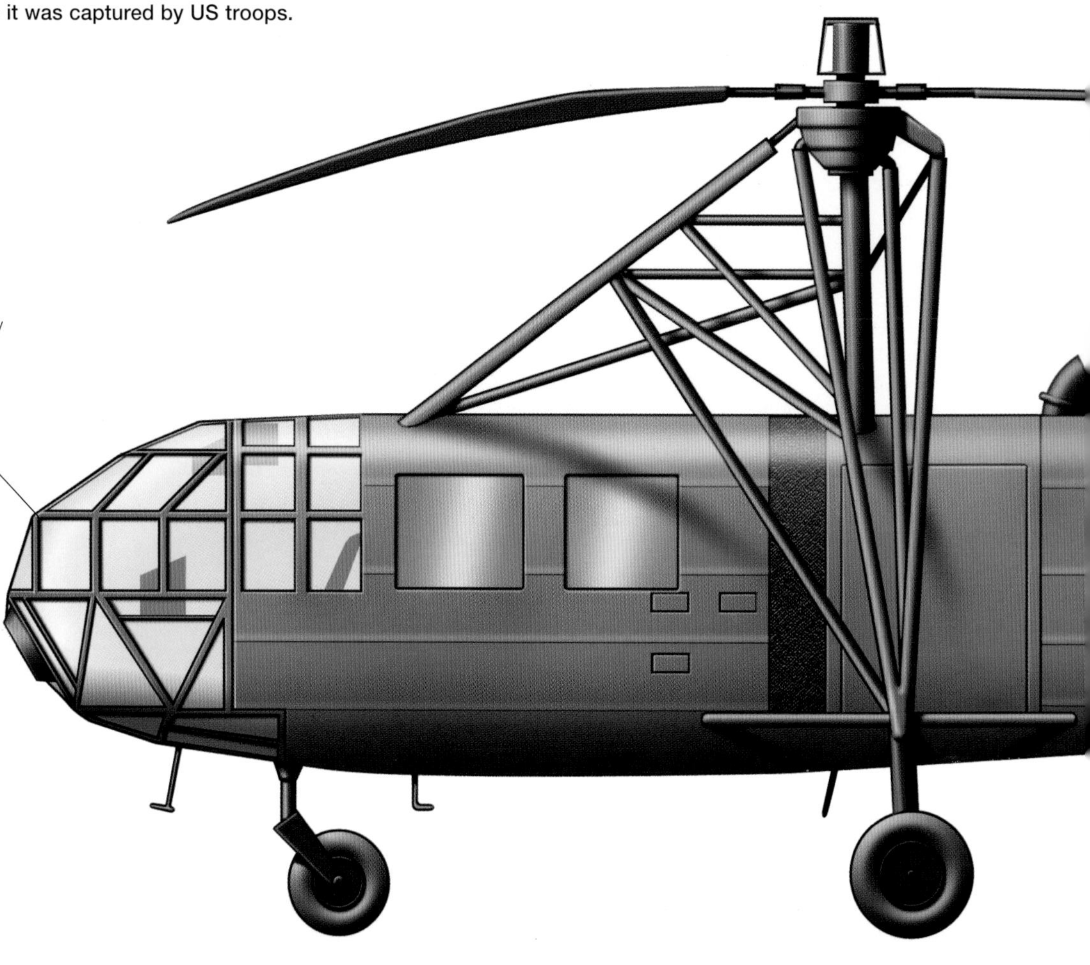

flight took place in August that year. Further development continued under military jurisdiction as the Fa 223 Drache ('kite'), with an order placed for 39 examples for evaluation in roles including training, transport, rescue and anti-submarine patrol. Different equipment included a 7.92mm (0.31in) MG 15 machine gun and two 250kg (551lb) bombs, a camera for reconnaissance, or a jettisonable 300-litre (66 gal) auxiliary fuel tank.

Production run

Ten of the planned 30 pre-production aircraft were completed in Bremen before the factory was bombed, and another seven were completed at Laupheim near Stuttgart. One more emerged from a factory in Berlin before the war came to an end. Most of those completed didn't ever fly, but at least two entered service with Lufttransportstaffel 40,

Specifications: Fa 223

Type:	Helicopter
Dimensions:	Length: 12.25m (40ft 2in); Height: 4.36m (14ft 4in)
Weight:	4,315kg (9,513lb) maximum take-off
Powerplant:	1 x 750kW (1,006hp) Bramo 323D-2 nine-cylinder radial engine
Maximum speed:	176km/h (109mph)
Range:	437km (272 miles)
Service ceiling:	4,875m (15,994ft)
Crew:	2 + 4 passengers
Armament:	1 x 7.92mm (0.31in) MG 15 machine gun; 250kg (550lb) bombs or 2 x depth charges

Rotors
The counter-rotating rotors were constructed of wood with a steel spar. Each rotor had three blades while the outriggers held them sufficiently far apart to avoid any overlap, which would have required synchronization of their rotation.

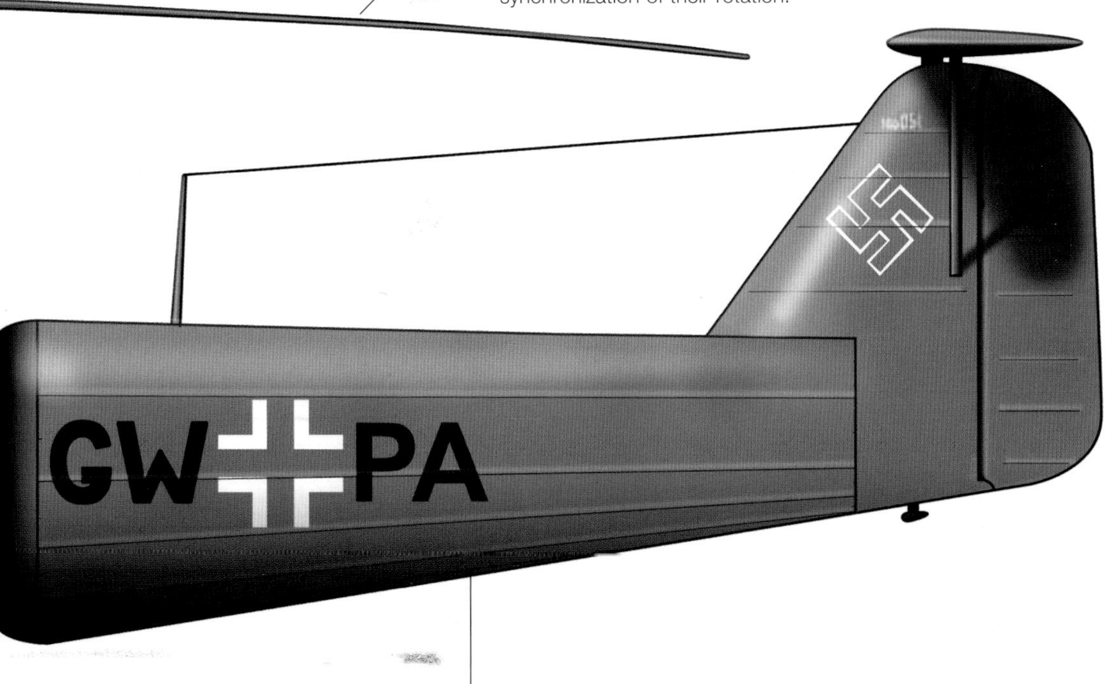

Construction
The fuselage was constructed of steel tubes covered in doped fabric to save weight. The rotors were mounted on outriggers also constructed from steel tubing.

The initial prototype, Fl 265 V1, registration D-EFLV, completed its maiden flight in May 1939 and subsequently undertook extensive military testing in conjunction with the Kriegsmarine before being removed from service in April 1940.

these being captured by US forces at Ainring, Austria, in May 1945.

One example became the first helicopter to cross the English Channel when it was transferred to the UK for evaluation in September 1945, although it was destroyed in an accident the following October. Two more aircraft were completed in Czechoslovakia after the war using German-made components. Development also continued in post-war France, where the Sud Est SE 3000 first took to the air in October 1948.

Flettner Fl 265

Rotary-wing aircraft pioneer Anton Flettner developed the two-seat Fl 184 autogyro powered by a 104kW (140hp) Siemens-Halske Sh 14 radial engine driving a tractor propeller, plus an auto-rotating three-blade rotor.

The aircraft was destroyed before evaluation could commence but it was followed by the Fl 185 autogyro/ helicopter, in which a similar engine drove not only the main rotor but also two variable-pitch propellers mounted on outriggers, one on each side of the fuselage. Once selected for helicopter mode, the outrigger propellers were set so one acted as a tractor and the other as a pusher, offsetting main rotor torque. When operating as an autogyro, the outrigger propellers were both set to act as pushers and the main rotor was left to auto-rotate.

This concept paved the way for the further-refined Fl 265, construction of which began in 1937. This utilized the same basic airframe design as the Fl 185, with a radial piston engine in the nose, but the previous variable-pitch propellers were deleted. Instead, it became a true helicopter. Power was supplied to two intermeshing and synchronized main rotors, which were counter-rotating to cancel out each other's torque. The tail unit featured an adjustable tailplane for trimming and a large fin and rudder to supplement the differential collective pitch change of the rotors in providing steering.

Active service

The first Fl 265 was lost in an accident when the rotors struck each other just three months after its first flight. The second prototype, the Fl 265 V2, was used for a variety of military trials. A total of six prototypes were built to meet a requirement of the German Navy and, although an order for quantity production followed in 1940, this was abandoned in favour of continued work on a more advanced Flettner design, the Fl 282 *Kolibri*.

Focke-Achgelis Fa 330 *Bachstelze*

The Fa 330 *Bachstelze* ('white wagtail') was a single-seat rotary-wing observation kite, developed in order to provide U-boat commanders with an aerial surveillance capability.

The surveillance was designed to extend around 8km (5 miles) from the submarine, primarily to locate naval targets. In 1942, Focke-Achgelis began work on a small, rotary-wing gyrokite that could be launched, towed and retrieved from a submarine. This was intended to be quickly assembled and disassembled and featured a free-turning, three-blade rotor mounted on a pylon above a simple framework. The aircraft carried a pilot/observer and the tail unit was fitted with a tailplane, fin and rudder at the end of a braced tubular boom. Fa 330s were built by Weser Flugzeugbau, which produced around 200 examples.

Surface run

When employed operationally, the *Bachstelze*'s rotor was spun up manually before auto-rotating in the wind, with the U-boat running on the surface. The observation kite then flew at the end of a cable, towed along by the submarine.

The pilot/observer was provided with a telephone to communicate with the submarine, the cable extending to a length of 120m (394ft). At the end of the observation sortie, the aircraft would be winched back to the U-boat's deck. While the *Bachstelze* was ablc to extend the commander's situational awareness by a factor of five, in practice it was found to be hazardous to employ the observation kite, as the U-boat was no longer able to make an emergency dive. Their use was limited as a result, with most activity taking place in the South Atlantic and Indian Ocean, where the presence of Allied warships was less likely.

Fl 282 *Kolibri*

The Fl 282 *Kolibri* ('hummingbird') was an improved development of the earlier Fl 265 and emerged as a two-seater primarily intended for naval use. In early 1940, before it had even been tested, an order was placed for no fewer than 30 prototypes and 15 pre-production examples for use by the Kriegsmarine. The basic fuselage design was inherited from the Fl 265, but an important change was the introduction of a Bramo Sh 14A engine mounted in the centre fuselage, with the pilot seated in the nose.

A total of 24 prototypes were eventually built, these featuring different cockpit arrangements, including enclosed, semi-enclosed and fully open configurations. Some Fl 282s were also completed as single-seaters, while the two-seaters carried an observer in a position aft of the main rotor pylon, offering a good view to the rear of the aircraft. German Navy trials of the *Kolibri* began in 1942 and the type proved to be very stable and highly manoeuvrable, with flying characteristics making it safe to fly in even poor weather conditions.

By the following year, around 20 of the 24 prototypes were operating on board warships in the Mediterranean and Aegean, mainly on convoy protection duties.

The successful operational trials prompted an order for 1,000 production examples, but these efforts were frustrated by the Allied bombing campaign targeting the BMW and Flettner factories responsible for the work. By the end of the war, only three of the prototypes survived, many more having been destroyed to prevent their capture.

Jet Aicraft

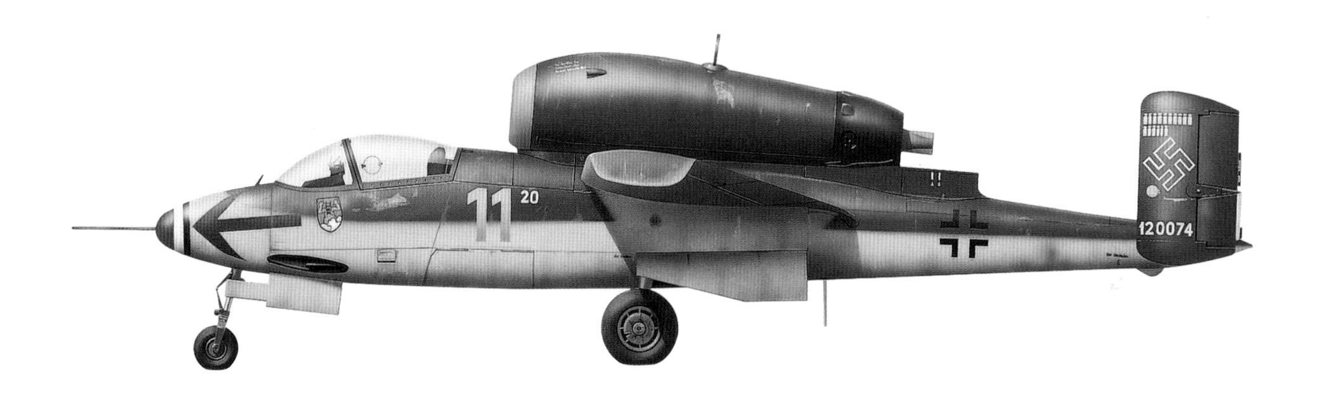

Far in advance of the Allies' application of jet and rocket technology, the Luftwaffe operated at the limit of what was possible with 1940s technology. Given the constraints imposed by Allied blockade and bombing campaigns, this led to some extremely high-risk operations. The Me 163 was powered by toxic chemicals that could dissolve human skin, despite which the pilot sat between two fuel tanks. Only the Me 262 came close to fulfilling the promise of jet technology and achieving results that justified the effort involved in its production. It was let down more by politics than performance, delays in production meaning it was never deployed in sufficient numbers to make a significant impact on the war.

Opposite: Messerschmitt Me 262 frames sit in woodland after the end of hostilities. More than 1400 of the high-performance jet fighter were built.

Messerschmitt Me 163B Komet (1941)

The fastest aircraft of World War II, the Me 163B had its roots in gliding. Outright speed would not, however, compensate for poor endurance and the hazards of exotic rocket fuels.

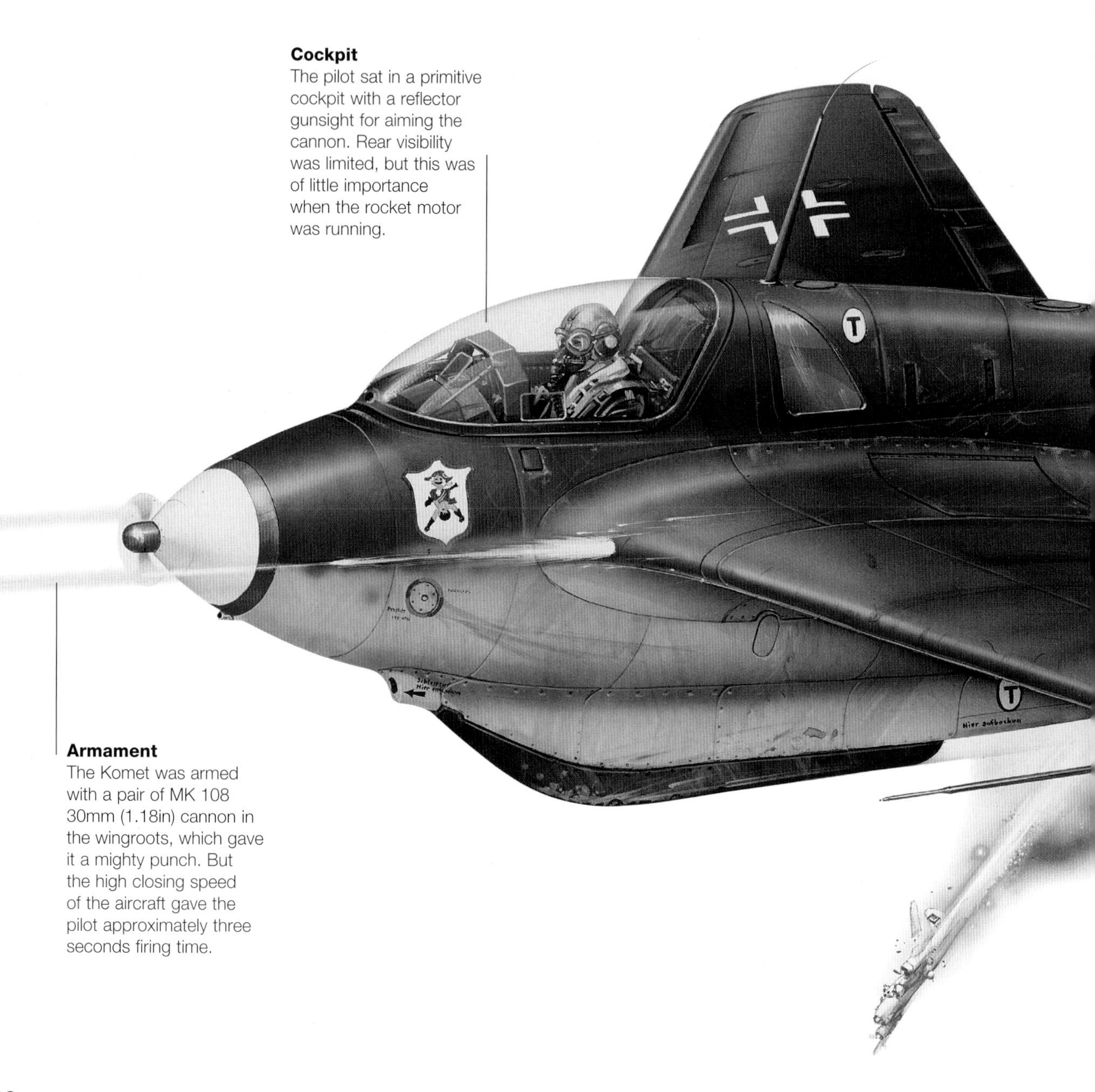

The Messerschmitt Me 163B Komet was a point defence interceptor and possibly the most specialized aircraft of the war. Its development can be traced to 1937, when glider and tailless aircraft proponent Dr Alexander Lippisch was asked by the German Air Ministry to develop an aircraft to test the new 3.92kN (882lb)-thrust Walter I-203 rocket motor. This led to the all-wood DFS 194, which first flew on 3 June 1940 at Peenemünde and eventually reached 547km/h (340mph) in level flight. Encouraged by this, Lippisch was directed by the RLM to design an interceptor using the 7.36kN (1,653lb)-thrust Walter II-203b engine in cooperation with Messerschmitt, with whom he had worked on the DFS 194.

The first Me 163A was ready for gliding trials in March 1941, the engine not yet having been installed. Although the handling was found to be excellent, it was almost too good as a glider, frequently floating just above the ground as the pilot rapidly ran out of airfield to land on. Powered

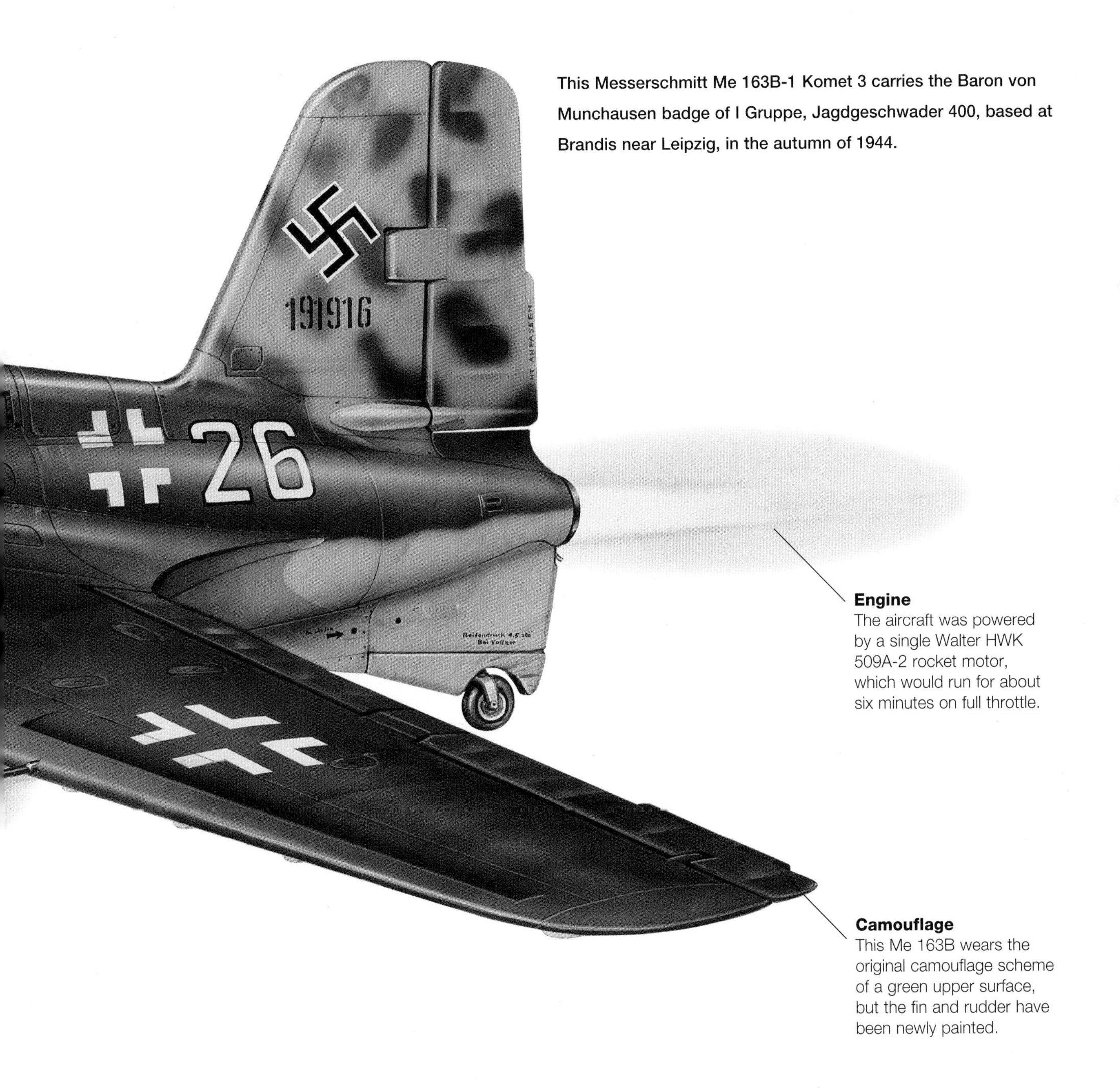

This Messerschmitt Me 163B-1 Komet 3 carries the Baron von Munchausen badge of I Gruppe, Jagdgeschwader 400, based at Brandis near Leipzig, in the autumn of 1944.

Messerschmitt Me 163B (V41). The first production interceptor designated Me 163B was the red-painted V41, first flown on 14 May 1944 by Major Wolfgang Späte.

trials commenced in August with the first flight exceeding 800km/h (497mph) while subsequent towed launches reached 1,000km/h (621mph) – an unofficial world record. This was not without its problems: the bounds of aeronautical knowledge were being broken as the aircraft approached the sound barrier and the nose of the Me 163A would drop violently without warning. This would result in the wing being redesigned for the Me 163B with a constant sweep on the leading edge and fixed slots on the outer 40 per cent. As well as resolving the issues with high-speed flight, this also made the aircraft unspinnable. An area that would not be so easily resolved was the landing gear.

New type

Due to its gliding pedigree, the Me 163 did not have a conventional undercarriage but took off from a detachable dolly and then landed on a sprung skid. At the same time, with no propeller slipstream over the rudder, directional control at low speed was poor.

Rough surfaces could lead to disaster, the aircraft bouncing uncontrollably or if the landing wasn't directly into wind, slewing round and overturning. In either case, there was a high chance that any remaining rocket fuel would explode.

Rocket motor

The Hellmuth Walter Kommanditgesellschaft HWK 109-509A-1 rocket motor was powered by two fuels, known as C-Stoff and T-Stoff. The first was a mix of hydrazine, methanol and water, with a trace of potassium copper cyanide. T-Stoff, meanwhile, was 80 per cent hydrogen peroxide, stabilizer and water. Neither is user-friendly. C-Stoff had to be stored in glass or enamel containers to prevent it reacting with anything, while T-Stoff was highly corrosive to organic material, such as cotton or human flesh. Given the dangers of handling the fuel, special protective clothing was required for everyone involved, and the tanks and engine were flushed through with fresh water after every flight.

With two T-Stoff tanks in the cockpit there was always a potential for injury to the pilot if leaks occurred due to damage, such as a heavy landing. To start the engine, the T-Stoff was passed through a catalyst, causing it to break down and produce steam at over 600°C (1,112°F). This then turned the turbine driving the fuel pump and exited through the exhaust nozzle. C-Stoff would be circulated by the pump around the combustion chamber and returned to the main tank. As the throttle was opened in the cockpit, C-Stoff and T-Stoff would be fed into the combustion chamber, where they would react to produce thrust with an exhaust temperature of 1,800°C (3,272°F).

Erprobungskommando 16 operated Me 163B-1 GH+IN (prototype V35) from Bad Zwischenahn, Germany, in 1944. EKdo 16 was the Me 163 test unit, formed at Bad Zwischenahn in summer 1942 and receiving its first Me 163B in May 1944.

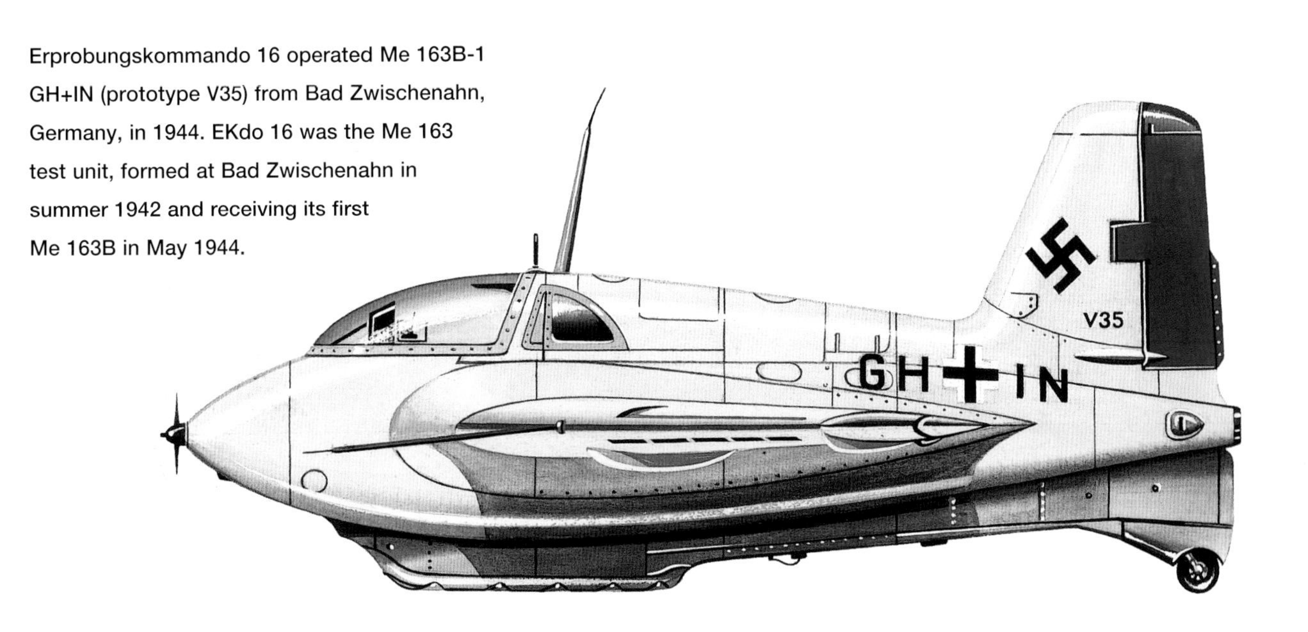

Work on the Me 163B began in December 1941 and would build on the lessons learnt with the Me 163A. The aircraft that emerged had wooden outer wings with a sweep of 23.3 degrees, the combined ailerons and elevators forming the outer trailing edge while the inner

An Me 163B-1a launching at Bad Zwischenahn, home of the trials unit Erprobungskommando 16, which accepted its first Me 163B during May 1944.

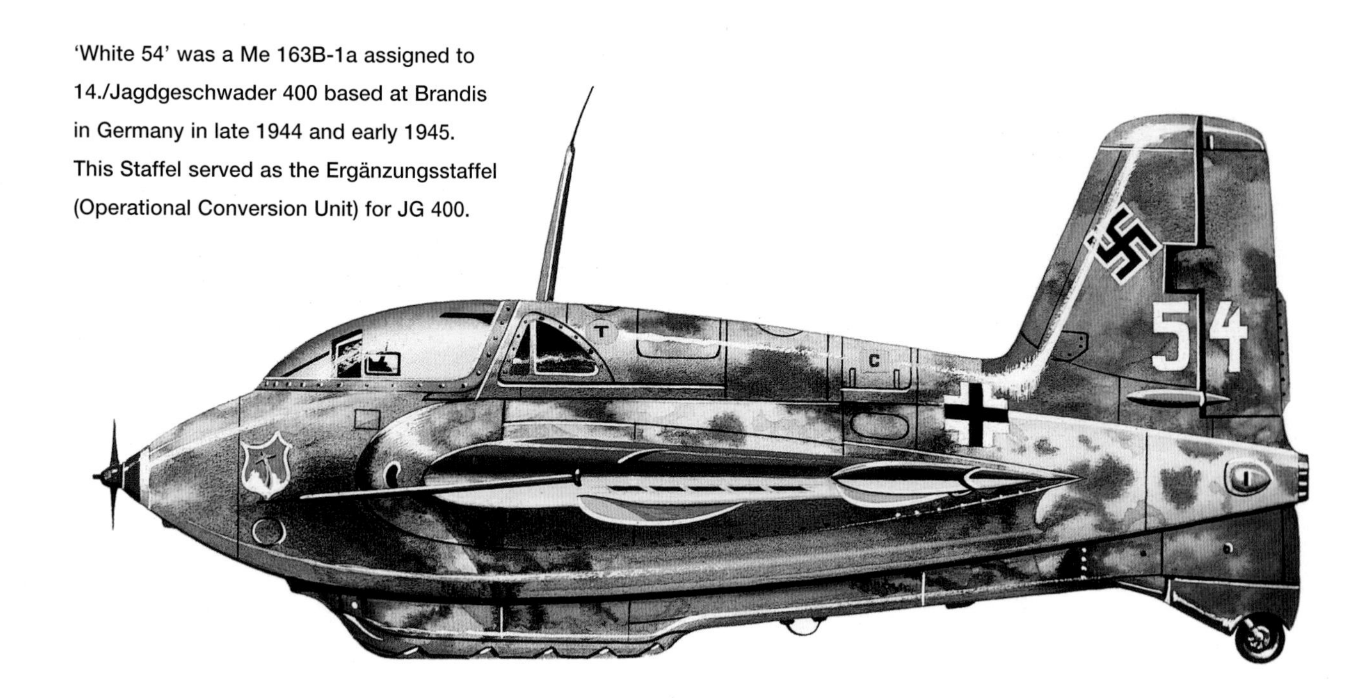

'White 54' was a Me 163B-1a assigned to 14./Jagdgeschwader 400 based at Brandis in Germany in late 1944 and early 1945. This Staffel served as the Ergänzungsstaffel (Operational Conversion Unit) for JG 400.

sections formed the flaps. The fuselage was constructed of alloy and contained the HWK Type 109-509A-1 motor, which was throttleable between 2kN (450lb) and 16.7kN (3,800lb) of thrust in four steps.

The engine itself weighed just over 100kg (220lb) and a fully loaded Me 163B weighed 4,310kg (9,502lb), giving a thrust-to-weight ratio of 0.42. Most of the rest of the fuselage behind the cockpit consisted of fuel tanks and pumps, with additional fuel in the cockpit either side of the pilot. The wing roots each held a cannon, early aircraft using the 20mm (0.79in) MG 151/20 while later production used the 30mm (1.18in) MK 108. The extreme nose held the avionics powered by a small battery, which was itself charged via the nose-mounted windmill generator.

Specifications: Me 163B-1

Type:	Point defence interceptor
Dimensions:	Length: 5.7m (18ft 8in); Wingspan: 9.3m (30ft 6in); Height: 2.5m (8ft 2in)
Weight:	4,310kg (9,502lb) maximum take-off
Powerplant:	1 x 14.71kN (3,307lb) HWK 109-509A-1 bi-propellant rocket motor
Maximum speed:	955km/h (593mph)
Range:	41km (25.5 miles)
Service ceiling:	12,100m (39,698ft)
Crew:	1
Armament:	2 x 20mm (0.79in) MG 151/20 or 2 x 30mm (1.18in) MK 108 cannon in the wing roots

Given the difficulty and danger involved in landing the Me 163, a graduated training programme was developed for pilots new to the type. After gaining initial gliding experience, they would fly DFS Habicht, or 'Hawk', gliders first with the original 14m (45ft 11in) wingspan, before moving on to the Stummel-Habicht, or 'Stumpy Hawk', with 8m (26ft 3in) and then 6m (19ft 8in) wingspans. These modified aircraft had landing speeds representative of the operational Me 163B. The final steps were unpowered flights either in an Me 163B or one of the handful of Me 163As, towed aloft by Bf 110 tugs and ballasted with water to an operational weight. Once this phase was complete, the pilots would be cleared to conduct powered flights.

First service

With interest in the Komet waxing and waning depending on how Germany was faring in the war, the first operational unit, I./JG 400, did not form until May 1944. Their first major engagement took place on 16 August

A Messerschmitt Me 163 Komet is retreived and moved by Luftwaffe ground crew using a Scheuschlepper tractor and trailer after landing.

when five aircraft engaged over 1,000 USAAF bombers. The results could only be described as disappointing, one Me 163 being lost to a B-17's tail gun, while a second fell to a P-51 after managing to at least hit a B-17. The difficulty of successfully engaging a target in the few seconds available while speeding through the enemy formation had been underestimated.

Attack profile

The typical attack profile involved a powered climb at 70 degrees to 12,000m (39,370ft) before cutting the engine and making an unpowered diving attack on the enemy bombers. The rocket would then be relit to reposition for another attack. The best results were obtained on 24 August when four heavy bombers were destroyed. However, this was something of a high point, only five other bombers falling to the Me 163B during the war while 14 Komets were lost in combat. At the same time, more than 200 were lost in take-off or landing accidents. Attempts were made to share the Komet design with the Japanese, but the two submarines carrying engines and blueprints were sunk before they reached the home islands.

Despite this, the Mitsubishi J8M was developed based on what knowledge had reached Japan. However, the first flight resulted in the loss of both the aircraft and the pilot and the type would not become operational before the war ended.

Flawed concept

Although technically impressive and with absolute performance far in excess of anything operated by the Allies, the Me 163B was a flawed concept. With around eight minutes of fuel for its engines, it could only spend a limited time engaging the enemy and although its inflight handling was rated highly, the difficulties in the take-off and landing phases were never resolved.

Junkers Ju 287 (1941)

The Junkers Ju 287 was a flying testbed that combined forward-swept wings, jet engines and a fixed undercarriage. Arriving too late to become a production aircraft, it would however point the way to aviation's future.

With development of the Ju 288 suffering continued delays, the spring of 1943 saw Junkers' chief designer exploring alternative configurations to rectify the performance-related issues. Wind tunnel testing showed that the most promising alternatives featured swept wings that allowed higher speeds for the same basic layout of fuselage and engines. Forward-swept wings permitted better performance with a smaller, lighter wing than aft-swept wings, although they suffered bending due to the aerodynamic forces at the tips. Around this time, the Luftwaffe issued a requirement for a fast bomber and by

The Ju 287 V1 was the only example to be completed and flown before the war ended. The Soviets would however use parts of the V2 and V3 for the essentially identical OKB-1 EF-131 which would fly post-war.

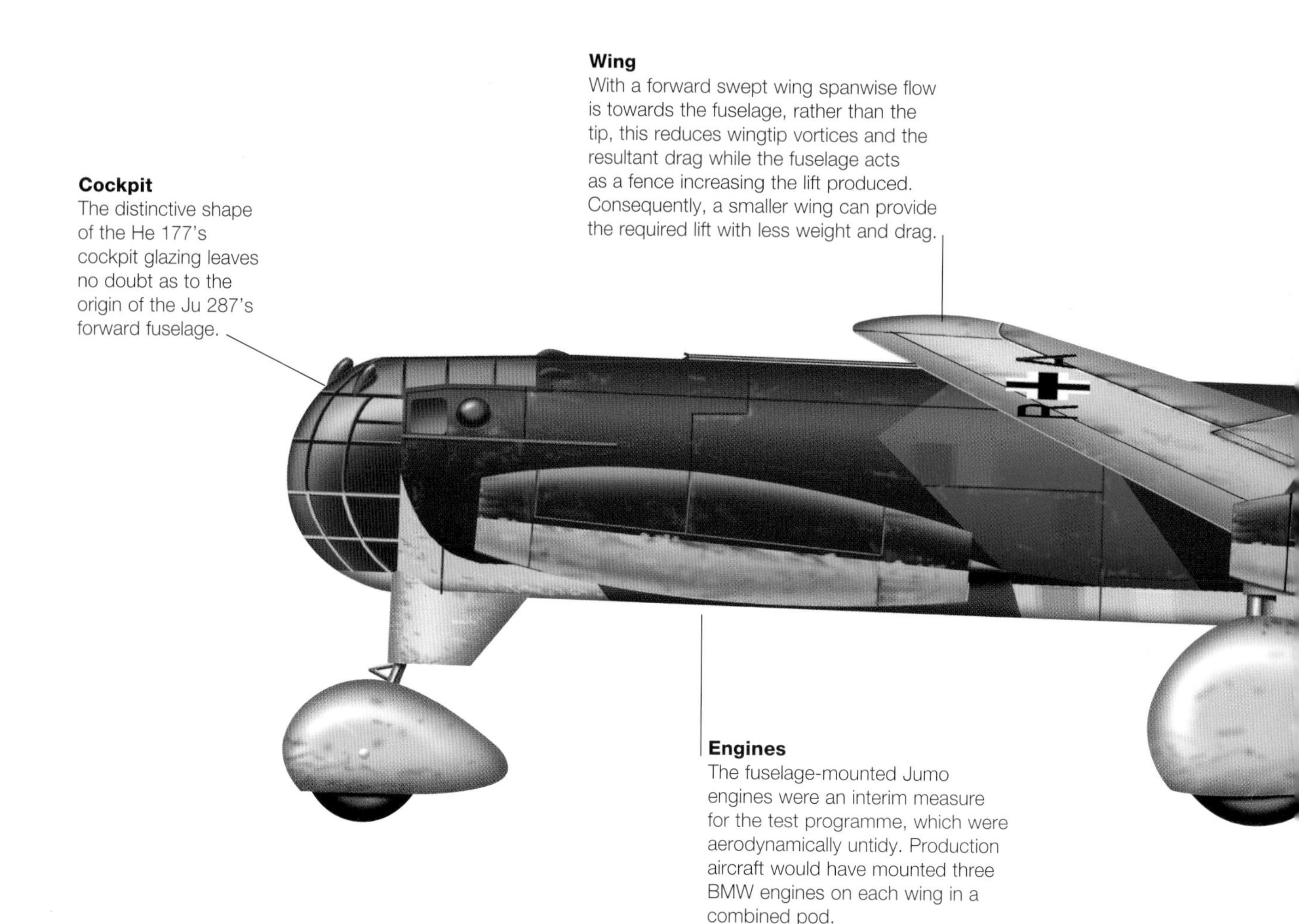

the end of 1943, Junkers were awarded a development contract based on a forward-swept wing design powered by four to six turbojet engines with the designation Ju 287, reusing that of a failed Stuka replacement programme. The number of engines used was dependent on the thrust produced, with a variety of configurations considered, including fuselage-mounted and separate and combined underwing pods. These layouts were subsequently refined with extensive wind tunnel testing.

Forward-swept wing

Given such a departure from conventional aerodynamics, two test aircraft were commissioned to fully explore the forward-swept wing's performance. To expedite development, these used the forward fuselage of the Heinkel He 177 and the tail section of a Ju 188G-2, the two parts being connected by a new centre section which mounted the forward-swept wings. The fixed landing gear, meanwhile, was salvaged from B-24 Liberators that had been shot down over the Reich, with two nosewheels placed beneath the cockpit and the mainwheels between the fuselage and engines. Two 8.83kN (1,984lb)-thrust Jumo 004Bs were placed either side of the nose just aft of the cockpit and two underwing units were mounted mid-span and extending aft behind the trailing edge.

Construction began in May 1944 and by August, Junkers' chief test pilot was carrying out fast taxi runs to become familiar with the controls. Finally, on 8 August, everything was ready for the first flight. With a take-off weight of 15.4 tonnes (17 tons), three HWK 109-501 rocket engines were used to assist the Ju 287 in getting off the ground. Two under the wings were fired during the ground

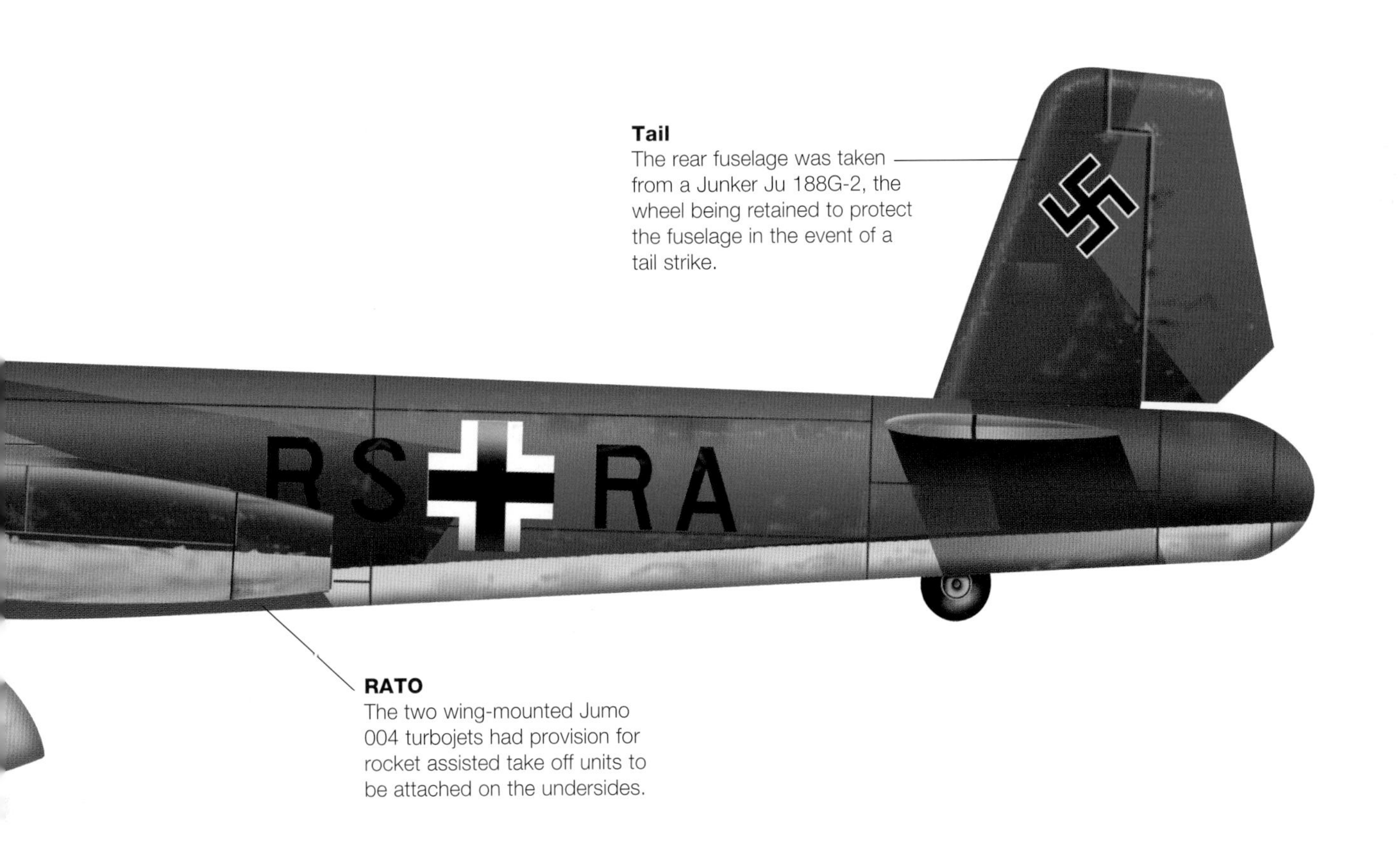

Tail
The rear fuselage was taken from a Junker Ju 188G-2, the wheel being retained to protect the fuselage in the event of a tail strike.

RATO
The two wing-mounted Jumo 004 turbojets had provision for rocket assisted take off units to be attached on the undersides.

run and the third under the right forward engine was used once while airborne.

Despite the Ju 287's unusual configuration, the aircraft handled normally for an aircraft of its size, reaching a top speed of 370km/h (230mph) during the 40-minute flight. The test programme revealed no serious flaws with the layout although the RLM had prohibited flight above 550km/h (342mph) due to concern over the increasing stress from aerodynamic bending as speed increased. This could be prevented with a stronger, stiffer structure, but the extra weight would eliminate the advantage of the forward-swept wing's smaller size.

Production aircraft would instead have a twist designed into the wing, giving extreme washout and a negative angle of attack at the tip. It was thought that this, together with mounting the engines on the leading edge, would counter the bending forces. Overall, the test programme correlated closely with the results found in Junkers' wind tunnel tests: the main problems encountered involved engine or boost rocket failures, none of which proved catastrophic. The only area that required real change was the design of the nosewheel fairings, which caused in-flight flutter and were eventually removed completely.

Flying the Ju 287

As a testbed, the Ju 287 was intended to prove the viability of a number of the intended operational aircraft's features, including the forward-swept wing, and the general handling of large jet aircraft. Ground movements were relatively easy although the lack of propeller slipstream meant the rudder was ineffective at low speeds and the brakes had to be used for directional control. The nosewheels pivoted freely, and some pilots found taxiing easier if the ground crew took charge of them via metal rods through the axles. Slow to accelerate under jet power alone, the prototype was effectively underpowered compared to the planned production model, which would have approximately 50 per cent more thrust.

Although initially limited to 370km/h (230mph), later flights would reach 650km/h (404mph) with no obvious issues experienced with wing bending or flutter. Stable in all axes, the Ju 287 could be slightly slow to respond, which was thought to be due to the influence of the non-retractable undercarriage.

There were no serious handling issues during the engine and rocket failures that were experienced, and it was found to be easy to maintain control of the aircraft and make a safe landing. Landings themselves were benign although there could be some wing-bending during the flare as the angle of attack was increased.

Different configurations

A variety of configurations were considered for the production Ju 287, which evolved towards an aircraft with broadly the same fuselage but with a Ju 388-style cockpit. Power would be provided by six BMW 003 turbojets – three in a combined pod under each wing – the fuselage-mounted engines being deleted to reduce turbulence. The second prototype would feature this engine layout and a retractable undercarriage as part of the development process.

Production was planned to begin slowly with single aircraft delivered during August, September and December 1944 before ramping up through 1945 to 100 a month by the end of the year. Even this would rapidly prove beyond the capability of German industry and with the Allied landings in Normandy turning the tide of the war by September 1944, all work on the programme was stopped on the orders of Hitler himself, and the second prototype would never actually fly.

Specifications: Ju 287 V1

Type:	Aerodynamic testbed
Dimensions:	Length: 18.3m (60ft); Wingspan: 20.11m (66ft); Height: 5.1m (16ft 9in)
Weight:	20,000kg (44,093lb) maximum take-off
Powerplant:	4 x 8.83kN (1,984lb)-thrust Jumo 004B turbojets
Maximum speed:	558km/h (347mph)
Range:	1,500km (932 miles)
Service ceiling:	10,000m (32,808ft)
Crew:	2
Armament:	None

The plan view of the Ju 287 clearly shows the degree of forward sweep and the limited clearance between the forward-mounted engines and the fuselage.

Although briefly considered for the Amerika Bomber project in early 1945, this would prove to be a dead end. The Ju 287 would, however, influence bomber design on both sides of the Cold War as American and Soviet engineers and scientists would plunder Junkers' research into swept-wing configurations. While Boeing would develop the swept-wing B-47 and Martin would even use a similar engine layout on the B-51, the Soviet Union took the next step and moved the Junkers facility and staff to Podberesje, north of Moscow. This would ultimately lead to the Type 150, in a continuation of Junkers' internal numbering system, an aft-swept wing bomber with podded turbojets under each wing. However, like the Ju 287, this too would never enter service.

Cutting-edge concept

The Ju 287 was a cutting-edge concept that helped accelerate post-war aviation, but it was the wrong aircraft at the wrong time for Germany and was ultimately a distraction from producing a successor to the Ju 88.

Messerschmitt Me 262 (1942)

Assured of its place in history as the first jet fighter to enter service, the Luftwaffe's Me 262 was the most advanced fighter to reach operational status during World War II, and ushered military aviation into the jet age.

Exploiting German research into gas turbines that had begun prior to the outbreak of the war, the Me 262 was the first turbojet-aircraft to achieve operational status. Its design origins date back to late 1938 and, after Messerschmitt had been requested to produce specifications for the new aircraft in January 1939, the first prototype airframes were available in 1941.

It was always planned that the new fighter would be powered by two examples of a new gas turbine engine then under development with BMW. Although some design work had been done on a single-engine twin-boom design, none of the available engines were considered

Tail
Control surfaces included fabric-covered elevators, replaced with stronger metal skins on later production aircraft. The powerful rudder was required to maintain directional stability.

Me 262A-1a 'Yellow 7' was on strength with the 11. Staffel of Jagdgeschwader 7, based at Prague in April 1945. The aircraft was eventually captured by the Allies at Lechfeld and is now preserved in the National Air and Space Museum in Washington, D.C.

Engines
Power was provided by a pair of Junkers Jumo 004B-1 axial-flow turbojets, which suffered from poor reliability and limited service life, primarily due to the effect of Allied bombing on production facilities, and the lack of certain materials required for the turbine blades.

powerful enough. The planned BMW P.3302 jet engines took a long time to refine, and it was not until April 1942 that the aircraft made its maiden flight, the initial prototype V1 being powered by a single 545kW (731hp) Jumo 210G piston engine mounted in the nose. A first flight under turbine power followed in March, the powerplant being the intended pair of 5.39kN (1,213lb)-thrust BMW 003 turbojets, in addition to the Jumo 210, should it be required in an emergency. This proved prophetic as during the maiden flight with jet power, both turbojets failed soon after take-off and the prototype was forced to land on piston power alone. In part at least, this was due to German engine manufacturers' determination to develop axial-flow engines. Although these are theoretically more powerful and fuel-efficient and had lower drag than the centrifugal jet engine being pursued in the UK, they are more difficult to build, requiring high-temperature alloys, and are far more susceptible to damage.

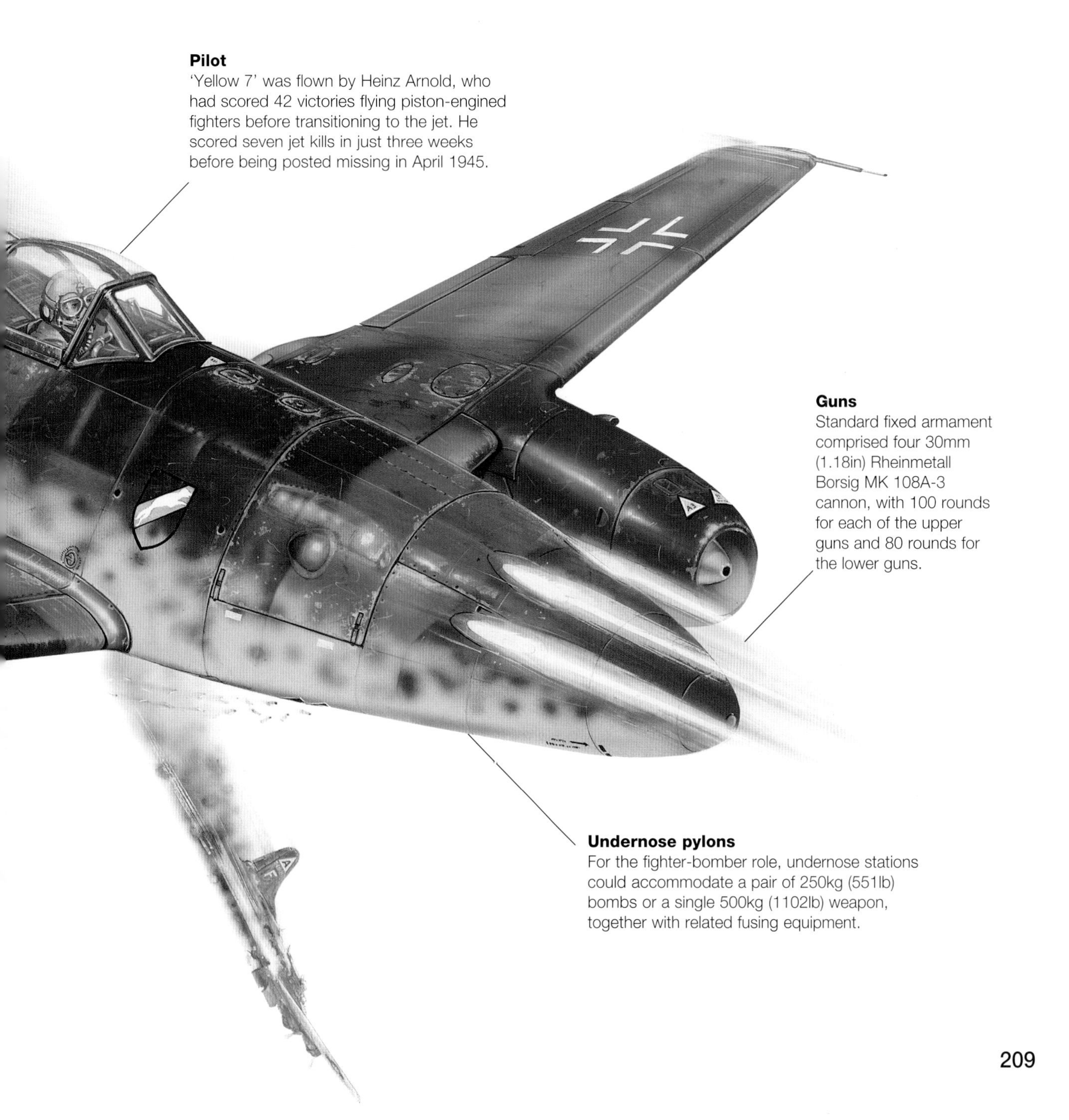

Flying with 10 Staffel, Nachtjagdgeschwader 11 (Kommando Walter), the Me 262B-1a/U1 was an interim conversion of a trainer into a night-fighter. Less than a dozen were produced.

Fixed undercarriage

While the first prototypes were equipped with a tailwheel undercarriage, Me 262 V5 had a fixed undercarriage with a nosewheel while V6, first flying in October 1943, had a fully retractable tricycle undercarriage. Meanwhile, the failure of the BMW 003 turbojets in the prototype had been attributed to the compressor blades, and work was underway to redesign the engine. Despite this setback, testing continued under the power of 5.88kN (1,323lb)-thrust Junkers Jumo 004 turbojets, which were larger and heavier than the BMW units. With the prototype appropriately redesigned, the Me 262 returned to the air on July 1942, now powered by two examples of the Jumo 004A. The improved Jumo 004B-1 turbojet came online in November 1943, when it was flown in prototype V6. These units were lighter than the interim Jumo 004As and became the standard for the production aircraft. After the completion of 10 prototypes, efforts switched to 20 pre-production machines.

Thanks to the engine tribulations, development of the jet fighter was not unsurprisingly slow. At the same time, Willy Messerschmitt worried that termination of his company's other programmes to concentrate on the Me 262 would be financially disastrous. Intervening directly with Hitler to have the cancellation of the Me 109 and 209 programmes revoked, Messerschmitt was effectively responsible for delaying the Me 262's entry to service purely for his own self-interest. Consequently, it was not until July 1944 that the Me 262 entered Luftwaffe service. The first to achieve operational status was the Me 262A-1 armed with four 30mm (1.18in) cannon that joined Erprobungskommando 262 at Lechfeld. It was this unit that gave the RAF its first encounter with the type on 25 July when a Mosquito of 544 (PR) Squadron was intercepted near Munich; the previously near-invulnerable British aircraft was damaged and barely managed to escape. Further units were formed as deliveries progressed, with Einsatzkommando Schenck forming at Lechfeld on the Me 262-2 fighter-bomber before being moved to France to oppose the Allied landings in Normandy. In addition to the baseline Me 262A-1, the Me 262A-1/U1 added another pair of cannon, while the Me 262A-1/U2 was equipped for poor-weather operations and the Me 262A-1/U3 was a reconnaissance version with the armament deleted.

Specifications: Me 262A-1

Type:	Single-seat fighter
Dimensions:	Length: 10.58m (34ft 9in); Wingspan: 12.5m (41ft 0in); Height: 3.83m (12ft 7in)
Weight:	6,387kg (14,080lb) maximum take-off
Powerplant:	2 x 8.83kN (1,984lb)-thrust Junkers Jumo 004B axial-flow turbojets
Maximum speed:	869km/h (540mph)
Range:	1,050km (652 miles)
Service ceiling:	12,100m (39,698ft)
Crew:	1
Armament:	4 x 30mm (1.18in) MK 108; plus up to 12 R4M air-to-air rockets under each wing

Me 262B-1/U1 night-fighter

The Me 262B-1/U1 night-fighter was created on the basis of the Me 262B-1 dual-control trainer, which differed from the basic fighter in having a second seat in the aft section of an elongated cockpit. Since this resulted in a reduction in internal fuel capacity, auxiliary fuel tanks were added below the forward fuselage.

The first trials of the night-fighter were undertaken at Rechlin in October 1944, using Lichtenstein SN-2 radar. The production version featured a radar operator in the rear seat and a FuG 218 Neptun V radar that featured a nose-mounted antenna array. These joined 10./NJG 11 in March of 1945, when the unit was already flying single-seat Me 262s as night-fighters attacking Allied aircraft caught in searchlights. Both types had some success, including against the previously invulnerable Mosquito, although there was always the danger of overshooting the target aircraft in the dark. At the end of the war, work was underway on an improved night-fighter, the Me 262B-2, with an extended fuselage, increased range and upwards-firing cannon.

Production delays

The Me 262's delay into service has often been described as a result of Hitler's insistence that the Me 262 take on an offensive bomber role. Although the Führer did favour development of this version, the Me 262A-2, the degree to which this hindered the type's overall progress and pace of deliveries is debatable. As well as the engine problems, Allied raids on the Me 262 production centre forced the evacuation of the manufacturing effort from Regensburg to Oberammergau, where problems were exacerbated by the lack of a suitable workforce. The Me 262A-2 was equipped to carry up to 500kg (1,102lb) of bombs in addition to the basic four-cannon armament and was also developed as a two-seater, the Me 262A-2/U2, with a bomb-aimer in a glazed nose. This was chosen in preference to the alternative of sitting the bombardier behind the pilot with a 2m (6ft 7in) periscope to see beyond the bombs. Ultimately, development on this version ceased after Hitler's 'bomber-only' edict was abandoned in November 1944.

In the fighter role, the Me 262 was regularly used against the USAAF's large day-bomber formations. Here, new tactics were required to make best use of the type's

The Me 262B-1a/U1 night-fighter was created on the basis of the Me 262B-1a dual-control trainer, which differed from the basic fighter in having a second seat in the aft section of an elongated cockpit. Since this resulted in a reduction in internal fuel capacity, auxiliary fuel tanks were added below the forward fuselage.

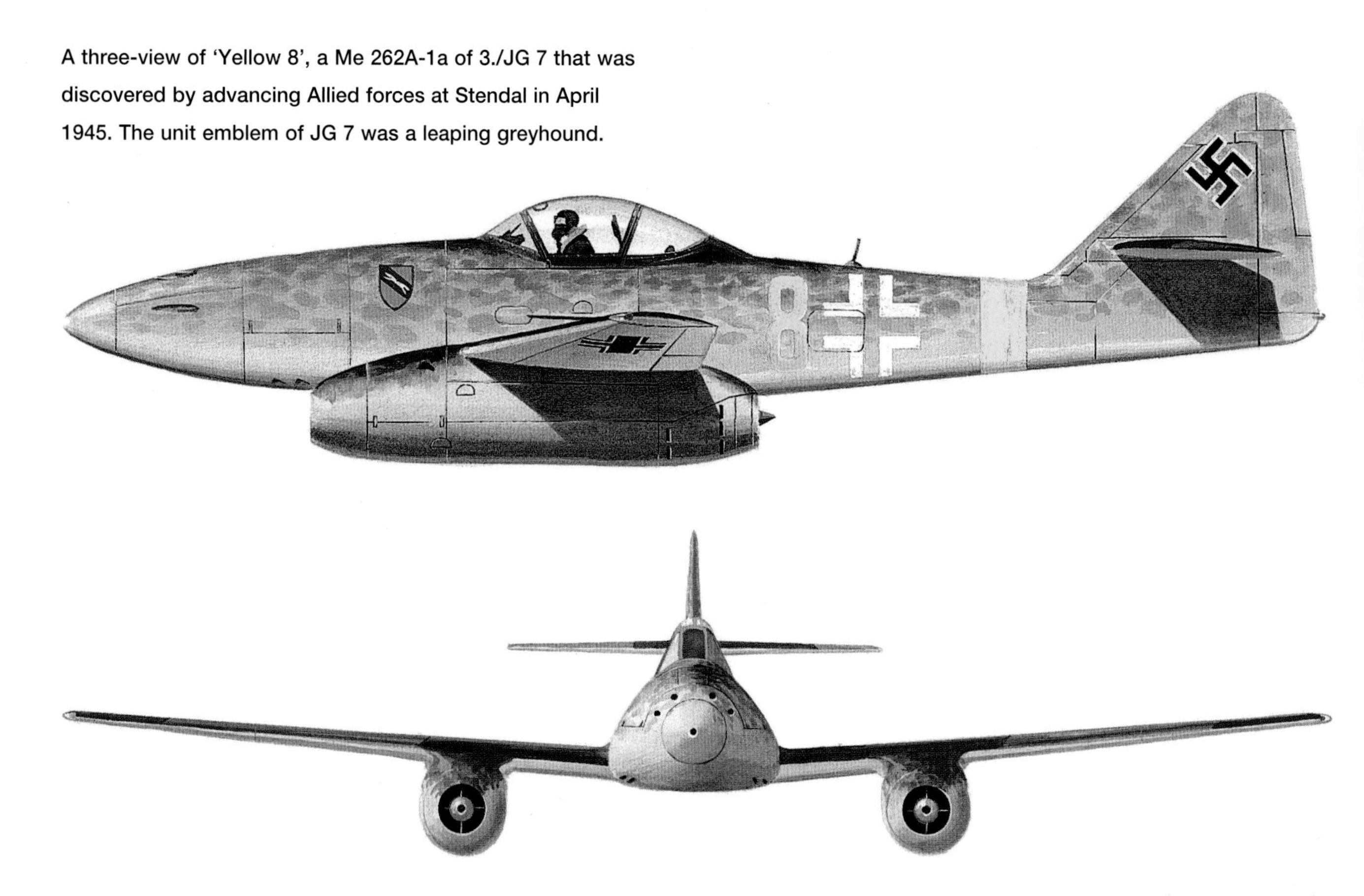

A three-view of 'Yellow 8', a Me 262A-1a of 3./JG 7 that was discovered by advancing Allied forces at Stendal in April 1945. The unit emblem of JG 7 was a leaping greyhound.

high speed and relatively slow-firing cannon. Positioning themselves around 5km (3 miles) behind and 2,000m (6,562ft) above the bomber stream, the Me 262s would dive to build up speed to 850km/h (528mph), allowing them to evade the escorting fighters. Once around 1,500m (4,921ft) behind and 500m (1,640ft) below the bombers, the German fighters would pull up to position themselves level with and behind their targets. This also reduced the Messerschmitt's speed to give a closure of around 150km/h (93mph), increasing the time spent in the firing window – which was from 500m (1,640ft) down to 200m (656ft) – to around seven seconds, after which the attack was broken off.

Escape was made low over the bomber formation to make it difficult for the gunners to hit them. Depending on fuel and ammunition state, the Me 262s could either disengage by diving away at high speed or reposition for a follow-on attack. In addition to the 30mm (1.18in) MK 108 cannon, some aircraft carried up to 24 R4M unguided rockets. These could be launched at around 600m (1,969ft) and only a couple of hits

A test aircraft rather than a dedicated single-seat night-fighter the second Me 262 V2 is seen here fitted with the antenna array for FuG 218 Neptun radar. The white lines on the tail are tufts of wool used to determine the airflow.

could down a B-24 or B-17. The Me 262 also saw some limited success in the night-fighter role. A dedicated night-fighter model appeared before the end of the war, in the form of the radar-equipped Me 262B-1/U1.

Against Allied fighters and fighter-bombers, the jet had to avoid getting into a turning fight. The slower-propeller aircraft could turn inside the Me 262 and in trying to match them, too much speed would be lost, making them vulnerable. Consequently, the Luftwaffe pilots would attempt to engage with a height and speed advantage, making a slashing attack through the Allies, ideally making no more than half a complete turn before climbing away.

The Me 262 pilots had the advantage of being able to pick and choose their engagements, as no Allied fighter was able to catch them if they maintained height and speed. Although its pilots thought the Me 262 should concentrate on enemy fighters, leaving the heavy bombers to the Fw 190 and Me 109, they entered service at a point when the shortage of fuel meant all available fighters were tasked with the anti-bomber role. Despite this, many aces shot down multiple fighters in the Me 262, including P-51s, Spitfires and in probably the last Luftwaffe air-to-air victory of the war, a Yak-9, which fell to a Me 262A-1 of 2./JG 7 over Czechoslovakia at 16:00 hrs on 8 May 1945.

As a dive-bomber, the Me 262 was able to make attacks relatively unhindered, diving at around 900km/h (559mph). The downside to this was that it made it difficult to strike small targets. Additionally, some aircraft were lost during the pull-out phase of the attack due to fuel in the rear tank. This left the centre of gravity too far aft and the resultant abrupt pitch-up could cause the wings to break off. Strafing attacks, meanwhile, were high-risk, low-reward activities, the MK 108 being poorly suited with its low muzzle velocity, and with only 360 rounds of ammunition, it was soon exhausted.

The Me 262 was most vulnerable at low speed, being unable to accelerate away rapidly due to the slow response of the early jets. Taking advantage of this, Allied fighters would attempt to catch them during take-off and landing by loitering near their bases. The Luftwaffe countered this with patrols of Fw 190s and by concentrating flak batteries along the approach lanes. On one occasion, six Me 262s were destroyed during their take-off by Mustangs of the 55th Fighter Group.

Ultimately, the Me 262 was a case of 'too little, too late' (a total of around 1,430 were completed) and despite its undoubted performance advantage, it was unable to do anything to alter the course of the war. Although it had been agreed that the Japanese would produce their own Me 262s, it proved impossible to get the necessary plans and material there, the two Messerschmitt employees dispatched being lost when U-864 was sunk by HMS *Venturer* off Norway in February 1945. Captured examples were examined by engineers from many Allied nations and they aided in the development of jet fighters in the USA, Britain and the Soviet Union. Meanwhile, Avia in Czechoslovakia produced nine single-seat and three two-seat examples after the war, and these served until 1951.

Heinkel He 162 (1944)

A desperate attempt to produce a winning weapon in the dying days of the war, the He 162 suffered from rushed development, which would make it almost as dangerous to its pilots as to the Allies' fighters.

As World War II approached its inevitable conclusion, the Third Reich increasingly pinned its hopes on wonder weapons that would overturn the odds and save the day. The requirement for one of these was set on 8 September 1944: a lightweight 2,000kg (4,410lb) single-engine jet fighter, armed with one or two 30mm (1.18in) cannon, and capable of being flown by members of the Hitler Youth with minimal training. Heinkel's proposal was selected, more due to successful lobbying than overwhelming superiority, Blohm und Voss's proposal being regarded as the better aircraft. Approval for production to begin was given at the end of the month and, in a sign of the continued mass delusion in the upper echelons of the Nazi regime, plans were made for an initial rate of production of 1,000 aircraft a month. The entire programme was code-named Salamander, while Heinkel called the aircraft itself the *Spatz*, or 'Sparrow'. The He 162 designation was chosen in an attempt to confuse Allied intelligence, the number having previously been used for a stalled pre-war Messerschmitt project.

The first of 10 prototype He 162s flew on 6 December, only 90 days after the original specification had been issued. The resulting aircraft had an alloy fuselage with a moulded plywood nose. The aft hinged canopy covered the rather cramped cockpit and was situated immediately in front of the intake for the dorsal-mounted engine. This was a 7.8kN (1,764lb)-thrust BMW 003A-1 turbojet, which was attached to the fuselage by three bolts and covered by front and rear cowls with central panels that could be rapidly removed for servicing. The small main wing was of plywood construction with alloy flaps and had space for 180 litres (40 gals) of fuel in the inter-spar volume, the main fuselage tank holding 695 litres (153 gals). A widely spaced H-tail plane was used, presumably to avoid the exhaust gases affecting the vertical stabilizer. The tricycle undercarriage was operated

Speed
The He 162A-2 had a maximum speed at normal thrust of 490mph (789km/h) at sea level or 837km/h (520mph) at 6000m (19,685ft). This speed could be increased for short periods with a burst of extra thrust. The range at full throttle was 620km (385 miles) at 6000m (19,685ft).

Engine
The He 162A-2 was powered by a single BMW 003E-1 or E-2 axial flow turbojet, rated at 7.8kN (1,764lb st) with a 9.02kN (2,028lb st) emergency rating available for periods of up to 30 seconds. The pre-production aircraft had been powered by the BMW 003A-1, while some prototypes flew with the BMW 003R, combined with the 7.8-kN (1,764lb st) BMW 718 liquid-fuel rocket.

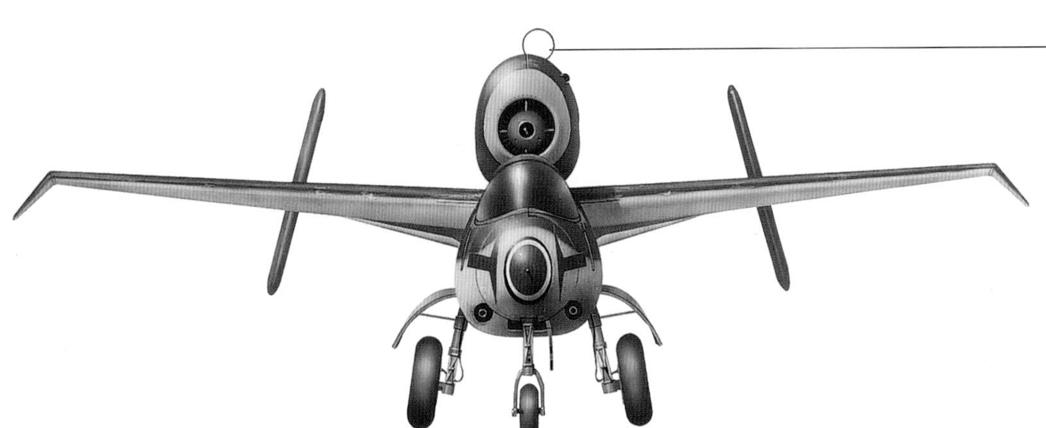

Configuration
Experts predicted that the He 162's unusual top-mounted engine would suffer airflow problems (which to a great extent did not occur), but did not foresee the pitch instability which made the aircraft so tricky to fly and fight in.

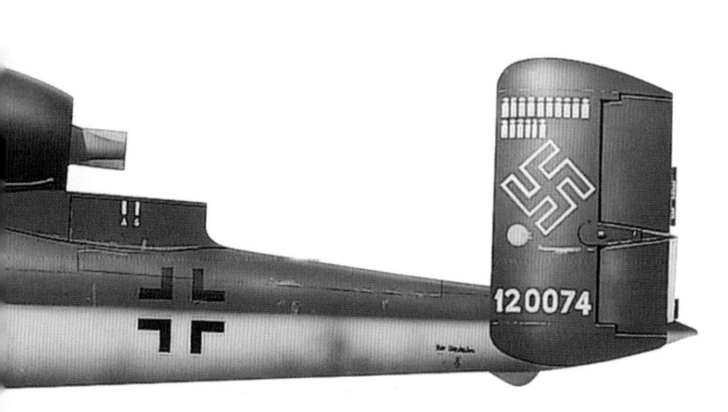

This Heinkel He 162A-2 was captured by British forces at Leck. It had previously served with 3. Staffel, Einsatz-Gruppe I/JG 1, and was the personal aircraft of the Staffelkapitän. It carried his 16 victory marks on the tail, although these had been gained on other aircraft.

Iain Wyllie

This He 162A-2 was allocated to 3. Staffel/JG 1 at its Leck base in May 1945. By this time, the 50 aircraft had been reorganised into one single Gruppe, Einsatz-Gruppe I./JG 1; many pilots from other fragmented units at Leck were absorbed by this new Gruppe.

by hydraulics, but the control surfaces were all manually operated. While the prototype carried two 20mm (0.79in) MG 151s, one either side of the nose gear bay, there would be some debate with the RLM as to the best weapon and later aircraft would trial the 30mm (1.18in) MK 108, although only 50 rounds could be carried compared to 120 for the smaller-calibre weapons.

During the first flight, a maximum speed of 840km/h (522mph) was achieved although the longitudinal stability was found to be poor. More alarmingly, one of the wooden main gear doors had detached in flight. This was due to the poor quality of adhesive used in their construction, the RAF having bombed the plant making the originally specified product and the substitute being too acidic in nature. Four days later, a second flight was made with an audience of Air Ministry officials. While conducting a high-speed pass in front of the VIPs the right leading edge broke off, taking the wing's upper skin with it. The He 162 rolled violently and crashed into the ground just outside the airfield. Once again, the poor quality of the glue used in the aircraft's construction was blamed. Despite this setback, the programme continued, with flights of the second prototype highlighting further stability issues. These were addressed with an enlarged tail and downwards-turned tips to the mainplane, named 'Lippisch ears' after the aerodynamicist who proposed them. Although this solved the immediate problem, the He 162 would continue to suffer mishaps during its short service life, undoubtedly due to the compressed development process.

Production programme

It was planned to produce the He 162 in vast numbers, to the extent that aircraft would simply be replaced with new ones if they couldn't be rapidly repaired. To this end, Heinkel were expected to build 1,000 a month at their Rostock Marienehe facility, while Junkers would produce a similar number at Bernburg. Meanwhile, a further 2,000 a month would be turned out by slave labour in an underground factory in the Harz mountains.

Later plans would involve another 1,000 coming from a chalk mine near Heinkel's factory in Vienna. To supply these assembly lines, components and assemblies would be built around the Third Reich, with a salt mine in Urseburg being used to assemble the BMW 003 engines. Although impressive on paper, this plan bore no relation to Germany's strategic situation and during the six months that the He 162 was built, only around 1,000 in total were completed.

Specifications: He 162A-2

Type:	Single-seat jet fighter
Dimensions:	Length: 9.05m (29ft 8in); Wingspan: 7.2m (23ft 7in); Height: 2.6m (8ft 6in)
Weight:	2,805kg (6,184lb) maximum take-off
Powerplant:	1 x 9.02kN (2,028lb)-thrust BMW 003E-1 axial-flow turbojet
Maximum speed:	905km/h (562mph)
Range:	620km (385 miles)
Service ceiling:	12,000m (39,370ft)
Crew:	1
Armament:	2 x 20mm (0.79in) MG 151 cannon

Erprobungskommando 162 was formed in January 1945 for operational test and evaluation of the small jet. This rapidly led to the abandonment of the proposal for the Hitler Youth and other low-experience pilots to fly the aircraft. Instead, I. and II./JG 1 were converted from the Fw 190A-8 to the He 162A-2. During their conversion, more aircraft and pilots were lost but by mid-April, I./JG 1 was operational and moved to Leck in Schleswig-Holstein. In the remaining three weeks of the war, the He 162 would be involved in a number of air combats, with mixed results – at least two falling to Hawker Tempests of the RAF while JG 1 would claim a Typhoon shot down two days before the end of the war. The lack of development time was, however, proving costly, with 10 aircraft lost due to engine flameouts and structural failures that had still not been resolved. There was some hope for the He 162's pilots, the type at least being fitted with an early ejection seat that appears to have been successfully used at least twice in late April.

As the Allies overran the He 162's airfields, many were captured for subsequent evaluation, while some units destroyed theirs to prevent this happening. Approximately 170 had been delivered to the Luftwaffe by VE Day, with 100 or so awaiting collection and 800 on the production lines. In another denial of reality, Heinkel was even planning new models, including ones with forward-swept wings and twin pulse-jet propulsion.

The He 162, when serviceable, was a decent jet fighter with post-war evaluation concluding it had the edge over the Meteor in manoeuvring. However, the speed of development and the limited resources available meant it was at least as dangerous to its operators as to the enemy during the war.

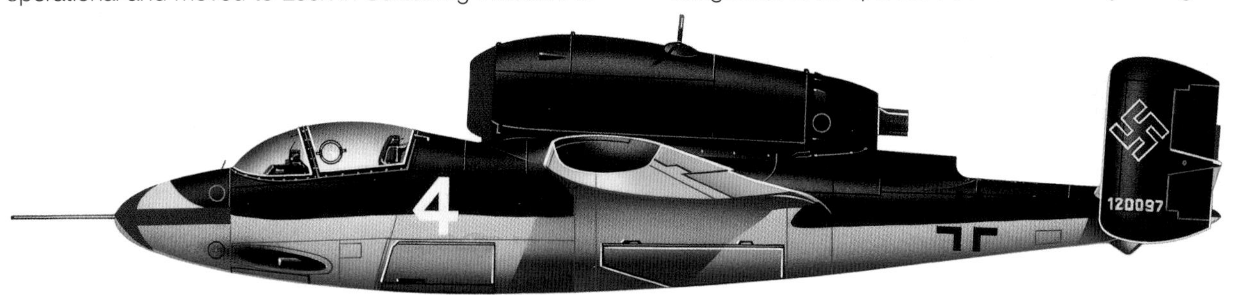

Above: Heinkel He 162A-1 'White 4' was operated by a frontline unit, 1./Jagdgeschwader 1, based at Leck. This particular aircraft, Werknummer 120097, was captured by the Allies and tested by the RAF after the war.

Below: The first prototype, He 162 V1, crashed at Schwechat Airfield, Vienna, on 10 December 1944, in front of an invited audience of high-ranking Nazi officials. The pilot, Flugkapitän Gotthold Peter, was killed.

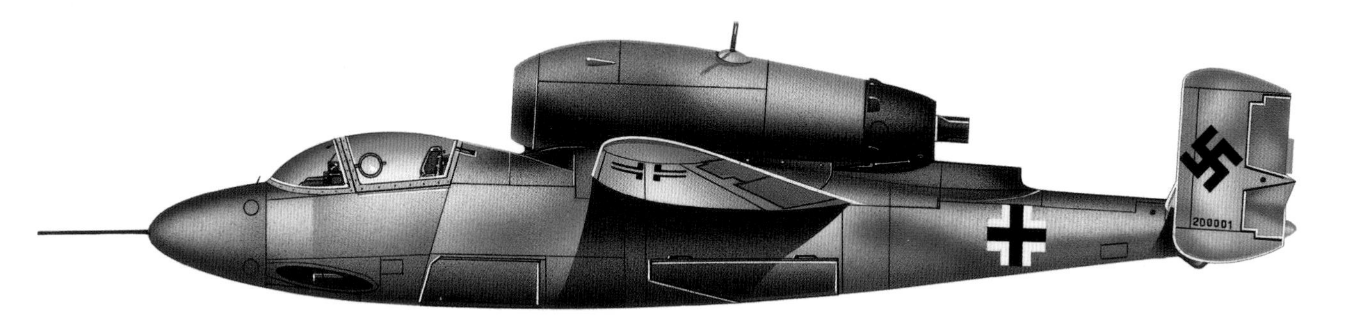

Glossary

Aufklärung	Reconnaissance
Ausbildungs-	Training
Befehlshaber	Commander
Behelfs	Auxiliary
Beobachter	Observer/Navigator
Bodenlafette	Ventral gun mount
Bordkanone	Fixed aircraft cannon
Bordfliegerstaffel	Shipborne aircraft squadron
B-Schule Advanced/Blind	Flying Training School
B-Stand	Dorsal gunner's position
Buna	Synthetic rubber (originally a trade name)
C-Schule	Advanced Flying Training School, multi-engine
C-Stand	Ventral gunner's position
DFS	*Deutsches Forschungsinstitut für Segelflug*
Einsatzkommando	Combat Operations Detachment
EJG	Ergänzungs-Jagdgeschwader
EKdo	*Erprobungs Kommando*
Elektrische Trägervorrichtung	Electrically-operated bomb racks
Entwicklungs-	Development-
Ergänzungs-	Replacement-
Ergänzungs-Jagdgeschwader	Fighter Replacement Training Group
Ersatz	Substitute or Replacement
Fallschirmjäger	Paratroopers
Fernaufklärung	Long-range Reconnaissance
Fernaufklärungsgruppe	Long-range Reconnaissance Gruppe
Fernaufklärungstaffel	Long-range reconnaissance Squadron
Fernnachtjagd	Long-range night fighter/intruder
Fernzielgerät	Remote aiming device or bombsight
FFS	*Flugzeugführerschule*
FHL	*Ferngerichtete Hecklafette*
Flak	*Fliegerabwehrkanone*
Flieger	Pilot (as description) or Airman (as rank)
Fliegerabwehrkanone	Anti-Aircraft Gun/Artillery
Fliegerdivision	Air Division
Fliegerkorps	Air Corps
Flugzeugführerschule	Pilot/Aircraft Commander School
FuG	*Funkgerät*
Funkgerät	*Radio or Radar set*
Führerkurierstaffel	Führer's courier squadron
Führungsstab	Operations Staff
General	Lieutenant General or Air Marshal
General der Jagdflieger	General of Fighters
General der Kampfflieger	General of Bombers
Generalfeldmarschall	General of the Air Force/Marshal of the RAF
Generalleutnant	Major-General/Air Vice Marshal
Generalmajor	Brigadier-General/Air Commodore
Generaloberst	General/Air Chief Marshal
Geschwader	Equivalent to Allied Group
Geschwaderkommodore	Geschwader commander
Gruppe	Equivalent to Allied Wing
Gruppenkommandeur	Group commander
Hauptmann	Captain/Flight Lieutenant
Heeres-	Army
Heeresaufklärungstaffel	Army or Tactical Reconnaissance Squadron
Jabo	*Jagdbomber*
Jabo-Rei	*Jagdbomber mit vergrosster reichweite*
Jagd-	Fighter (Hunt, Chase, Pursuit)
Jagdbomber	Fighter bomber
Jagdfliegerführer	Fighter Command
Jagdfliegerschule	Fighter Training School
Jafü	*Jagdfliegerführer*
Jagdgeschwader	Fighter Group
Jagdgruppe	Fighter Wing
Jagdstaffel	Fighter Squadron
JG	*Jagdgeschwader*
JGr	*Jagdgruppe*
JFS J	*agdfliegerschule*
Jumo	*Junkers Motoren Werke*
Kampf	Battle (Bomber, when applied to aircraft)
Kampfbeobachter	Artillery Observer
Kampfgeschwader	Bomber Group
Kampfgeschwader zur	Special Duty/Transport Group
Kampfgruppe	Bomber Wing
Kdo	*Kommando*
Kette	Flight of three aircraft
KG	*Kampfgeschwader*
KGr	*Kampfgruppe*
KGzbV	*Kampfgeschwader zur besonderen Verwendung*
Koluft	*Kommander der Luftwaffe bei einen AOK*
Kommando	Detachment
Kü.Fl	*Küsten Flieger*
Küsten Flieger	Coastal Aviation
Langstrecken-	Long-range
Lehr-	Instruction
Lehrgeschwader	Demonstration/Operational development Group
Luftflotte	Air Fleet
Lufttorpedo	Air-dropped Torpedo

Lufttransportstaffel	Air Transport Squadron
Luftwaffe	Air Force
Luftwaffenführungsstab	Luftwaffe Operations Staff
Luftwaffengeneralstab	Luftwaffe Air Staff
Major	Major/Squadron Leader
Maschinengewehr	Machine Gun
Maschinenkanone	Machine Cannon
MG	*Maschinengewehr*
Minensuchgruppe	*Minehunting/sweeping wing*
Mistel (Mistletoe)	combination aircraft
MK	*Maschinenkanone*
Nachtjagd-	Night Fighter
Nachtjagdgeschwader	Night Fighter Group
Nachtschlacht-	Night Harassment
Nachtschlachtgruppe	Night Harassment Wing
NAGr	*Nahaufklärungsgruppe*
Nahaufklärungs-	Short-range reconnaissance
Nahaufklärungsgruppe	Short-range reconnaissance group
NJG	*Nachtjagdgeschwader*
NSGr	*Nachtschlachtgruppe*
Ob.d.L	*Oberbefehlshaber der Luftwaffe*
Ob.d.M	*Oberbefehlshaber der Marine*
Oberbefehlshaber der Luftwaffe	Commander-in-Chief of the Luftwaffe
Oberbefehlshaber der Marine	Commander-in-Chief of the Navy
Oberfeldwebel	Master Sergeant/Flight Sergeant
Oberkommando des Heeres	Army High Command
Oberkommando der Luftwaffe	Air Force High Command
Oberkommando der Marine	Navy High Command
Oberkommando der Wehrmacht	High Command of the Armed Forces
Oberleutnant	First Lieutenant/Flying Officer
Oberst	Colonel/Group Captain
Oberstleutnant	Lieutenant Colonel/Wing Commander
OKH	*Oberkommando des Heeres*
OKL	*Oberkommando der Luftwaffe*
OKM	*Oberkommando der Marine*
OKW	*Oberkommando der Wehrmacht*
Rauchgerät	Rocket-booster unit
R-Gerät	*Rauchgerät*
Rotte	*A* flight of two aircraft
R-Stoff	Rocket fuel (57% Monoxylidene, 43% triethylamine)
Rüstatz	Field conversion kit
SAGr	*See-Aufklärungsgruppe*
Sanitätsstaffel	Air Ambulance Squadron
Sch.G S	*chlachtgeschwader*
Schlacht-	Close-support/Assault
Schlachtgeschwader	Close Support Group
Schlepp-	Towing
Schnellkampfgeschwader	High-speed Bomber/Attack Group
Schwarm	Flight of four fighters
Schnellbomber	Fast bomber
Schräge Musik	'Slanting' or 'Jazz Music' – cannon firing obliquely upwards
Sd.Kdo	Special Detachment
Seeaufklärungsgruppe	Maritime Reconnaissance Wing
Seenotsdienst	Air Sea Rescue Service
Seenotsstaffel	Air Sea Rescue Squadron
SG	*Schlachtgeschwader*
SKG	*Schnellkampfgeschwader*
Sonder-	Special purpose
Spanner-Anlage	Early infra-red sensor system
S-Stoff	Rocket fuel (97% Nitric Acid, 3% Sulphuric Acid)
Stab-	Staff
Stabschwarm	Staff flight in a Gruppe
Staffel	Squadron
Staffelkapitan	Squadron commander
St.G	*Sturzkampfgeschwader*
Störkampstaffel	Night Harassment Squadron
Stuka	*Sturzkampfflugzeug*
Sturm-	Assault
Sturmgruppe	Assault Wing
Sturzkampfflugzeug	Dive bomber
Sturzkampfgeschwader	Dive bomber Group
Sturz-visier	Dive Bombing Sight
Trägergeschwader	Aircraft Carrier Group
Troika-schlepp	Triple tow (of large gliders by three aircraft)
Umbau	Reconstruction
Umrüst-Bausatz	Factory conversion kit
V	*Versuchs* (Experimental)
Verband	Formation
Verstellschraube	Variable pitch propeller
VS	*Verstellschraube*
Werfer-Granate	Grenade projector/rocket propelled shell
Wettererkundungsstaffel	Meteorological squadron
X-Gerät	Electronic blind-flying/bombing aid
Y-Gerät	Electronic blind-flying/range-finding aid
Zwilling	Twin or coupled
Zerstörer	Destroyer, or heavy fighter
Zerstörergeschwader	Heavy Fighter Group
ZG	*Zerstörergeschwader*

Index

Note: page numbers in **bold** refer to information contained in captions.

Index

Picture credits

Photographs

Alamy: 37 (Pictorial Press), 175 (Sueddeutsche Zeitung Photo)

AirSeaLandPhotos: 7, 23, 27, 32, 35, 45, 56, 58, 62, 65, 67, 80, 91, 101, 106, 110, 114, 119, 124, 125, 128, 130, 132, 138, 139, 141, 143, 149, 151, 156, 157, 161, 163, 164, 171, 179, 181, 183, 186, 187, 189, 203, 213

Amber Books: 5, 6, 9, 10, 17, 33, 49, 51, 72, 84, 97, 121, 137, 147, 191, 201, 211, 216

Getty Images: 12 (ullstein bild Dtl.), 75 (De Agostini Picture Library), 107 (Central Press)

National Air and Space Museum: 129 (Dane A. Penlan)

National Museum of the United States Air Force: 196

Artworks

Amber Books: 14–16, 20/21, 26, 28/29, 41–43, 46–47, 52–55, 59, 64–66, 68–71, 73, 78/79, 81–83, 91–94, 98–100, 102–105, 108/109, 111, 113, 115, 120, 126/127, 131, 133–135, 140/141, 144–146, 148–150, 155, 158–160, 162/163, 166/167, 172/173, 176–178, 184–185, 188–191, 197–201, 202, 204–210, 212, 214–215, 217

Ronny Bar: 11, 13, 30/31, 38/39, 48, 50, 51, 56/57, 118

David Bocquelet: 9, 24/25, 27, 34–36, 74–77, 85–89, 112/113

Vincent Bourguignon: 192–195

Ed Jackson - artbyedo.com: 60/61, 63, 116/117, 121–123, 136, 152/153, 157, 165, 168–169, 174, 180–181

Murilo Martins: 8, 18–19, 22, 95–97